Experimental Hematology Today—1988

S.J. Baum   K.A. Dicke   E. Lotzová   D.H. Pluznik
Editors

Experimental Hematology Today—1988

Selected Papers from the 17th Annual
Meeting of the International Society for
Experimental Hematology August 21–25,
1988, Houston, Texas, USA

With 71 Illustrations

 Springer Science+Business Media, LLC

S.J. Baum
Treasurer, International Society for Experimental Hematology, Bethesda, Maryland 20817, USA

K.A. Dicke
Bone Marrow Transplantation Section, Department of Hematology, University of Texas M.D. Anderson Cancer Center, Houston, Texas 77030, USA

E. Lotzová
Section of Natural Immunity, Department of General Surgery, University of Texas M.D. Anderson Cancer Center, Houston, Texas 77030, USA

D.H. Pluznik
Laboratory of Immunology, National Institute of Dental Research, National Institutes of Health, Bethesda, Marlyand 20892, USA

Library of Congress number 79-641222

Printed on acid-free paper

Camera-ready copy provided by the editors.

9 8 7 6 5 4 3 2 1

ISBN 978-1-4613-8864-7   ISBN 978-1-4613-8862-3 (eBook)
DOI 10.1007/978-1-4613-8862-3

Preface

Experimental Hematology Today—1988 represents a selection of the outstanding papers presented at the 17th Annual Meeting of the International Society for Experimental Hematology in Houston, Texas, August 21–25, 1988. The manuscripts were selected after a careful review by the local program committee and finally by the editors of the book. The book is divided into six parts dealing with molecular involvement in hematopoietic precursor proliferation, growth factors, specific cell regulators, genetic manipulation, regulation of leukemogenesis, and bone marrow transplantation.

The first part, chaired by Dr. Bradley, presents recent interesting findings on "Hematopoietic Regulation by Cytokines." Dr. Zipori heads part II, entitled "Hematopoietic Cellular Growth Regulation." Reports in this chapter discuss the interaction of stromal cells with hematopoietic stem and progenitor cells. Part III deals with "Granulopoietic Regulators" and is chaired by Dr. Pluznik. Papers in part IV, introduced by Dr. Evinger-Hodges, relate to findings dealing with means of gene transfers into hematopoietic progenitor cells. Dr. Ruscetti chairs part V, which deals primarily with "Regulation of Leukemogenesis." Finally, part VI comprises papers dealing with recent discoveries in "Bone Marrow Transplantation." Dr. Dicke is the very able leader of this section.

The present yearbook of experimental hematology reflects the diverse interests of basic and clinical hematologists. As such, it should be of considerable value to all biomedical scientists.

Contents

Contributors

P. Aegerter, INSERM U88 (Public Health and Social and Economic Epidemiology), Paris, France

Kunihisa Akai, Researcher, Research Institute of Life Science, Snow Brand Milk Products Co., Ltd., Tochigi, Japan

Zuhair S. Al-Lebban, Research Associate, Department of Environmental Practice, College of Veterinary Medicine, University of Tennessee, Knoxville, Tennessee 37901-1071, USA

Julian L. Ambrus, Jr., Senior Investigator, Laboratory of Immunoregulation, National Institute of Allergy and Infectious Diseases, National Institutes of Health, Bethesda, Maryland 20892, USA

W. French Anderson, Laboratory of Molecular Hematology, National Heart, Lung and Blood Institute, National Institutes of Health, Bethesda, Maryland 20892, USA

B. Auvert, INSERM U88 (Public Health and Social and Economic Epidemiology), Paris, France

Siegmund J. Baum, Treasurer, International Society for Experimental Hematology, Bethesda, Maryland 20817, USA

A.D. Bearpark, Leukemia Research Fund Center, Institute of Cancer Research, London, United Kingdom

Th.M. Berkvens, Department of Medical Biochemistry, University of Leiden, Leiden, The Netherlands

Ivan Bertoncello, Senior Research Fellow, Peter MacCallum Cancer Institute, Melbourne, Victoria, Australia

Matthias Bickel, Laboratory of Immunology, National Institute of Dental Research, National Institutes of Health, Bethesda, Maryland 20892, USA

F. Birg, Unit 119, INSERM, Marseille, France

Thomas R. Bradley, Head, Cell Biology Unit, Peter MacCallum Cancer Institute, Melbourne, Victoria, Australia

Anna Butturini, University of Parma, Parma, Italy

C.T. Caskey, Howard Hughes Medical Institute, Institute for Molecular Genetics, Baylor College of Medicine, Houston, Texas 77030, USA

Mridula Chandan, Tumor Immunology Laboratory, Department of Therapeutic Radiology and Bone Marrow Transplantations Program, University of Minnesota Twin Cities Medical School, Minneapolis, Minnesota 55455, USA

Laura Chesky, Laboratory of Immunoregulation, National Institute of Allergy and Infectious Diseases, National Institutes of Health, Bethesda, Maryland 20892, USA

D. Clarke, Leukemia Research Fund Center, Institute of Cancer Research, London, United Kingdom

I. Cox, University of Texas M.D. Anderson Cancer Center, Houston, Texas 77030, USA

Mary K. Cullen, University of Florida College of Medicine, Gainesville, Florida 32605, USA

R. Andrew Cuthbertson, The Howard Florey Institute for Experimental Physiology and Medicine, University of Melbourne, Parkville, Victoria, Australia

T. Michael Dexter, Department of Experimental Hematology, Christie Hospital and Paterson Institute for Cancer Research, Manchester, United Kingdom

Karel A. Dicke, Bone Marrow Transplantation Section, Department of Hematology, University of Texas M.D. Anderson Cancer Center, Houston, Texas 77030, USA

Ashley R. Dunn, Head, Molecular Biology Program, Ludwig Institute for Cancer Research, Royal Melbourne Hospital, Victoria, Australia

F. Dunphy, University of Texas M.D. Anderson Cancer Center, Houston, Texas 77030, USA

Martin A. Eglitis, Laboratory of Molecular Hematology, National Heart, Lung and Blood Institute, National Institutes of Health, Bethesda, Maryland 20892, USA

M.P.W. Einerhand, Radiobiological Institute TNO, Rijswijk, The Netherlands

Larry R. Ellingsworth, Collagen Corporation, Palo Alto, California 94303, USA

Carol L. Epstein, Director of Clinical Affairs, Immunex Research and Development Corporation, Seattle, Washington 98101, USA

M.J. Evinger-Hodges, University of Texas M.D. Anderson Cancer Center, Houston, Texas 77030, USA

Anthony S. Fauci, Laboratory of Immunoregulation, National Institute of Allergy and Infectious Diseases, National Institutes of Health, Bethesda, Maryland 20892, USA

C. Fay, Unit 119, INSERM, Marseille, France

Robert Peter Gale, University of California, Los Angeles School of Medicine; Armand Hammer Center for Advanced Studies in Nuclear Energy and Health, Los Angeles, California, USA

John T. Gallagher, Department of Medical Oncology, Christie Hospital and Paterson Institute for Cancer Research, Manchester, United Kingdom

Myrtle Y. Gordon, Leukemia Research Fund Center, Institute of Cancer Research, London, United Kingdom

Norbert C. Gorin, Professor of Hematology, Saint Antoine Hospital; Associate Professor, Department of Hematology-Bone Marrow Transplant Unit, Saint Antoine University Hospital Center and National Blood Transfusion Institute; President, Working Party on Autologous Bone Marrow Transplantation of the European Bone Marrow Transplantation Group (EBMTG), Paris, France

Anton Hagenbeek, Chairman, Leukemia Research Department, Radiobiological Institute TNO, Rijswijk, The Netherlands

L.E. Healy, Leukemia Research Fund Center, Institute of Cancer Research, London, United Kingdom

George S. Hodgson, Director, Peter MacCallum Cancer Institute, Melbourne, Victoria, Australia

P.M. Hoogerbrugge, Radiobiological Institute TNO, Rijswijk, The Netherlands

L.J. Horwitz, University of Texas M.D. Anderson Cancer Center, Houston, Texas 77030, USA

S. Jagannath, Department of Hematology, University of Texas M.D. Anderson Cancer Center, Houston, Texas 77030, USA

J.B. Jones, Department of Environmental Practice, College of Veterinary Medicine, University of Tennessee, Knoxville, Tennessee 37901, USA

Michael Kalai, Department of Cell Biology, The Weizmann Institute of Science, Rehovot, Israel

Gosei Kawanishi, Research Institute of Life Sciences, Snow Brand Milk Products Co., Ltd., Tochigi, Japan

Armand Keating, Director, University of Toronto, Autologous Marrow Transplant Program, Toronto General Hospital, Toronto, Ontario, Canada

R.E. Kellems, The Verna and Marrs McLean Department of Biochemistry, Baylor College of Medicine, Houston, Texas 77030, USA

Jonathan R. Keller, Biological Carcinogenesis Development Program, Program Resources, Inc., National Cancer Institute – Frederick Cancer Research Society, Frederick, Maryland 21701, USA

Stephen M. Kobb, University of Florida College of Medicine, Gainesville, Florida 32605, USA

Anthony B. Kriegler, Scientific Officer, Peter MacCallum Cancer Institute, Melbourne, Victoria, Australia

William W. Kwok, Fred Hutchinson Cancer Research Center, Seattle, Washington 98104, USA

Richard A. Lang, Ludwig Institute for Cancer Research, Royal Melbourne Hospital, Victoria, Australia

M. Lopez, Unit 119, INSERM, Marseille, France

Clinton D. Lothrop, Jr., Department of Environmental Practice, College of Veterinary Medicine, University of Tennessee, Knoxville, Tennessee 37901, USA

Eva Lotzová, Section of Natural Immunity, Department of General Surgery, University of Texas M.D. Anderson Cancer Center, Houston, Texas 77030, USA

Patrice Mannoni, Head, Department of Cell Biology, Regional Cancer Center, Marseille, France

N. Maroc, Unit 119, INSERM, Marseille, France

Anton C.M. Martens, Radiobiological Institute TNO, Rijswijk, The Netherlands

E.A. McCulloch, Head, Division of Biological Research, The Ontario Cancer Institute; University Professor, The University of Toronto, Toronto, Ontario, Canada

Patrick McFarland, Laboratory of Immunoregulation, National Institute of Allergy and Infectious Diseases, National Institutes of Health, Bethesda, Maryland 20892, USA

Stephan E. Mergenhagen, Laboratory of Immunology, National Institute of Dental Research, National Institutes of Health, Bethesda, Maryland 20892, USA

A. Dusty Miller, Fred Hutchinson Cancer Research Center, Seattle, Washington 98104, USA

Alan M. Miller, Assistant Professor, Department of Medicine, University of Florida College of Medicine, Gainesville, Florida 32610, USA

Yasusada Miura, Division of Hematology, Department of Medicine, Jichi Medical School, Tochigi, Japan

Howard Mostowski, Laboratory of Immunoregulation, National Institute of Allergy and Infectious Diseases, National Institutes of Health, Bethesda, Maryland 20892, USA

Richard Nash, Fred Hutchinson Cancer Research Center, Seattle, Washington 98104, USA

T. Philip, Centre Léon Bérard, Lyon, France

Dov H. Pluznik, Laboratory of Immunology, National Institute of Dental Research, National Institutes of Health, Bethesda, Maryland 20892, USA

D. Razanajaona, Unit 119, INSERM, Marseille, France

Francis W. Ruscetti, Head, Lymphokine Section, Laboratory of Molecular Immunoregulation, National Cancer Institute, Frederick, Maryland 21701-1013, USA

C.A. Savary, Section of Natural Immunity, Department of General Surgery, University of Texas M.D. Anderson Cancer Center, Houston, Texas 77030, USA

Maurizio Scarpa, Postdoctoral Fellow, Institute for Molecular Genetics, Baylor College of Medicine, Houston, Texas 77030, USA

Friedrich G. Schuening, Associate in Clinical Research, Fred Hutchinson Cancer Research Center, Seattle, Washington 98104, USA

Frank W. Schultz, Radiobiological Institute TNO, Rijswijk, The Netherlands

Garwin K. Sing, Laboratory of Molecular Immunoregulation, Biological Response Modifiers Program, National Cancer Institute – Frederick Cancer Research Facility, Frederick, Maryland 21701, USA

Larry Souza, AMGEN Inc., Thousand Oaks, California 91320, USA

Jorge A. Spinolo, Assistant Internist, Department of Hematology, University of Texas M.D. Anderson Cancer Center, Houston, Texas 77030, USA

Gary Spitzer, University of Texas M.D. Anderson Cancer Center, Houston, Texas 77030, USA

Richard B. Stead, Fred Hutchinson Cancer Research Center, Seattle, Washington 98104, USA

Rainer Storb, Professor of Medicine, University of Washington; Member, Fred Hutchinson Cancer Research Center, Seattle, Washington 98104, USA

Toshio Suda, Division of Hematology, Department of Medicine, Jichi Medical School, Tochigi, Japan

A. Tabilio, University of Perugia, Perugia, Italy

Kerry McD. Taylor, Director, Department of Hematology, Mater Public Hospital, South Brisbane, Queensland, Australia

Frances Toneguzzo, Research Chemist, E.G. & G. Biomolecular, Natick, Massachusetts 01760, USA

H. Torres, Unit 119, INSERM, Marseille, France

Faith M. Uckun, Tumor Immunology Laboratory, Department of Therapeutic Radiology and Bone Marrow Transplantations Program, University of Minnesota Twin Cities Medical School, Minneapolis, Minnesota 55455, USA

Masatsugu Ueda, Research Institute of Life Science, Snow Brand Milk Products Co., Ltd., Tochigi, Japan

D. Valerio, Research Scientist, Radiobiological Institute TNO, Rijswijk, The Netherlands

D.W. van Bekkum, Radiobiological Institute TNO, Rijswijk, The Netherlands

V.W. van Beusechem, Radiobiological Institute TNO, Rijswijk, The Netherlands

H. van der Putten, The Salk Institute for Biological Studies, San Diego, California 92138, USA

I.M. Verma, The Salk Institute for Biological Studies, San Diego, California 92138, USA

P.M. Wamsley, The Salk Institute for Biological Studies, San Diego, California 92138, USA

Roy S. Weiner, University of Florida College of Medicine, Gainesville, Florida 32605, USA

K. Randall Young, Jr., Laboratory of Immunoregulation, National Institute of Allergy and Infectious Diseases, National Institutes of Health, Bethesda, Maryland 20892, USA

J. Yau, University of Texas M.D. Anderson Cancer Center, Houston, Texas 77030, USA

Dov Zipori, Department of Cell Biology, The Weizmann Institute of Science, Rehovot, Israel

Part I.   Hematopoietic Regulation by Cytokines

Chairperson: T.R. Bradley

1   Interleukin 1 in Hematopoiesis

C.L. Epstein

This paper will review some of the important events in the development of our understanding of Interleukin-I (IL-1), and will cover discovery and characterization, in vitro and in vivo animal studies, and will speculate on the role of IL-1 in the clinic.

IL-1 was cloned at Immunex using two separate techniques - 1) hybrid selection from macrophage cDNA library using sized mRNA, and 2) using oligonucleotide probes derived from N-terminus amino acid sequence information obtained from purified IL-1 protein. Both techniques yielded proteins which possessed activities attributed to IL-1[1,2]. However, these proteins proved not to be identical.

The two gene products, termed IL-1α and IL-1β, are both 30KD molecular weight in their primary gene transcript form, and, in vivo, are processed to a 17.5 KD length. They share homology at only 25% of their amino acid sequence[2]. The homology is scattered rather evenly throughout the sequence leading to the conclusion that there might have been a mutual forbearer gene. IL-1α and IL-β bind competitively to the same receptor. IL-1α in its full length gene transcript (30KD molecular weight form) binds equally well as the 17.5KD form, as measured by percent inhibition of binding of radiolabeled IL-1α. However, it shows that IL-1β is active only in its processed (17.5KD) form. This IL-1β competes with IL-1α for the binding site[3]. In side-by-side equal-dose studies to date, there have been no demonstrated differences between IL-1α and IL-1β. Therefore, this paper will consider them as equivalent and will refer to both proteins as IL-1.

The receptor for IL-1 is an 80KD glycosylated cell surface protein[4] which we have recently cloned[5]. It binds both IL-1α and IL-1β, and is expressed in all cells responding to IL-1. Very recent work by Sims and colleagues has demonstrated that the N-terminus 319 amino acids have a structure reminiscent of the immunoglobulin superfamily with three external domains. There is a region consisting of 20 uncharged amino acids which is presumed to be the transmembrane region[5].

Within the past two years, investigations have led to the conclusion that IL-1 is identical to hematopoietin-1. Mochizuki and co-workers[6] found, using hematopoietin-1 from medium conditioned by the human bladder tumor cell line 5637 and an in vitro 5FU bone marrow proliferation assay, that hemopoietin-1 activity was indistinguishable from that of IL-1. HBT 5637 was shown to express mRNA coding for IL-1. Hematopoietin-1, co-purified with IL-1, and monoclonal antibodies directed against IL-1α neutralized the activity of hemopoietin-1.

In general, IL-1 was then believed to place stem cells in cycle, thus protecting against the effects of radiation and chemotherapy, to induce CSF production, to synergize with CSFs, and to induce CSF receptor expression.

IL-1 also has multiple other effects which may be significant in clinical applications. Among these are synergism with IL-2 to generate LAK activity (Grimm, personal communication). Work at Immunex has demonstrated that it induces fibroblast proliferation, acts as a chemo-attractant to leukocytes, and demonstrates angiogenic activity. In addition, we have shown that it potentiates the primary humoral response to a variety of antigens.

Figure 1 is a diagram showing the sites of action of IL-1 and other cytokines. The activity on IL-1 on lymphopoiesis appears to be limited to the generation of the lymphoid stem cell. The activities on the generation of other hemopoietic cells is much more complicated, and it appears to have effects in generating cells of varying maturity.

Some of the earliest work on IL-1 was done by Neta and co-workers[7,8] who investigated IL-1 as a radioprotectant. They found that when IL-1 was given 20 hours prior to irradiation, the mice became more radioresistant. The effect was dependent on the interval between IL-1 therapy and the dose of radiation. When IL-1 was given 45 hours or 4 hours preradiation, the effect was diminished, and dosing 1 hour after radiation did not show a protective effect.

These observations were confirmed in Talmadge's laboratory by Castelli and co-workers[9] and extended to treatment with cyclophosphamide. Mice given sublethal doses of cyclophosphamide were treated with IL-1 at varying times relative to the dose of cyclophosphamide. Optimal CFU-C recovery in the femur occurred when cyclophosphamide was given as a single dose 20 hours prior to radiation, or 48 hours after. Single doses at 2 hours and 24 hours post-cyclophosphamide showed decreased CFU-C on day 4, but increased CFU-C by day 8, when compared to the cyclophosphamide-treated controls. Treatment schedules beginning on day 1 or day 2 and extended over several days, resulted in increased numbers of CFU-C. These results supported the theory that IL-1 has protective (optimally when given 20 hours pre-cyclophosphamide or radiation) or restorative (when given once or more post-cyclophosphamide or radiation) effects on bone marrow progenitor cells.

In order to study whether the effects of pretreatment with IL-1 could be explained by the induction of other cytokines, Neta's group[7,10] compared treatment with IL-1 with treatment with other cytokines. IL-1 given 20 hours prior to radiation improved survival better than GM-CSF, gamma interferon or IL-2. Subsequent work[11] has demonstrated that in vivo, administration of IL-1 20 hours preradiation results in placing GM-CSF-responsive cells into cycle. This supports that theory that the pretreatment of cells with IL-1 prior to radiation (or chemotherapy) may confer its protective effect by placing the cells into cycle, as cells in S phase are said to be less sensitive to radiation than normal cells[12].

Morrissey and co-workers[13] have studied CFU-GM in mice which were sublethally irradiated, then treated b.i.d. with IL-1 intraperitoneally, b.i.d. IL-1 increased the number of CFU-GM over control animals which were untreated. In addition, on day 4, there were more cells in the IL-1-treated animals, which were responsive to GM-CSF, IL-3, and G-CSF, than seen in the control animals treated with these same cytokines. Peripheral polymorphonuclear leukocytes were increased on days 5 through 20. However, in this model, chronic IL-1 administration (postirradiation through day 20) led to hypoplasia of the thymus. MSA-treated control mice rapidly recovered thymic cellularity following radiation, unlike the IL-1-treated mice. In addition, IL-1 treatment resulted in diminishing the number of pre-B cells in the marrow. Responses of the spleen cells to T cell and B cell mitogens were decreased, suggesting that chronic IL-1 therapy may have limited clinical drawbacks.

Unpublished extension of this work (Morrissey, personal communication) has shown that in normal mice given IL-1 for 4 days, thymic cellularity was decreased by 90%, but recovered rapidly following cessation of treatment. In addition, serum corticosteroids increased. The decrease in thymocytes occurred in the CD4+/CD8+ population. They propose that the decrease in thymic cellularity is a result of the demonstrated increase in serum corticosteroids which is know to cause cell intrathymic death, presumably by activation of a calcium-dependent endonuclease. Finally, Neta and co-workers[14] studied the effect of combining CSFs on the survival of mice after lethal radiation. GM-CSF alone was ineffective, but in combination with IL-1 was more effective than IL-1 alone or G-CSF alone. The combination of G-CSF and IL-1, although better than either of the two alone, was not as good as IL-1 plus GM-CSF.

A number of other combination studies are currently being performed in order to clarify the effects of the cytokines, and how they may best be used in the clinic. Preliminary reports by Moore and co-workers (personal communication) indicate that IL-1 treatment may improve the hematologic reconstitution in the 5FU-treated mouse, and that combination with G-, GM-, and IL-3 may be better than single-agent therapy.

There is speculation that IL-1 may be useful in the setting of bone marrow transplantation. Recent studies with GM-CSF indicate that, at least in some clinical trials, it provides enhanced engraftment[15]. However, Blazar and co-workers (personal communication) in Minnesota have had less striking results. We speculate that this is due to the 4HC purging treatment that the center uses. 4HC dramatically decreases the population which is responsive to GM-CSF. Therefore, the use of IL-1 in this setting, which may help to increase the population of cells ranging from the pluripotent stem cell down to committed cells which can respond to GM-CSF, may be appropriate in this situation.

However, the use of IL-1 in the setting of malignancies will require studying various malignant cells to examine their potential for responding by proliferating to IL-1. Referring back to Figure 1, it would appear that the lymphoid malignancies might be particularly appropriate for IL-1 therapy as cells beyond the lymphoid stem cell are not believed to proliferate in response to IL-1. Thus, one possibility would be to treat bone marrow cells in culture with IL-1, either preceding or following treatment with an agent such as 4HC, or both. This might a) protect the stem cell from the effects of 4HC, and b) encourage the replication and differentiation (perhaps in combination with other cytokines) of these cells to enhance engraftment. A variety of other clinical settings can be imagined. However, given the time course of beneficial IL-1 effect we have seen in vivo in animal models it may be necessary to develop new chemotherapeutic or radiotherapeutic schedules to result in optimal effects.

References

1. March CJ, Mosley B, Larsen A, Cerretti DP, Breadt G, Price V, Gillis S, Henney CS, Kronheim SR, Grabstein K, Conlon PJ, Hopp TP, Cosman D (1985) Cloning, sequence and expression of two distinct human interleukin-1 complementary DNAs. Nature 315:641.

2. Hopp TP, Dower SK, March, CJ (1986) The molecular forms of interleukin 1. Immunol. Res. 5: 271.

3. Dower SK, Kronheim SR, Hopp TP, Cantrell M, Deeley M, Gillis S, Henney CS, Urdal DL (1986) The cell surface receptors for interleukin-1α and interleukin-1β are identical. Nature 324:266.

4. Dower SK, Kronheim SR, March CJ, Conlon PJ, Hopp TP, Gillis S, Urdal D, (1985) Detection and characterization of high affinity plasma membrane receptors for human interleukin-1. J. Exp Med 162:501.

5. Sims JE, March CJ, Cosman D, Widmer MB, MacDonald HR, McMahan J, Grubin CE, Wignall GM, Jackson JL, Call SM, Friend D, Alpert AR, Gillis S, Urdal DL, Dower SK (1988) cDNA expression cloning of the IL-1 receptor, a member of the immunoglobulin superfamily. Science 241: 585.

6. Mochizuki DY, Eisenman JR, Conlon PJ, Larsen AD, Tushinski RJ, (1984) Interleukin 1 regulates hematopoietic activity, a role previously ascribed to hemopoietin 1. PNAS USA 84:5267.

7. Neta R (1986) Cytokines in radioprotection. Comparison of the radioprotective effects of IL- 1 to IL-2, GM-CSF, G-CSF, and IFNλ Lymphokine Research 5: S105.

8. Neta R, Douches S, Oppenheim JJ (1986) Interleukin 1 is a radioprotector. J. Immunol. 136: 2483.

9. Castelli P, Black PL, Schneider M, Pennington R, Abe F, and Talmadge JE (submitted for publication)

10. Neta R, Oppenheim JJ, Douches SD, Giclas PC, Imbra RJ, Karin M (1986) Radioprotection with interleukin-1:comparison with other cytokines. Progress in Immunol. VI:900.

11. Neta R, Sztein MB, Oppenheim JJ, Gillis S, Douches SD (1987) The in vivo effects of interleukin 1. J. Immunol 139: 1861.

12. Boggs SS, Boggs DR (1975) Earlier onset of hematopoietic differentiation after expansion of the endogenous stem cell pool. Radiat. Res. 63:165.

13. Morrissey PJ, Charrier K, Bressler L, Alpert A (1988) The influence of IL-1 treatment on the reconstitution of the hemopoietic and immune systems after sublethal radiation. J. Immunol 14: 4204.

14. Neta R, Oppenheim JJ, Douches SD (1987) Interdependence of hrIL-1α, hr TNFα, HRG-CSF, and mr GM-CSF in radioprotection and in induction of the acute phase reactant fibrinogen. (abstr) Internat. Workshop on Monokines and other Non-Lymphocytic Cytokines. Hilton Head, SC, December, 1987.

15. Appelbaum F. (1988) Recombinant human granulocyte macrophage colony stimulating factor (GM-CSF) following autologous marrow transplantation in man. Abstr #892 Proc ASCO 7:231.

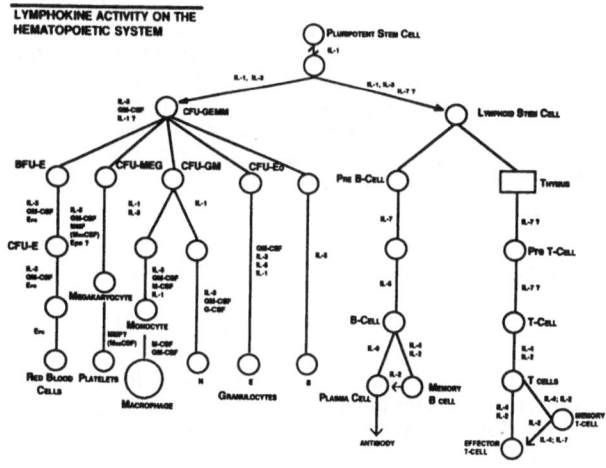

Fig. 1. Effects of various lymphokines and colony stimulating factors (CSFs) on the hematopoietic system. (Courtesy Immunex Corporation)

2 Regulation of Human B Cell Growth

J.L. Ambrus, Jr., A.S. Fauci, K.R. Young, Jr., P. McFarland, L. Chesky, H. Mostowski, M. Chandan, and F.M. Uckun

INTRODUCTION

The study of human B cell function involves examination of the events leading to maturation of B cells from B lineage lymphoid progenitor cells as well as the stages of activation, proliferation and differentiation which mature resting B lymphocytes must undergo to become immunoglobulin producing plasma cells [1-3]. In human B cell physiology (as opposed to murine B cell physiology) there appear to be molecules which act only at specific stages to induce either activation, proliferation or differentiation. However, as in murine B cell physiology, there are also molecules which can act at various stages. Antigen or anti-Ig will activate cells while B cell growth factors (BCGFs) [4-8] will induce their proliferation and B cell differentiation factors including interleukin-6 (IL-6) [9-11] will induce their differentiation. IL-2 in high concentrations can induce activation [12], proliferation, and differentiation [13] of human lymphocytes. In the next section, we will discuss some of our work examining the action of BCGFs in the regulation of normal human B cell proliferation.

Human B cell tumors are often clonal in nature and retain some of the maturational characteristics of the cell from which they were derived [14]. A similar observation has been made for Epstein Barr virus (EBV) transformed B cell lines [15]. Thus, these clonal B cell populations can be used to examine some of the regulatory elements involved in the earlier maturational stages of human B cell development. In the final section of this paper, we will discuss some of our work examining the role of human BCGFs in the regulation of growth of different human B cell tumors.

NORMAL HUMAN B CELL PROLIFERATION

The in vitro assays for normal human B cell proliferation are done using BCGFs either with anti-μ co-stimulation of resting B cells [3] or with cells which have been pre-activated either in vitro with Staphylococcus aureus cowan (SAC) [16] or in vivo and separated by counterflow centrifugation elutriation [17]. The use of pre-activated cells offers the advantage that the proliferative signal can be studied in the absence of the activating signal. Many lymphokines and other molecules have been described to enhance human B cell proliferation. A partial list is shown in Table 1.

Table 1. Molecules Described to Enhance Human B Cell Proliferation

Molecule	Comments
IL-1	minimal effects
IL-2	acts on ~50% activated B cells
IL-4	only effective with insoluble anti-μ; may inhibit proliferation induced by other lymphokines
IFN-α	some subtypes only
IFN-β	may have minimal effects
IFN-γ	reported to both enhance and suppress
TNF-α	
lymphotoxin	mostly enhances IL-2 induced proliferation
neuroleukin	does not work on SAC-activated B cells
B-BCGFs	heterogeneous; from normal SAC-activated and EBV transformed B cells
LMW-BCGF	apparently B cell specific
HMW-BCGF	apparently B cell specific; may be involved in expansion of memory B cells
complement component Bb	Antigenic and functional similarity to HMW-BCGF
complement component C3d	or any molecule properly binding CR2

Clearly, many of these molecules are noted for working on multiple cell types. We have focused our attention on high molecular weight-BCGF (HMW-BCGF) and low molecular weight (LMW-BCGF) because they appear to be not only B cell but also proliferation stage specific. HMW-BCGF and LMW-BCGF appear to act preferentially on different subpopulations of activated human B cells. HMW-BCGF is more active than LMW-BCGF on large in vivo activated B cells [17], long term in vitro grown B cells [18] and cells which are IgD negative before activation [unpublished data]. LMW-BCGF is more active than HMW-BCGF on small elutriated B cells [17], B cells from patients with common variable immunodeficiency [18], and cells which are IgD positive before activation (unpublished data).

The receptor for HMW-BCGF is a 90 kd heterodimer recognized by the monoclonal antibody BA5 [19]. The LMW-BCGF has not been structurally characterized and no antibodies to it have yet been published. Recent studies have shown that the HMW-BCGF receptor is phosphorylated on a tyrosine residue when bound by HMW-BCGF while no phosphorylation of the HMW-BCGF receptor is seen with either LMW-BCGF or IL-2 [20]. Multiple non-receptor proteins are also phosphorylated, however when activated B cells are stimulated with HMW-BCGF. We are currently identifying phosphorylated proteins to dissect the steps necessary for the proliferation event to occur.

While SAC-activation of human B lymphocytes leads to production of inositol phosphates and release of intracellular calcium [21], by 72 hours after SAC-activation, these cells have returned to a resting level of inositol phosphates and intracellular calcium. At this point, stimulation with either HMW-BCGF or LMW-BCGF results in the new generation of inositol phosphates and release of intracellular calcium [22]. Interestingly, calcium flux can also be demonstrated to occur in larger cells with HMW-BCGF than LMW-BCGF, to occur more in response to LMW-BCGF when IL-2R positive cells are chosen, and to be fully blocked by BA5 when HMW-BCGF is utilized (but not affected when LMW-BCGF is utilized) [23]. These experiments point out that calcium is reutilized for the proliferative event after being used for the activation event. Furthermore calcium flux occurs in response to HMW-BCGF or LMW-BCGF in different subpopulations of B cells. Thus HMW-BCGF and LMW-BCGF both provide signals capable of inducing proliferation but use different receptors (present on different subpopulations of B cells) and may also utilize some different intracellular signals to achieve different cell functions. One such difference found so far is the ability of HMW-BCGF but not LMW-BCGF to cause the accumulation of intracellular cAMP [22]. Additional means of approaching differences between the actions of HMW-BCGF and LMW-BCGF has been the evaluation of their effects on different malignant B cell populations.

MALIGNANT B CELL PROLIFERATION

Much of the work done by other laboratories with malignant B cells has been done with B cell lines. By definition, these lines represent clones of B cells which have developed whatever is necessary for completely autonomous growth. Most B cells from patients with B cell tumor will not, however, grow autonomously in culture and still require growth promoting signals to proliferate [24]. We have chosen to study bone marrow derived B lineage cells from patients with B cell precursor acute lymphocytic leukemia (BP-ALL), B lineage chronic lymphocytic leukemia (CLL), and B lineage malignant lymphoma to better understand the role of HMW-BCGf and LMW-BCGF in the regulation of their growth.

Fresh BP-ALL bone marrow blasts sorted with the CD19 antibody B43 and the CD10 antibody 24.1 were studied for proliferative responses, colony blast formation, and radiolabeled growth factor binding with HMW-BCGF and LMW-BCGF [25]. Seventy-eight percent of the 28 cases studied responded to LMW-BCGF while 57% responded to HMW-BCGF. All HMW-BCGF responsive cells also responded to LMW-BCGF. Many HMW-BCGF responsive cells had a structural chromosomal abnormality involving the 12p11-13 region of the short arm of chromosome 12. This may suggest a role for the K-ras2 oncogene in HMW-BCGF induced proliferation. Many patients also responded to IL-1 but responsiveness to multiple growth factors including IL-1, IL-2, HMW-BCGF and LMW-BCGF was found primarily in patients studied at relapse [26].

Fresh blood or bone marrow B cells from patients with CLL were studied in a similar manner [27]. Sixty percent of 10 patients formed in vitro colonies in response to HMW-BCGF while only 10% formed colonies in response to LMW-BCGF. HMW-BCGF responsive cells all stained with BA5 and demonstrated a similar 90 kd receptor to that seen with normal activated B cells. Lymphokines which failed to induce responses in these cells included IL-1, IL-2, IL-4, IL-5, and IFN-γ. Thus, HMW-BCGF appeared to be the major growth factor capable of expanding this population of B cells in the absence of other signals.

Five patients with B lineage malignant lymphoma were studied using a colony forming assay [25] in the presence or absence of HMW-BCGF or LMW-BCGF. The results are summarized below:

Table 2

Patient	Media	HMW-BCGF(ng/ml)				LMW-BCGF (ng/ml)			
		1	5	10	20	1	5	10	20
1(RS)	0	31	69	482	19	0	0	0	0
2(JR)	0	0	0	0	0	69	156	566	275
3(VF)	717	607	553	566	528	271	321	958	1008
4(DC)	343	571	597	630	672	270	222	371	394
5(LH)	0	0	0	0	0	0	0	0	0

Clearly, in these patients there is a heterogeneous response to HWM-BCGF and LMW-BCGF. Neither growth factor seems to predominate in enhancing the proliferation of these cells. Further, only one patient responded to both HMW-BCGF and LMW-BCGF. Additional lymphokines will have to be examined in this patient population.

When viewed as a whole many important points can be made from the study of these malignant B cell populations. In contrast to normal cells, malignant cells responding to BCGFs do not require an activation signal. In each disease category, the response to a given BCGF is heterogeneous and diferent populations tend to respond to different BCGFs. Furthermore, BCGFs which were described to act on mature activated human B lymphocytes can clearly stimulate pre-B cell as well as mature B cell derived malignant B cell populations. Analysis of the normal counterparts of these malignant B cell populations will have to be studied as well. The fact that certain groups such as CLL demonstrate such a strong response to HMW-BCGF (and contain apparently normal receptor for HMW-BCGF) suggests that HMW-BCGF and BA5 may be useful as carrier transport proteins for selective toxin or drug delivery to these malignant cells. Each patient would have to be studied individually before these therapeutic maneuvers were attempted, however.

CONCLUSIONS

While many lymphokines have been described to enhance human B cell proliferation in different assay systems, little is known about the true physiologic relevance of any of them. Two lymphokines which appear to be B cell specific and specific to the proliferative phase, HMW-BCGF and LMW-BCGF, were used to identify different subpopulations of normal and malignant human B lymphocytes. This will ultimately lead to a greater understanding of the role of each of these growth factors as well as the populations which they expand. Preliminary work utilizing HMW-BCGF and LMW-BCGF has also suggested both similarities and differences in the intracellular signals utilized by these growth factors in normal human B lymphocytes. This will help to dissect which signaling pathways are necessary for particular B cell functions. Clearly calcium plays a role in the induction of proliferation while cAMP plays a role in inhibiting terminal differentiation. Further studies will be required to determine where and how these signals are acting. Ultimately, expansion of these studies to malignant B cell populations will help to dissect the events leading to autonomous growth.

The authors would like to acknowledge the editorial assistance of Mary Rust. This work was supported in part by R29 CA 42111, R01 CA 42633 and P01 CA 21737 awarded by the National Cancer Institute. F.M. Uckun is recipient of a FIRST award from NCI and Special Fellow of the Leukemia Society of America.

REFERENCES

1. Dutton RW, Falkoff R, Hurst JA, Hoffman M, Kappler JW, Ketmann JR, Lesley JR (1971) Is there evidence for a non-antibody specific diffusable chemical mediator from the thymus-derived cell in the initiation of the immune response? Prog Immunol 1:355.

2. Kishimoto T, Miyake T, Nishizawa Y, Watanabe T, Yamamura Y (1975) Triggering mechanism of B lymphocytes. I. Effect of anti-immunoglobulin and enhancing soluble factor on differentiation and proliferation of B cells. J Immunol 1555:1179.

3. Ambrus JL Jr, Jurgensen CH, Bowen DL, Tomita S, Nakgawa T, Nakagawa N, Goldstein H, Witzel NL, Mostowski HS, Fauci AS (1987) The activation, proliferation, and differentiation of human B lymphocytes. Adv Exp Med Biol 213:163.

4. Ambrus JL Jr, Jurgensen CH, Brown EJ, Fauci AS (1985) Purification to homogeneity of a high molecular weight human B cell growth factor, demonstration of specific binding to activated B cells, and development of a monoclonal antibody to the factor. J Exp Med 162:1319.

5. Mehta SR, Conrad D, Sandler R, Morgan J, Montagna R, Maizel AL (1985) Purification of human B cell growth factor. J Immunol 135:3298.

6. Gordon J, Guy GR (1987) The molecules controling B lymphocytes. Immunol Today 8:339.

7. Delfraissay JD, Wallen C, Vazquez A, Dugas B, Dormant J, Galanaud P (1986) B cell hyperactivity in systemic lupus erythematosus: selectively enhanced responsiveness to a high molecular weight B cell growth factor. Eur J Immunol 16:1251.

8. Butler JL, Ambrus JL Jr, Fauci AS (1984) Characterization of monoclonal B cell growth factor (BCGF) produced by a human T-T hybridoma. J Immunol 133:251.

9. Kishimoto T, Yoshizaki K, Kimoto M, Okada M, Kuritani T, Kikutani H, Shimizu K, Nakagawa T, Nakagawa N, Miki Y, Kishi H, Fukunaga K, Yoshikubo T, and Tag T (1984) B cell growth and differentiation factors and mechanism of B cell activtion. Immunol Rev 78:97.

10. Goldstein H, Volkman DJ, Ambrus JL Jr, Fauci AS (1985) Characterization of a T4+/Leu8+ T cell clone that directly helps B cell Ig production by secreting B cell differention factor. J Immunol 135:339.

11. Hirano T, Yasukawa K, Horada H, Taga T, Watanabe Y, Matsuda T, et al (1986) Complementary DNA for a novel interleukin (BSF-2) that induces B lymphocyts to produce immunoglobulin. Nature 324:73-6.

12. Le thi Bich Thuy, Lane HC, Fauci AS (1986) Recombinant interleukin-2 induced polyclonal proliferation of in vitro unstimulated human peripheral blood lymphocytes. Cell Immunol 98:396-410.

13. Waldmann TA, Goldman CK, Robb RJ, Depper JM, Leonard WJ, Sharrow SO, Bongiovanni KF, Korsmeyer SJ, Greene WG (1984) Expression of interleukin-2 receptors on activated human B cells. J Exp med 160:14-50.

14. Greaves MF (1986) Differentiation-linked leukoemogenesis in lymphocytes. Science 234:697-704.

15. Aman P, Ehln-Henriksson B, Klein G (1984) Epstein Barr virus susceptibility of normal human B lymphocyte populations. J Exp Med 157:208-20.

16. Muraguchi A, Fauci AS (1982) Proliferative responses of normal human B lymphocytes. Development of an assay for human B cell growth factor (BCGF). J Immunol 129:1104.

17. Ambrus JL Jr, Fauci AS (1985) Human B lymphoma cell line producing B cell growth factor. J Clin Invest 75:732-9.

18. Ambrus JL Jr, Fauci AS (1987) Current studies examining regulation of the human B cell cycle. In: Webb DR, Pierce CW, Cohen S (eds) Molecular basis of lymphokine action, Clifton, New Jersey: Humana Press, 137-148.

19. Ambrus JL Jr, Jurgensen CH, Brown EJ, McFarland P, Fauci AS (1988) Identification of a receptor for high molecular weight B cell growth factor. J Immunol 141:660.

20. McFarland P, Ambrus JL Jr, Mostowski H, Fauci AS (1988) Phosphorylation of membrane proteins by activated human B lymphocytes in response to a high molecular weight B cell growth factor. Fed proc 2:A165 (7869).

21. Young KR Jr, Ambrus JL Jr, Chesky L, Chused T, Fauci AS (1988) Heterogeneity of transmembrane signaling pathways triggered by human B lymphocyte activating agents. Fed Proc 2:A1230 (5429).

22. Ambrus JL Jr, Chesky L, McFarland P, Young KR, Jr, Brown EJ, Peters M, Uckun F, Ledbetter J, Fauci AS (1988) High molecular weight B cell growth factor (HMW-BCGF) activtes two distinct pathways of inracellular signaling in SAC-activated human B lymphocytes. Fed Proc 2:A1466 (6796).

23. Ambrus JL Jr, Chused T, Uckun FM, Young KR Jr, Fauci AS (1988) Human B cell growth factors induce calcium fluxes in different subpopulations of activated human B lymphocytes. Manuscript in review.

24. Furth J (1953) Conditoned and autonomous neoplasms: a review. Cancer Res 13:477-92.

25. Uckun FM, Fauci AS, Heerema NA, Song CW, Mehta SR, Gajl-Peczalska KJ, Chandan M, Ambrus JL Jr (1987) B-cell growth factor receptor expresion and B cell growth factor response of leukemic B cell precursors and B lineage lymphoid progenitor cells. Blood 70:1020-34.

26. Uckun FM, Meyers D, Fauci AS, Ambrus JL Jr (1988) Leukemic B cell precursors constitutively express function receptors for human IL-1. Manuscript in review.

27. Uckun FM, Fauci AS, Chandon M, Ambrus JL Jr (1988) Detection and characterization of human B cell growth factor receptors on leukemic B cells in chronic lymphocytic leukemia. Manuscript in review.

3 Mouse Bone Marrow Cells Which Require Multiple Growth Factors for Proliferation: Examination of the Effects of Interleukin 1, Serum Factors and Inhibitor(s)

T.R. Bradley, I. Bertoncello, A.B. Kriegler, and G.S. Hodgson

INTRODUCTION

Bone marrow (BM) cells from 5-fluorouracil treated mice have been useful as model populations in hemopoietic studies because (a) early after FU treatment (2 day) the in vitro colony-forming-cells (CFC) responsive to single lineage growth factors (GF) are drastically depleted and only reappear late in the regeneration of the marrow, and (b) multiple GF responsive CFC are less depleted, develop large colonies in vitro (high proliferative potential, HPP-CFC) and exhibit early regeneration to supranormal levels at 8-10 days post FU [9,10,11]

The growth factors necessary for proliferation of HPP-CFC are (a) a basic requirement for the macrophage GF (M-CSF: CSF-1) [28] together with (b) synergistic factor(s), which have been derived from various sources eg. mouse, rat or human spleen conditioned medium (CM) [9,10] human placenta CM [21] and CM from the human bladder carcinoma cell line, 5637. CM from 5637 cells was used to purify hemopoietin-1 [20] which, with CSF-1, generated cells from FUBM which were used in a radioreceptor assay for H-1 [2]. The same assay system was used to show synergism of CSF-1 and IL-3 [3].

Highly enriched cell populations are necessary for assays of GF's in order to establish whether they are acting directly to produce proliferation or by induction of other factors from accessory cells. Bertoncello et al [6] purified HPP-CFC from regenerating FU8day BM by fluorescent activated cell sorting (FACS) using monoclonal antibody to the cell surface differentiation marker Qa-m7 [27]. In that study the HPP-CFC were enriched six fold resulting in a plating efficiency of 21%. For optimal proliferation the HPP-CFC required the presence of at least three growth factors in the agar colony assay, namely CSF-1 plus interleukin-3 plus 5637 CM.

Mochizuki et al [23] have produced evidence, in studies of purification of 5637 CM, that interleukin-1 (IL-1) copurified with hemopoietin -1 (H-1). From these two studies the questions were posed (a) could HPP-CFC be enriched from BM sampled early after FU treatment, (eg. FU_2day) analogous to those from FU8day BM? and (b) could interleukin-1 replace the 5637 CM used in those assays?

In initial studies Bartelmez et al [4] have used FU2day BM cells enriched by FACS with monoclonal antibodies to cell surface antigens 7/4, B220 and L3T4 to demonstrate optimal large colony development in response to CSF-1 plus IL-3 plus IL-1 when plated at low cell densities. In the present studies target cell populations suitable for testing the GF and inhibitor activities in mouse sera at low cell densities in nutrient agar cultures have been prepared by negative immunomagnetic bead sorting of FU2day BM.

METHODS

Enrichment of HPP-CFC from bone marrow

FU_2day BM cells were sampled from femurs of $(C_{57}Bl_6xDBA_2)F-1$ male mice after 200 mg/kg intravenous FU. The mice were reared and maintained in specific pathogen free conditions. The BM cells were enriched for HPP-CFC using an immunomagnetic bead purification procedure as described by Bertoncello et al [7]. Briefly the procedure was as follows: The marrow cells were dispersed and the low density (<1.085 gm/cm^3) cells isolated on discontinuous density gradient centrifugation using Nycodenz (Nyegaard Co. Oslo, Norway). The cells were then labelled with an anti-7/4, anti B-220 antibody cocktail using rat monoclonal antibody against the 7/4 murine differentiation antigen [8] (obtained from Dr. S. Gordon, Sir William Dunn School of Pathology, U. of Oxford England) and the RA3.6B23 rat monoclonal antibody against B220 antigen [13] (from Dr. W. Langdon, Walter and Eliza Hall Institute, Melbourne). Unconjugated affinity purified goat anti-rat IgG was purchased from Kpl Inc. (Gaithersberg, Md. USA) and covalently bound to uncoated monodisperse magnetic polystyrene beads (M-450 Dynabeads, Dynal AS, Norway). Negative selection was carried out by incubation of the cells with beads at 4 C for 45 minutes and the rosetted cells separated with a Dynal MPC-1 magnetic particle separator for 2 minutes. Non-rosetted cells were harvested, washed and counted.

In vitro assay: HPP-CFC

Nutrient agar cultures were made using the double layer technique (1.0 ml underlay, 0.5 ml overlay) modified as described by Bradley et al [12]. All GF's and serum samples were incorporated in the underlays and the target cells in the overlay. Cultures were incubated in 5% O_2, 10% CO_2: 85% N_2 for 14 days and colonies scored as <0.5 or >0.5 mm diameter.

Growth Factors

Recombinant human interleukin-1_α (r-h-IL-1) was supplied by courtesy of Dr. P. Lomedico (Hoffman La Roche) at a specific activity of 2.5×10^9 units per mg. protein and containing less than 0.3 endotoxic units per mg (LAL). Recombinant human IL-2 (r-h-IL-2) was supplied by Hoffman La Roche. Recombinant murine interleukin-3 (r-mu-IL-3) [15] was kindly supplied by Drs. A. Hapel and I. Young. (Australian National University, Canberra) at 2.3×10^6 units per mg protein. CSF-1 was purified from pregnant mouse uterus extract [8] by specific immuno-sorbent column chromatography column [31] by E.R. Stanley (Albert Einstein College of Medicine, New York) at 8×10^7 units per mg. protein. Murine GM-CSF (r-mu GM-CSF), [16,26] human G-CSF (r-hu-G-CSF) [25] at 3×10^4 and 1×10^5 units per μg. protein respectively and anti-IL-1_α antibody were kindly supplied by Immunex Corporation, Seattle, Washington. Anti-CSF-1 antibody [30,31] was kindly supplied by E.R. Stanley.

Sera

Sera from groups of at least 10 mice were collected after mild penthrane anaesthesia and sacrifice. Where post IL-1_α sera were collected the mice were routinely injected intravenously with 1×10^6 units in 0.9% saline plus 0.1% BSA (Sigma A3156). Control mice received vehicle only. C3H/HeJ male mice were used routinely for in vivo treatment experiments.

RESULTS

The ability of the FU2day 7/4, B220 negative BM cells to form colonies in nutrient agar cultures in the presence of CSF-1, IL-3 and IL-1_α was tested. As shown in Table 1 none of the individual GF's alone were effective in stimulating large colony formation. Except for an occasional large colony, combination of

any two factors was not effective for HPP-CFC colony development, however IL-1_α plus IL-3 grew some small colonies. The optimal development of large colonies required the presence of the three factors, CSF-1 plus IL-3 plus IL-1 (designated triple GF) at the optimal concentration of CSF-1 1000 units, IL-3 25 units, IL-1_α 5000 units. Substitution of r-mu-GM-CSF (1000 units per dish) for CSF-1 resulted in 50% of the large colony development obtained with CSF-1. Addition of r-hu-G-CSF (250 ng) or r-mu-GM-CSF or both to the triple GF did not influence colony development numerically. Addition of the high concentration of r-hu-G-CSF tended to spread out the colonies but did not result in increased numbers of cells per colony. Addition of r-hu-IL-2, (1500 units per dish) depressed large colony formation by 86%. Incubation of the FU2day BM cells with triple GF plus anti-IL-1_α negated 97% of the large colony development whilst anti-CSF-1 completely stopped the large colony development except for the colonies developed with IL-1_α plus IL-3. Increases in the concentrations of each of the three GF's to 32 fold the optimal concentration did not result in further significant colony development although the colonies were often slightly larger.

Since the in vitro data suggest that IL-1_α is involved in the proliferation of early cells of the hemopoietic system the effects of IL-1_α in vivo have been tested. Initially attempts were made to estimate the half-life of IL-1_α in the sera of mice. Normal C3H/HeJ mice were injected intravenously with 1×10^6 units (400 ng) of IL-1_α and sera sampled at various times after the injection. The sera were then tested in vitro for IL-1-like activity using the immunomagnetic bead enriched FU2day BM cells in the presence of optimal concentrations of CSF-1 plus IL-3. Standard curves of optimal CSF-1 plus IL-3 together with six concentrations of IL-1_α (62.5 to 2000 units per dish) were tested for comparison with the sera. The results are shown in Fig. 1.

Normal mouse serum contained a just detectable concentration of IL-1-like activity and exhibited complete lack of colony formation at the highest dose tested (0.2 ml) per 1.5 ml volume per dish). Five minutes after IL-1_α injection the sera contained IL-1-like activity equivalent to 120,000 units per ml. of serum; at 15 minutes 35,000 units; 60 minutes 37,500 units; 120 minutes 70,000 units; 240 minutes 120,000 units; 360 minutes 58,000 units; 480 minutes 35,000 units and 24 hours 30,000 units compared with the optimal concentration in normal serum of 2500 units per ml of serum. A second observation from the results was that at 1 hour partly and 2, 4 and 6 hours completely, the highest concentration of the sera, 0.2 ml, showed lack of inhibitory activity and large colonies were formed.

Inhibitory activity was again observed at 8 and 24 hours after the IL-1_α treatment. Also noteable is the fact that the 4 and 6 hour post IL-1_α sera not only lacked inhibitory activity but also increased colony formation compared with the optimal concentration of triple GF (CSF-1 1000 units; IL-3 25 units; IL-1_α 5000 units).

Table 1. Clonal growth of two day post 5-fluorouracil mouse bone marrow cells with growth factor combinations.

Growth factors	Colonies (per 1000 cells plated)*	
	< 0.5 mm	> 0.5mm
CSF-1	0.2 + 0.1 **	0
IL-1$_\alpha$	0	0
IL-3	0.3 + 0.2	0
IL-1$_\alpha$ + IL-3	5.1 + 0.6	0.1 + 0.1
CSF-1 + IL-3	2.8 + 0.7	0.5 + 0.3
CSF-1 + IL-1$_\alpha$	0.8 + 0.4	0.3 + 0.1
CSF-1 + IL-3 +IL-1$_\alpha$	5.0 + 0.7	19.5 + 2.7

*   (C57B1₆xDBA₂) F1 male mice: BM 7/4, B220 negative: (immuno-bead enrichment).
** Means $\pm$ SE: 9 replicate experiments.   CSF-1, 1000 units: IL-3, 25 units: IL-1$_\alpha$, 5000 units.

The sera were retested at four concentrations, 0.2, 0.1, 0.05, 0.025 ml per dish, specifically to measure CSF-1, IL-3 and IL-1-like activities on the FU2day BM target cells in the presence of optimal concentrations of the other two GF's. Thus, CSF-1-like activity in serum was measured in the presence of IL-1$_\alpha$ + IL-3, IL-3 like activity in the presence of CSF-1 + IL-1$_\alpha$ and IL-1 in the presence of CSF-1 + IL-3. In addition, in replicate dishes assaying for CSF-1 like activity an optimal concentration of specific anti-CSF-1 antibody was incorporated, likewise in replicate dishes assaying for IL-1-like activity anti-IL-1$_\alpha$ was incorporated, so these cultures specifically tested for CSF-1 and IL-1$_\alpha$ respectively. Since no specific neutralizing antiserum was available for testing IL-3 like activity it was tested using FDCP-1 cells and 32D cells for clonal growth in agar. The concentrations of the factors were

determined by comparison with standard curves for each of the growth factors. The results are shown in Fig. 2.

Antibody substantiated CSF-1 decreased slightly at 5 minutes, rose to a peak at 2 hours and then decreased steadily to near normal at 8 hours post IL-1$_\alpha$ injection. The concentration of IL-1$_\alpha$, substantiated with anti-IL-1$_\alpha$, peaked at 5 minutes after the IL-1$_\alpha$ injection and decreased steadily until it was undetectable at 2 hours, a serum half life of approximately 12 to 15 minutes. However, from 1 hour after the IL-1$_\alpha$ injection, IL-1-like activity was generated, which was not negated by anti-IL-1$_\alpha$, peaking at 4 hours and then decreasing steadily after this. None of the sera exhibited any ability to grow FDCP-1 or 32D cells despite the fact that a steadily increasing value for IL-3 like activity up to 5 hours was detected in the sera

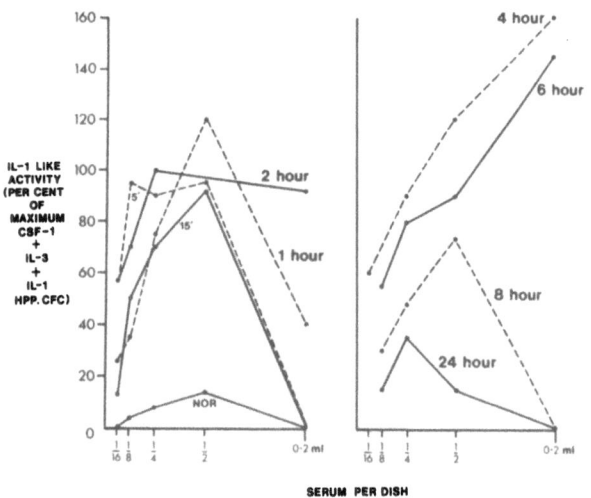

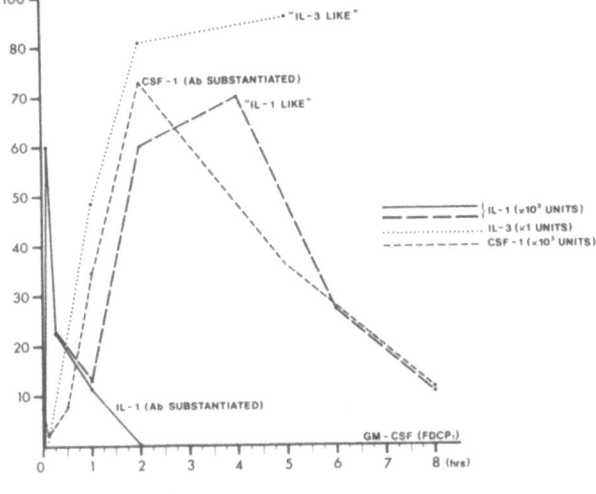

Fig. 1. The effects of addition of graded concentrations of C3H/HeJ sera sampled at various times after injection of IL-1$_\alpha$ on HPP-CFC colony development in the presence of CSF-1 plus IL-3.

Fig. 2 Assays of growth factors in the sera of C3H/HeJ mice at various times after injection of IL-1$_\alpha$

using the triple GF clonal FU2day BM assay. Control incubation of FDCP-1 cells with GM-CSF (5 units) and 32D cells with IL-3 (12.5 units) resulted in 74% and 25% plating efficiencies respectively in the same experiment.

The same sera were assayed in the presence of optimal concentrations of all three GF's, an assay to detect inhibitory activity or any further synergistic activity in the sera. Fig. 3 shows the results obtained using the concentration of 0.2 ml of the sera per dish. Confirmatory of the results shown in Fig. 1, inhibitory activity was present in normal serum, was unchanged at 5 minutes, decreased at 1 and 2 hours, absent or overriden at 4 and 6 hours and reappeared at 8 hours and 24 hours post IL-1_α injection.

The inhibitory activity was tested to see whether it could be overriden in vitro. By increasing the concentration of all three growth factors 32 fold, the inhibitory activity of a sample isolated from normal serum on Sephacryl S200 chromatography was overriden completely when a concentration, which normally exhibited 50% inhibition of HPP-CFC colony formation, was used. At the higher concentration of inhibitor (100% inhibition) even the large increase in triple GF's was not effective. (Table 2).

DISCUSSION

Using an enriched FU2day BM cell population as an agar culture assay system two principal changes were detectable in sera following the treatment of normal mice with one injection of IL-1_α. These changes were (a) increases in GF activities and (b) decreases in inhibitory activity, both of which exhibited approximately the same time relationship to the administration of the IL-1_α.

The in vitro cloning assay has detected two IL-1-like activities present in mouse serum after injection of IL-1_α. The first with a rapid peak and rapid disappearance is substantiated by the use of specific anti-IL-1_α to be IL-1_α. In this assay the estimate of the half life was approximately 12 - 15 minutes, which agrees reasonably with that of Dinarello et al[14] who estimated the half life of radiolabelled recombinant human IL-1_α in rabbits as 5.5 minutes. The second IL-1 like activity is evident at 1 hour after the IL-1_α injection, rises to a peak at 4 hours and then decreases progressively until the sample taken at 8 hours. A similar phenomenon was studied in rabbits by Dinarello et al [14] with a second IL-1-like activity being detected in plasma 3 hours after IL-1_α injection. Chromatographic analysis of the plasma suggested that the second activity was a newly synthesized IL-1 of rabbit origin. From further in vitro experiments these workers have substantiated a hypothesis that the second IL-1 activity is IL-1_β. Such an explanation is possible in our experiments since IL-1_β can effectively substitute for IL-1_α in the cloning assay [4]. However formal proof that the second IL-1-like activity generated in mouse serum is IL-1_β is still required.

Antibody substantiated CSF-1 in normal serum contained about 5000 units per ml, an estimate which agrees with that of Bartocci et al [5] using a radioimmunoassay. The concentration decreased at 5 minutes, then rose steadily to a peak at 2 hours and decreased to a near normal serum value at 8 hours post injection. Other workers have shown increases of colony-stimulating factor(s) in sera of mice.

Table 2. Antagonism of inhibitor(s) by growth factors in vitro.

Growth Factors (units)		%Inhibition* Inhibitor concentration (µl)				
		100	50	25	12.5	6.25
CSF-1 1000⎫ IL-3 25 ⎬x1 IL-1_α 5000⎭		100	100	61	50	7
"	x2	100	–	–	10	–
"	x4	100	–	–	32	–
"	x8	100	–	–	22	–
"	x16	100	–	–	25	–
"	x32	100	–	–	0	–

*   Inhibition of FU2day BM HPP-CFC. Control value triple GFx1=19 colonies per 1000 cells plated.

**   Inhibitor from normal serum: S200 Sephacryl. µl. per dish.

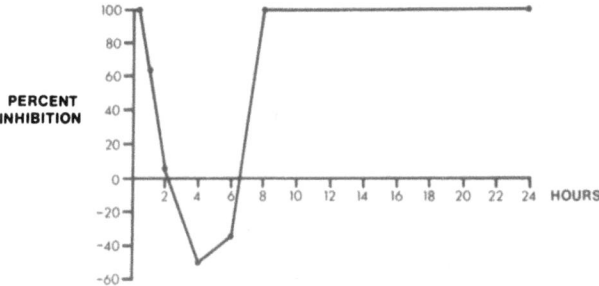

Fig. 3. Inhibition of FU2day BM (enriched) HPP-CFC colony development in vitro in response to CSF-1 + IL-3 + IL-1_α by sera (0.2 ml) from normal and IL-1_α treated mice (hrs. after IL-1_α) Control value;15 colonies;1x10³ cells plated.

following IL-1α injection. Vogel et al [33] detected a peak of activity at 2 - 3 hours which then decreased steadily over 24 hours after the injection of recombinant murine or human IL-1α. No criteria were applied to the assay system (colony assay using normal mouse BM cells at a cell density of 1x10^5 per ml) which permit an unequivocal proof of the type of GF induced. Stork et al [32] also demonstrated a colony stimulating factor in serum of mice following single injection of recombinant human IL-1α with a peak of activity at 6 hours but no characterization of the activity was carried out.

The third activity disclosed in sera following IL-1 injection was an IL-3 like activity as judged by the triple GF agar colony assay. However the results with FDCP-1 cells (responsive to both GM-CSF and IL-3) and with 32D cells (responsive to IL-3) showed neither GM-CSF nor IL-3 activities in the sera. One possible explanation was considered that these cells were inordinately sensitive to the inhibitor present in normal, 5 minute, 15 minute and 1 hour serum samples, but the 2, 4 and 6 hour samples do not exhibit inhibitory activity and it seems unlikely that a specific inhibitor for FDCP-1 and 32D cells exists in the 2, 4 and 6 hours sera at high concentrations. Thus the inference at present is that the IL-3 like activity is in fact another GF which is capable of acting with CSF-1 plus IL-1 in the cloning assay. The time course of production of IL-3 like activity was that it was absent in normal serum, rose steadily to 2 hours post IL-1 and was still slightly increasing at 6 hours. Since Ikebuchi et al [19] have demonstrated that recombinant IL-6 can act as a synergistic factor with IL-3 in colony growth on unpurified FU2day bone marrow cells but was not capable of inducing colony growth by itself, it is possible that the IL-3 like activity generated post IL-1α may be IL-6 or a similar factor. As far as we are aware IL-6 hasnot been shown to stimulate FDCP-1 or 32D cells. The identity of this post IL-1α serum factor(s) remains to be established.

The second facet of changes in the sera of mice following IL-1α injection is the decrease and then apparent absence of inhibitor at just over 3 hours. One possible explanation of this would be that an increase of GF's in the sera antagonize the inhibitory activity and hence exhibit overall stimulating activity in vitro. The question of whether this could be the mechanism depends of the concentration of GF's generated. The actual increases of all of the factors concerned in serum cannot, at the present time, be calculated in terms of the standard cocktail of triple GF's. Certainly the concentration of inhibitory activity partially purified from normal serum by Sephacryl S200 chromatography, which exhibits 50% inhibition of the triple GF colony forming ability, was antagonized by increasing the three factors to 32 times their normal optimal concentrations. However initial studies of the chromatographic isolation of inhibitory activity from normal versus 5 hours post IL-1α sera show that the 5 hour sera do not contain fractions exhibiting inhibitory activity, which suggests an active disappearance of inhibitor(s) rather than GF

antagonism. The point requires further investigation but at present the data have been interpreted to indicate that IL-1α antagonised the production of inhibitor by some cell type(s) as yet not identified.

The nature of the inhibitor has not yet been investigated beyond initial investigations identifying active fractions from Sephacryl S200 chromatography as having molecular weights in the range of 40,000 - 100,000 dalton The fact that the inhibitory activity disappears after IL-1α treatment of mice may be useful to helping ascertain whether its chemical nature is similar to those known to inhibit hemopoietic proliferation in vitro e.g. interferons, transforming growth factor prostaglandins, acidic isoferritins, or inhibitory pentapeptides [see 17 for a brief review].

Other workers have produced evidence of involvement of IL-1α in large colony development. Moore and Warren [24] found some HPP-CFC colony development with unfractionated FU$_2$day BM using IL-1 plus CSF-1, IL-3 or GM-CSF but the values quoted are for the totals of marrow plus spleen colonies and the actual numbers of colonies for BM alone are not described. Zsebo et al [34] have used low density (Ficoll-Paque) non-adherent cells of FU2day BM (BALB/C mice), which were plated at high density (1x10^5 cells/ml) to demonstrate the synergism of CSF-1 or IL-3 with IL-1α or β. The HPP-CFC enriched population from FU2day BM used in the present studies and in the work of Bartelmez et al [4] demonstrate a population which unequivocally requires at least three growth factors (CSF-1, IL-3, IL-1) for proliferation in vitro and provides an accurate in vitro biological assay system using low cell densities for (a) investigation of GF actions and for (b) studies of the hierarchy in marrow cells. The immunomagnetic bead enrichment is suitable for any mouse strains bearing the 7/4 antigen (eg. C57B1, AKR, DBA2, NZB, SJL) but only for hybrids with those (BALB/c, C3H,CBA) on which it is either poorly expressed or not detectable.

The timing of GF induction and inhibitor suppression after in vivo administration of IL-1α together with the fact that an increased GF concentration (x32 times optimal) can overrride a low concentration of inhibitory activity suggests that the normal control of hemopoiesis depends on the prevailing ratios of GF's: inhibitor(s) and that both of these factors may have to be taken into consideration in attempts to stimulate hemopoiesis in situations such as post-cytotoxic drug administration or transplantation of BM.

ACKNOWLEDGEMENTS

We gratefully acknowledge technical asssistance from L. Barber, S. Verschoor B. Williams and P. Graham in marrow cell preparation and plating, cell enrichment by immunomagnetic beads and partial preparation of CSF-1 and Allison Corry for preparation of this manuscript.

Supported by grants from National Cancer

Institute, U.S.A,. Ca32551; National Health
& Medical Research Council of Australia,
870868; Anti-Cancer Council of Victoria;
Integrated Genetics, Framingham, Massachusetts,
U.S.A

REFERENCES

1. Auron PE, Webb AC, Rosenwasser LJ, Mucci CF,
   Rich A, Wolf SM, Dinarello CA: 1984.
   Nucleotide sequence of human monocyte
   interleukin-1 precursor cDNA. Proc. Natl.
   Acad. Sci. (USA) 81:7907

2. Bartelmez SH, Stanley ER: 1985. Synergism
   between hemopoietic growth factors (HGF's)
   detected by their effects on cells bearing
   receptors for a lineage specific HGF: Assay
   of Hemopoietin-1. J. Cell Physiol. 122:370.

3. Bartelmez SH, Sacca R. and Stanley ER:1985.
   Lineage specific receptors used to identfy
   a growth factor for developmentally early
   hemopoietic cells: Assay for hemopoietin-2.
   J. Cell Physiol. 122:362.

4. Bartelmez SH, Bradley TR, Bertoncello I,
   Mochizuki DY, Tushinski RJ, Stanley ER,
   Hapel AF, Young IG, Kriegler AB, Hodgson GS:
   1988. Interleukin-1 plus interleukin-3 plus
   colony stimulating factor-1 are essential for
   clonal proliferation of primitive myeloid
   bone marrow cells (submitted for publication).

5. Bartocci A, Pollard JW, Stanley ER: 1986.
   Regulation of colony-stimulating-factor-1
   during pregnancy. J. Exp. Med 164:956.

6. Bertoncello I, Bartelmez SH, Bradley TR,
   Hodgson GS: 1987. Increased Qa-m7 antigen
   expression is characteristic of primitive
   hemopoietic progenitors in regenerating marrow.
   J. Immunol. 139:1096.

7. Bertoncello I, Bradley TR, Hodgson GS: 1988 The
   concentration and resolution of primitive
   hemopoietic cells from normal mouse bone
   marrow by negative selecti on using monoclonal
   antibodies and dynabead immunodisperse
   magnetic microspheres (Submitted for
   publication).

8. Bradley TR, Stanley ER, Sumner MA: 1971.
   Factors from mouse tissues stimulating colony
   growth of mouse marrow cells in vitro. Aust.
   J. Exp. Biol. Sci., 49:595.

9. Bradley TR, Hodgson GS: 1979 Detection of
   primitive macrophage progenitor cells in
   mouse bone marrow blood, 54,1446.

10. Bradley TR, Hodgson GS, Bertoncello I: 1980.
    Characteristics of primitive macrophage
    progenitor cells with high proliferative
    potential: their relationship to cells with
    narrow repopulating ability in 5-fluorouracil

treated mouse bone marros. In Exptl.
Hematology Today (1979) Ed. Baum S.J.
Ledney, G.D, Van Bekkum D. S. Karger, New
York, p 285.

11. Bradley TR, Hodgson GS, Kriegler AR, McNiece
    IK: 1985. Generation of CFU-S13 in vitro.
    In progress in Clinical and Biological
    Research.' "Hemopoietic Stem Cell Physciology"
    Ed. Cronkite, Dainiak, McCaffrey, Polek
    and Queensberry. Alan R. Inc., New York.
    184:39.

12. Bradley TR, Hodgson GS, Rosendaal M: 1978.
    The effect of oxygen tension on haemopoietic
    and fibroblast cell proliferation in vitro.
    J. Cell Physiol. 94; 517.

13. Coffman RL: 1982. Surface antigen
    expression and immunoglobin gene
    rearrangement during mouse pre-B cell
    development. Immunol. Rev. 69:5.

14. Dinarello CA, Ikejima T, Warner SJC,
    Orencole SF, Lonneman G, Cannon JG, Libby P:
    1987. Interleukin 1 induces interleukin 1.
    1. Induction of circulating interleukin
    1 in rabbits in vivo and in human
    mononuclear cells in vitro. J. Immunol.
    139: 1902.

15. Fung MC, Hapel AJ, Ymer S, Cohen DR,
    Johnson RM, Campbell HD, Young ID: 1984.
    Molecular cloning of cDNA for murine
    interleukin-3. Nature (London) 307:233.

16. Gough NM, Metcalf D, Gough J, Geach D,
    Dunn AR: 1985. Structure and expression of
    the mRNA for murine granulocyte-macrophage
    colony stimulating factor. EMBO J., 5:1193.

17. Guignon M, Najman A: 1988. The inhibitors of
    hematopoiesis. Int. J. Cell Cloning 6:69

18. Hirsh S, Gordon S: 1983. Polymorphic
    expression of a neutrophil differentiation
    antigen revealed by monoclonal antibody 7/4.
    Immunogenetics, 18:229.

19. Ikebuchi K, Wong GG, Clark SC, Ihle JN,
    Hirai Y, Ogawa M: 1987. Interleukin 6 enhance-
    ment of interleukin 3 -dependent proliferation
    of multipotential hemopoietic progenitors. Proc
    Nat Acad Sci USA 84:9035.

20. Jubinsky PT, Stanley ER: 1985. Purification
    of hemopoietin-1, a multilineage hemopoietic
    growth factor. Proc. Nat. Acad. Sci.,
    82:2764.

21. Kriegler AB, Bradley TR, Janusceiwicz E,
    Hodgson GS, Elms EF: 1982. Partial
    purification and characterization of a
    growth factor for macrophage progenitor cells
    with high proliferative potential in mouse
    bone marrow. Blood 60:503.

22. March CJ, Moseley B, Larsen A, Ceretti DP, Braedt G, Price V, Gillis S, Henney CS, Kronheim SR, Grabstein K, Conlon PJ, Hopp TP, Cosman D. 1985: Cloning, sequence and expression of two distinct human interleukin-1 complementary DNA's. Nature (London) 315:641.

23. Mochizuki DY, Eisenman JB, Conlon PJ, Larsen AD, Tushinki RJ: 1987 Interleukin-1 regulates hematopoietic activity, a role previously ascribed to hemopoietin-1. Proc. Nat. Acad. Sci., 84:2567.

24. Moore MAS, Warren DJ: 1987. Synergy of interleukin-1 and granulocyte colony-stimulating factor. In vivo stimulation of stem-cell recovery and hematopoietic regeneration following 5- fluorouracil treatment of mice. Proc. Natl. Acad. Sci 84:7134.

25. Nagata S, Tsuchinya M, Asano S, Kaziro Y, Yamazaki T, Yamamoto O, Hirata Y, Kubota N, Ikeda M, Nomura H, Ono M: 1986. Molecular cloning and expression of cDNA for human granulocyte colony-stimulating factor. Nature, 319:415.

26. Price V, Mochizuki DY, March DJ, Cosman D, Deely MC, Kinke R, Clevenger W, Gillis S, Baker P, Urdal D. 1987. Expression , purification and characterization of recombinant murine granulocyte-macrophage colony-stimulating factor and bovine interleukin-2 from yeast. Gene, 66:287.

27. Sandrin MS, Hogarth PM, McKenzie IFC. 1983. Two Qa specificities: Qa-m7 and Qa-m8 defined by monoclonal antibodies. J. Immunol. 131: 546.

28. Stanley ER, Heard PM. 1977. Factors regulating macrophage production and growth. Purification and some properties of the colony stimulating factor from medium conditioned by mouse L. cells. J. Biol. Chem., 252:4305.

29. Stanley ER. 1979. Colony stimulating factor (CSF) radioimmunoassay. Detection of a CSF subclass stimulating macrophage production. Proc. Natl. Acad. Sci, USA. 76:2969.

30. Stanley ER, Guilbert LJ. 1981. Methods of purification, assay, characterization and target cell binding of a colony stimulating factor. (CSF-1). J. Immunol. Meth. 42:253.

31. Stanley ER. 1985. The macrophage colony stimulating factor, CSF-1. Methods in Enzymology, 116: 564.

32. Stork LC, Peterson VM, Rundus CH, Robinson WA. 1988. Interleukin-1 enchances murine granulopoiesis in vivo. Exp. Hematol. 16:163.

33. Vogel SN, Couches SD, Kaufman EN, Neta R, 1987. Induction of colony stimulating factor in vivo by recombinant interleukin-1 and recombinant tumour necrosis factor- . J. Immunol. 138:2143.

34. Zsebo KM, Wypych J, Yuschenkoff VN, Lu H, Hunt P, Dukes PP, Langley KE. 1988. Effects of hematopoietin-1 and interleukin-1 activities on early hematopoietic cells of the bone marrow blood. 71:962.

4 Role of NK Cells in Tumor Cell Growth and Eradication

E. Lotzová, C.A. Savary, K.A. Dicke, and S. Jagannath

INTRODUCTION

We have been interested in dissecting the population of interleukin-2 (IL-2) activated lymphocytes involved in the growth inhibition or destruction of malignant cells from the population reactive against normal hematopoietic tissues. This question became even more important after the observation indicating that adoptive transfer of IL-2 activated lymphocytes was effective in therapy of some human cancers [1], and after our own observation demonstrating that IL-2 activated natural killer (NK) cells represent the major cell type involved in destruction of leukemic cells in vitro [2-5].

In this communication, we will demonstrate that NK cells display preferential reactivity against leukemic cells as opposed to normal hematopoietic tissues.

RESULTS

Initially we investigated the role of NK cells in resistance against leukemia in both normal donors and leukemic patients. These studies showed that the peripheral blood NK cell cytotoxicity of patients with various types of leukemia (measured in a 3 hr or 16 hr ^{51}Cr release cytotoxicity assay against highly NK-sensitive line, K-562) was inferior to that of healthy donors (Table 1). Subsequently, we showed that this low cytotoxicity in a majority of patients was not due to the dilution or suppression of NK cell function by leukemic cells, but could be attributed to multiple defects in the NK cell lytic machinery [4,5].

We also demonstrated that the NK cell cytotoxic defect of leukemic patients could be corrected after culture of effector cells with IL-2 (Table 2).

Table 1. Defective NK Cell Cytotoxicity of Patients with Leukemia

Time of Assay[a]	E:T Ratio[b]	Percent Lysis	
		Normal Donors	Leukemic Patients
3 hrs	12	17.0 ± 5.1	1.0 ± 0.4
	25	29.2 ± 8.3	1.6 ± 0.8
	50	39.3 ± 7.5	3.0 ± 1.4
16 hrs	12	38.5 ± 5.3	2.3 ± 0.8
	25	45.1 ± 3.7	3.5 ± 1.2
	50	58.7 ± 3.3	6.2 ± 2.2

[a] Peripheral blood mononuclear cells, prepared by Ficoll-Hypaque gradient [2] were tested against K-562 in a ^{51}Cr release assay. Values represent mean ± S.E. of normal donors and patients with various types of leukemia.

[b] E:T = effector:target ratio

However, in contrast to cancer patients with solid tumors, the effector cells of leukemic patients required frequently two to three weeks of culture with IL-2 in order to acquire the optimal levels of cytotoxicity.

It is of interest to note that activation of effector cells with IL-2 resulted not only in correction, but also in augmentation of all parameters of the lytic mechanism (Table 3). Specifically, the level of the binding efficiency of NK cells, their frequency, speed of lysis, recycling activity and production of cytotoxic factor (NKCF) were comparable

Table 2. Correction of Cytotoxic Defect
of Leukemic Patients with Interleukin-2

Leukemic Patients	Percent Lysis[a]			
	Weeks in Culture			
	0	1	2	3
AML	0.5	0.8	64.9	56.0
AML	1.2	14.5	29.3	28.6
AML	0.9	22.4	58.2	76.2
CML	1.4	47.9	74.3	53.8
CML	0.8	3.8	16.2	24.4
CML	1.8	nt[b]	61.3	72.9
CML	0.9	nt	1.9	76.1
ALL	0.1	4.3	15.2	50.0
ALL	0.5	0.5	nt	44.1[c]

[a]Cytotoxicity of peripheral blood mononuclear cells was tested against K-562 in a 3 hr ^{51}Cr release assay at 12 E:T ratio. Effector cells were cultured with 10^3 U/ml of IL-2.

[b]nt = not tested

[c]Cells of ALL patient were cultured for 4 weeks.

to those manifested by IL-2 activated effector cells from healthy individuals [2].

Importantly, IL-2 activated cytotoxic cells manifested the lytic activity also against fresh autologous and allogeneic leukemia targets (Table 4). Of note was the observation that these cells were also effective in the down-regulation of the in vitro growth of clonogenic leukemia, the population of leukemic cells which may be responsible for expansion of leukemia in vivo (Table 5).

All of these data suggested the possible therapeutic relevance of IL-2 activated lymphocytes in the treatment of patients with leukemia. In order to design the most effective therapeutic modality, it was important to dissect the most aggressive population of antileukemia-directed killer cells.

Since both NK cells and T cells proliferate and can be activated in cultures with IL-2, we examined the involvement of either of these effector cells in antileukemia activity. NK cells can be easily distinguished from T cells based on their cell surface markers (NK cells display CD3$^-$, CD5$^-$, CD16$^+$ and NKH1$^+$ phenotype, and T cells CD3$^+$, CD5$^+$ and CD16$^-$ phenotype). Thus, using the

Table 3. Correction of Several Steps of Lytic Mechanism of Leukemic Patients with Interleukin-2

Parameter Tested[a]	Mean ± SE	
	-IL-2	+IL-2
% Binding	4.2 ± 1.0	17.9 ± 2.1
% Killing	25.7 ± 8.1	39.1 ± 8.0
Freq. of Killer Cells	1.4 ± 1.2	6.7 ± 1.2
Rate of Lysis	1.9 ± 1.2	30.0 ± 1.7
Recycling	1.2 ± 0.5	4.0 ± 0.1
% Lysis by Factor	2.5 ± 1.6	27.0 ± 3.0

[a]LGL of patients with various types of leukemia were analysed before and 1-5 weeks after culture with IL-2 for various parameters of cytotoxic mechanism, as described before [2].

monoclonal antibodies (MoAb) and a complement-dependent cytotoxicity assay, we depleted the IL-2 activated killer cells of either NK cells or T cells, and tested the activity of such depleted populations in destruction or inhibition of the growth of fresh leukemic cells.

Table 6 demonstrates that the anti-autologous leukemia-directed lytic activity was not affected by the removal of T cells, but was either totally abolished or significantly reduced after depletion of NK cells. Similarly using the clonogenic assay, we showed that the growth of clonogenic leukemic cells was inhibited by NK cells; in contrast, NK-depleted populations (comprised primarily of T cells) were not effective

Table 4. Lysis of Autologous Leukemia by Interleukin-2 Activated Lymphocytes

Leukemia Patients	Percent Lysis[a]	
	-IL-2	+IL-2
AML	0.1	31.5
AML	-1.7	15.5
AML	-0.9	32.4
CML	0.5	11.8
ALL	0.6	30.1

[a]Cytotoxicity of peripheral blood mononuclear cells of leukemic patients was tested in a 3 hr ^{51}Cr release assay at 50 to 100 E:T ratio, before and 14-17 days after culture with IL-2.

Table 5. Inhibition of Clonogeneic
Leukemia (CL) by Activated NK Cells

Leukemia Target	Lymphocyte Treatment[a]	% Inhibition of CL
CML	None	77
AML	None[b]	50
	Anti-CD5	90
	Anti-CD16	2
AML	None[b]	58
	Anti-CD5	83
	Anti-CD16	17

[a]Peripheral blood mononuclear cells of
normal donors were cultured for 3-7 days
with IL-2, and tested for inhibition of
leukemia in a clonogenic assay, at an ef-
fector:leukemia ratio of 10. Effector
cells were treated with MoAb against NK
cells (anti-CD16 and anti-NKH1) or T
cells (anti-CD5), using a complement-
dependent lysis technique [6]. Values
represent percent inhibition of leukemia
colony formation; the number of colonies
formed by 10^5 leukemic cells ranged from
95-149.

[b]Effector cells were treated with complement alone.

in manifesting the leukemia-inhibitory
activity (Table 5). These observations
clearly established the role of NK cells
in resistance against leukemia.

Even though it has been reported that the
IL-2 activated killer cells do not neg-
atively affect normal tissues, we considered it important to examine this
phenomenon more thoroughly. Using a ^{51}Cr
release cytotoxicity assay, we tested the
cytotoxic reactivity of IL-2 activated
killer cells against normal bone marrow,
blood lymphocytes and monocytes. Table 7
shows that the IL-2 generated effector
cells displayed low, but significant
levels of lytic activity against all of
these normal tissues.

We considered the possibility that a ^{51}Cr
release assay may not be optimal for
these studies, since it measures only the
cytotoxic, and not the cytostatic ability
of effector cells and furthermore, it may
not reflect realistically the lysis of
the cell population which is present in
the target cell suspension in a minority
(e.g. the hematopoietic progenitors).
Thus, we analysed the regulatory effect
of IL-2 activated T or NK cells on
clonogenic hematopoietic progenitors and

Table 6. Lysis of Autologous Leukemia is
Mediated by Interleukin-2 Activated NK
Cells

Leukemia Patients	Treatment of Effector Cells[a]	Percent Lysis[b]
AML	None	23.2
	Anti-CD5	28.2
	Anti-CD3[c]	36.2
	Anti-NKH1	9.0
AML	None	16.6
	Anti-CD16	3.2
	Anti-NKH1	3.0

[a]Peripheral blood mononuclear cells of 2
AML patients were cultured for 14-19 days
with IL-2. MoAb treatment is described
in the legend of Table 5.

[b]Cytotoxicity was tested in a 3hr ^{51}Cr release assay at 100 E:T ratio.

[c]Anti-CD3 MoAb is directed against T
cells.

bone marrow cells, using the GM-CFC assay
and ^{3}H-thymidine incorporation assay,
respectively.

As illustrated in Table 8, we observed a
significant inhibition of autologous as
well as allogeneic GM-CFC clonogenic
activity after incubation of bone marrow
cells with IL-2 activated lymphocytes.
Such inhibitory activity was totally
removed or significantly diminished after
depletion of T cells. On the contrary,
removal of NK cells did not affect in any

Table 7. Sensitivity of Normal Tissues
to Lysis by IL-2 Activated Killer Cells

Target Cells	Percent Lysis (Mean ± S.E.)[a]		
	25[b]	50	100
BM	4.7 ± 1.3	6.4 ± 2.6	11.8 ± 2.8
PB	4.2 ± 0.8	6.5 ± 0.9	9.6 ± 1.3
MO	5.5 ± 0.6	6.1 ± 0.5	10.9 ± 1.3

[a]IL-2 activated peripheral blood mononuclear cells of normal donors were tested
against allogeneic bone marrow (BM), peripheral blood (PB) and monocytes (MO) in
a 3 hr ^{51}Cr release assay. Monocytes
were obtained by centrifugal elutriation
[7].

[b]E:T ratio

Table 8. Inhibition of GM-CFC by Inter-leukin-2 Activated T Cells

Lymphocyte Treatment[a]	% Inhibition of GM-CFC[b]	
	Exp 1	Exp 2
None[c]	58	64
Anti-CD5	8	37
Anti-CD16	40	70
Anti-NKH1	40	72

[a]Peripheral blood mononuclear cells of normal donor (Exp 1) and breast cancer patient in remission (Exp 2) were cultured with IL-2 for 5-7 days.

[b]Bone marrow and effector cells (1:1) were incubated for 18 hr, and cultured for 7 days in agar [6]. Values represent percent of inhibition of colony formation; colony-forming ability of 10^5 bone marrow cells was 106 ± 8.0 and 122 ± 5.0 (mean ± S.E.) for exp 1 and 2, respectively.

[c]Effector cells were treated with complement alone.

significant way, the inhibitory activity. Similar observations were made in a ^{3}H-thymidine incorporation assay (Table 9); again, T cells, but not NK cells, were found to be inhibitory for bone marrow cell proliferation.

These studies clearly demonstrated that the highest antileukemia activity was manifested by IL-2 activated NK cells,

Table 9. Inhibition of Growth of Bone Marrow Cells by IL-2 Activated T Cells

Lymphocyte Treatment[a]	% Inhibition of Growth[b]	
	Exp 1	Exp 2
None[c]	59	56
Anti-CD5	29	0
Anti-CD16	72	52
Anti-NKH1	76	32

[a]Peripheral blood mononuclear cells were cultured for 7 days with IL-2.

[b]Growth of bone marrow cells was measured by ^{3}H-thymidine incorporation assay. Effector and bone marrow cells (1:1 ratio) were cultured for 7 days. Values represent percent inhibition of bone marrow growth.

[c]Effector cells were treated with complement alone.

Table 10. Comparison of the Growth and Cytolytic Activity of Interleukin-2 Activated LGL and Unfractionated Lymphocytes

Parameter Tested[a]	Weeks in Culture	Type of Effectors	
		LGL	LY
% Lysis	0	41.5	16.7
	1	74.0	56.0
	2	77.7	61.6
	3	74.2	54.5
Expansion Index	1	2.1	1.7
	2	9.2	4.2
	3	57.0	7.0

[a]Cytotoxicity of LGL or unfractionated lymphocytes (LY) was tested at 12 E:T ratio in a 3 hr ^{51}Cr release assay; expansion index reflects the cell growth.

while the most effective reactivity against normal bone marrow was displayed by IL-2 activated T cells. Consequently, these observations suggest that adoptive therapy of leukemic patients with in vitro activated NK cells (depleted of T cells) may result in maximal antileukemia activity and no or minimal interference with normal hematopoiesis.

In light of this possibility, it was important to determine whether the highly enriched NK cell population displayed an ability to proliferate in vitro. In subsequent experiments, we compared the growth kinetics and cytotoxicity levels of NK cell (LGL)-enriched population and the population of unseparated lymphocytes, during culture with IL-2 (Table 10).

The results of these investigations showed that NK cell highly-enriched population grew effectively in vitro and maintained the high levels of cytotoxicity for two to three weeks in IL-2 culture. In fact the LGL population was superior to the unseparated IL-2 activated lymphocytes in both the growth ability and cytotoxic activity. These experiments provided the evidence that highly cytotoxic NK cells can be propagated in vitro for therapeutic purposes.

SUMMARY

1.   IL-2 activated NK cells represent the primary effector cell population involved in destruction and regulation of the growth of leukemic cells.

2.   IL-2 activated T cells exhibit minimal if any antileukemia activity,

but are effective in down-regulation of normal hematopoietic tissues.

3. IL-2 activated NK cells may provide a new therapeutic modality for treatment of leukemia.

4. Two therapeutic approaches in leukemic patients with IL-2 activated NK cells are plausible:

   a. Adoptive transfer with in vitro propagated IL-2 activated autologous NK cells.

   b. Use of NK cells (autologous and perhaps, allogeneic) as a tool for removing the residual leukemic cells from bone marrow before autologous bone marrow transplantation therapy.

ACKNOWLEDGEMENT

This research was supported by the grant CA 39632 from National Cancer Institute.

REFERENCES

1. Rosenberg SA, Lotze MT, Muul LM, Leitman S, Chang AE, Ettinghausen SE, Matory YL, Skibber JM, Shiloni E, Vetto JT, Seipp CA, Simpson C, Reichert CM (1985) Observations on the systemic administration of autologous lymphokine-activated killer cells and recombinant interleukin-2 to patients with metastatic cancer. N Engl J Med 313:1485-1492

2. Lotzová E, Savary CA, Herberman RB (1987) Induction of NK cell activity against fresh human leukemia in culture with interleukin 2. J Immunol 138:2718-2727

3. Lotzová E (1987) Human natural killer cells: their role and possible therapeutic application in leukemia. Clinl Immunol Newsltr 8:56-60

4. Lotzová E, Savary CA, Herberman RB (1987) Impaired NK cell profile in leukemia patients. In: Lotzová E (ed), Herberman RB (assoc ed)Immunobiology of Natural Killer Cells. CRC Press, Inc Vol II 3:29-53

5. Lotzová E (1987) Interleukin-2-generated killer cells, their characterization and role in cancer therapy. Cancer Bull 39:30-38

6. Lotzová E, Savary CA, Herberman RB (1987) Inhibition of clonogenic growth of fresh leukemia cells by unstimulated and IL-2 stimulated NK cells of normal donors. Leuk Res 11:1059-1066

7. Lotzová E (1980) Centrifugal elutriation allows enrichment of natural killing and separates xenogeneic and allogeneic reactivity. In: Herberman RB (ed) Natural Cell-Mediated Immunity Against Tumors. Academic Press, Inc, New York, 131-137

Part II. Hematopoietic Cellular Growth Regulation

Chairperson: D. Zipori

5 A Quantitative Assay for Stroma Dependent Hemopoiesis

M. Kalai and D. Zipori

INTRODUCTION

Hemopoiesis is strictly dependent upon the self renewal of pluripotent stem cells in the bone marrow. A small fraction of these cells differentiates in a series of steps and continuously supplies mature cells to the myeloid and immune systems. Ultrastructural studies of the bone marrow have revealed close physical associations of stromal cells with hemopoietic stem and progenitor cells [1, 2]. Continuous hemopoiesis and long term maintenance of stem cells *in vitro* can be obtained only in coculture with bone marrow stromal cells [3]. Functional significance has been ascribed mostly to stromal-adipose cells [4]. A number of laboratories have reported on the establishment of continuous cell lines from mouse bone marrow stroma. These cell lines have been characterized morphologically and functionally and some were shown to support both myelopoiesis and pre-B lymphopoiesis in long term cultures [5, 6, 7, 8, 9]. Although these cell lines have significantly simplified the model systems used for the study of hemopoiesis *in vitro*, the cultures are very complex and are hard to use when one is trying to evaluate the molecular basis of the mechanisms that control the process. Extracellular matrix components were proposed to be involved in the regulation of hemopoiesis. It was found that stromal cells deposit high amounts of extracellular matrix components, some of which were detected at attachment sites between marrow derived adherent cells and developing granulocytes and macrophages [10, 11]. So far the exact role of the extracellular matrix is unknown.

This paper further describes the ability of an endothelial-adipose cell line (14F1.1) to support the growth of hemopoietic stem cells and offers a rapid and simple coculture system for the study of the interactions between hemopoietic progenitor cells and bone marrow stroma.

MATERIALS AND METHODS

Cell Lines

A mouse bone marrow derived cell line, designated 14F1.1, which was previously classified as endothelial-adipocyte, and a human foreskin fibroblast cell strain designated ASJ-1 were used [6, 7, 8]. The cells were grown in 60 mm tissue culture plates (Falcon), in DMEM (Gibco) containing 4.5 g glucose per liter supplemented with 10% fetal calf serum (FCS) and were passaged by rubber policeman scraping and dilution in fresh growth medium. In all cases experiments were carried out using confluent cell layers at the sixth week of culture.

Primary Cells

Bone marrow cells used in liquid culture experiments were obtained from the femur and tibia of 8-10 week old male BALB/c mice (Olac), while bone marrow cells used in semisolid medium experiments (see below) were obtained from 6-7 day old BALB/c mice. Cell suspensions were prepared in DMEM supplemented with 10% FCS and were depleted from adherent cells by two cycles of 1 h incubation in tissue culture plates (Falcon).

Conditioned Media (CM)

Medium was removed from confluent layers of 14F1.1 cells and was replaced by fresh growth medium. The cultures were incubated for 7 days (previous experiments indicated that this procedure yields the highest titers of factors a given stromal cell would secrete [12]). The medium was then collected, millipore filtered and examined for biological activity.

Extracellular Matrix (ECM) and Dry Stromal Cell Layers

Plates coated with ECM of 14F1.1 cells were prepared from confluent cell cultures. The cultures were washed with phosphate buffered saline (PBS), exposed to 0.5% (v/v) Triton-X-100 in PBS for 30 min and then washed three times with PBS [13,14]. Dry 14F1.1 cell layers were prepared by washing confluent 14F1.1 cultures with PBS and then removing all medium and allowing the cells to dry in a laminar flow hood.

Liquid Cultures of Bone Marrow Cells

Bone marrow cells were suspended in DMEM supplemented with 10% FCS or horse serum (HS), as specified, at a final concentration of 10^5 cells per ml. Five-milliliter aliquots were seeded onto 14F1.1 cell

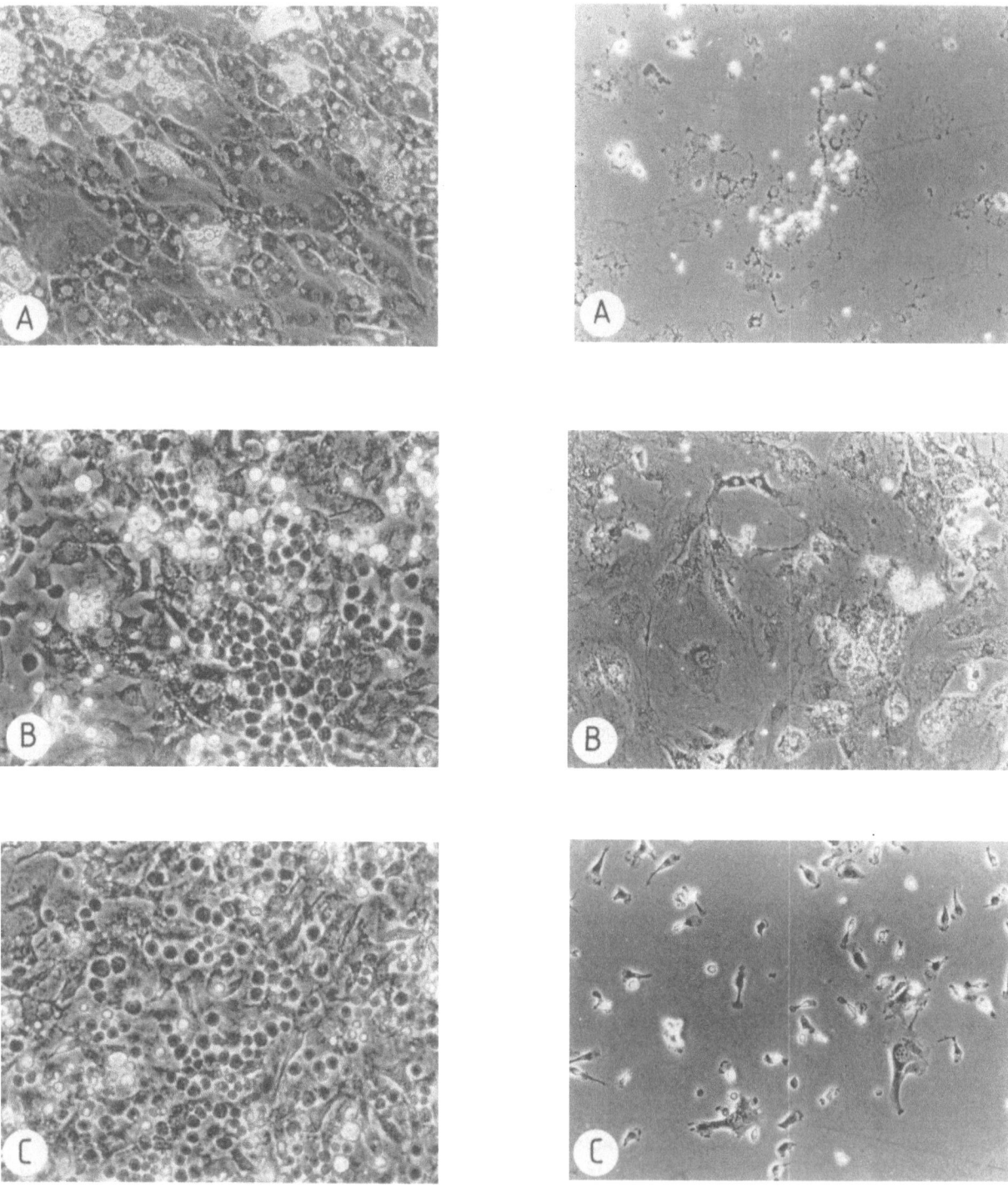

Fig. 1: Phase contrast micrographs of (A) endothelial-adipose cell line-14F1.1 (x128) and appearance of cobblestone-like areas in bone marrow cocultures, (B) in medium supplemented with HS at the second week (x260) and (C) in medium supplemented with FCS at the seventh week (x260).

Fig. 2: Phase contrast micrographs of bone marrow cells cultured with subcellular fractions of 14F1.1 cells: (A) 14F1.1 ECM (x128); (B) dry 14F1.1 cells (x128); (C) medium conditioned by 14F1.1 cells (x170).

layers, 14F1.1-ECM coated plates and dry 14F1.1 cell preparations in 60 mm tissue culture plates (Falcon). Control cultures containing bone marrow cells only were also prepared. Cultures were incubated at $37^{o}C$ in an atmosphere of 10% CO_2 in air. Once a week, all of the growth medium containing nonadherent cells was removed, and replaced by fresh medium supplemented with the corresponding serum. Total nucleated cell counts were performed and the cells were further examined, as detailed below.

Semisolid Cultures of Bone Marrow Cells

Freshly isolated nonadherent bone marrow cells were embedded in methylcellulose medium (0.8% w/v) supplemented with 20% HS or FCS, as specified, at a concentration of $5x10^4$, $2x10^4$ or 10^4 cells per ml. Five-milliliter aliquots were seeded onto confluent layers of 14F1.1 or ASJ-1 cells, as specified. After one week of incubation in a humidified atmosphere of 10% CO_2 in air, at $37^{o}C$, the semisolid medium was removed from the culture plates. The plates were washed once with PBS and the adherent cell compartment was fixed and stained with May-Grunwald Giemsa.

Myeloid Progenitors (GM-CFC)

Cells recovered from long-term cultures or cells picked from colonies of hemopoietic cells growing in semisolid bone marrow cocultures were embedded in methylcellulose medium (0.8% w/v) supplemented with 20% FCS and 14FIL3.CB6 conditioned medium as a source of CSF (14FIL3.CB6 is a clone of 14F1.1 cells transfected with an IL3 cDNA;. this cell clone was previously shown to secrete CSF-1 and IL3 [12]). One-milliliter aliquots were seeded into 35-mm tissue culture plates (Nunclon). Macrophage and granulocyte colonies were counted following 7 days of incubation in a humidified atmosphere of 10% CO_2 in air, at $37^{o}C$. Only aggregates containing 40 cells or more were counted.

Immunofluorescence

B220 antigen was detected in cells recovered from long-term cultures by use of 6B2 rat antibodies to mouse B220 and a second fluorescein isothiocyanate (FITC) conjugated rabbit anti-rat antibody [15]. Fluorescence profiles were analyzed with a fluorescence-activated cell sorter.

RESULTS

Correlation Between Long-Term Hemopoiesis and Cobblestone-like Area Formation

The 14F1.1 cell line was shown to be capable of supporting the continuous proliferation of myeloid progenitor cells and pre-B lymphocytes in long-term bone marrow cocultures [6, 7, 8, 12, 14]. We examined the appearance of cobblestone-like areas and evaluated the incidence of GM-CFC and pre-B lymphocytes in these cultures. Myeloid progenitor cells could be recovered from bone marrow cocultures supplemented with HS up to 6 weeks in culture. The mature cell types accumulating in the suspension were mostly myeloid cells at various stages of differentiation. Similar cocultures supplemented with FCS, showed mostly pre-B lymphopoiesis. The incidence of pre-B cells was determined by the use of a specific antibody. Cells exhibiting membrane B220 antigen could be recovered continuously from such long term cultures

and the incidence of these cells was 50% at the sixth week. These results could be directly correlated with the formation and maintenance of cobblestone-like areas. As long as the latter were observed in the culture plates, GM-CFC and pre-B cells could be recovered. The morphological appearance of the cobblestone-like areas containing cells with blast-like appearance is shown in Figure 1. The above results suggested that such blast-like colonies are essential for the maintenance of long term bone marrow cocultures.

Inability of ECM and Conditioned Medium from 14F1.1 Stroma to Support Long-Term Hemopoiesis

Nonadherent bone marrow cells seeded onto plates coated with ECM from 14F1.1 cells formed minute colonies of small cells (Fig. 2A), while cultures of cells seeded onto dry 14F1.1 cell layers (Fig. 2B) yielded merely small clusters of macrophage-like cells. Cells harvested and pooled from 10 bone marrow cultures seeded and incubated for 7 days in plates coated with 14F1.1-ECM did not yield even a single GM-CFC. The same results were obtained from bone marrow cells cultured on dry 14F1.1 cell layers or from plates containing bone marrow cells only. Addition of medium conditioned by 14F1.1 did not change these results significantly and at the most allowed the survival of some more macrophages. This was also observed in control plates in which medium conditioned by 14F1.1 cells was added to bone marrow cells cultured by themselves (Fig. 2C). It seems that while intact living 14F1.1 cells could support long-term hemopoiesis, the functional components were not found in medium conditioned by these cells or in their ECM. In order to study further the properties of the molecules involved in the induction of blast-like colony formation, it was necessary to develop a simpler system.

Stroma-Dependent Hemopoietic Colony Formation

When nonadherent bone marrow cells were embedded in methylcellulose-semisolid medium supplemented with HS and seeded onto 14F1.1 cells, macroscopically distinct dense hemopoietic colonies were formed within the stromal cell layer (Fig. 3, top left). This phenomenon occurred without the addition of external

Table 1: Incidence of hemopoietic colonies

Colony type	Colonies per plate ± SE (%)
Blast-like	33.90 ± 0.87
Granulocyte	13.50 ± 1.42
Macrophage	52.60 ± 1.43
Megakaryoid	Infrequent

Colonies from semisolid bone marrow cocultures at one week of incubation were stained with May-Grunwald and Giemsa. Result averages were obtained from 10 individual experiments (a total of 24 culture plates). The incidence of each type of colony in percent was constant and independent of the cell concentration used. Results from three different cell concentrations $5x10^4$, $2x10^4$ and 10^4 bone marrow cells per ml, were pooled.

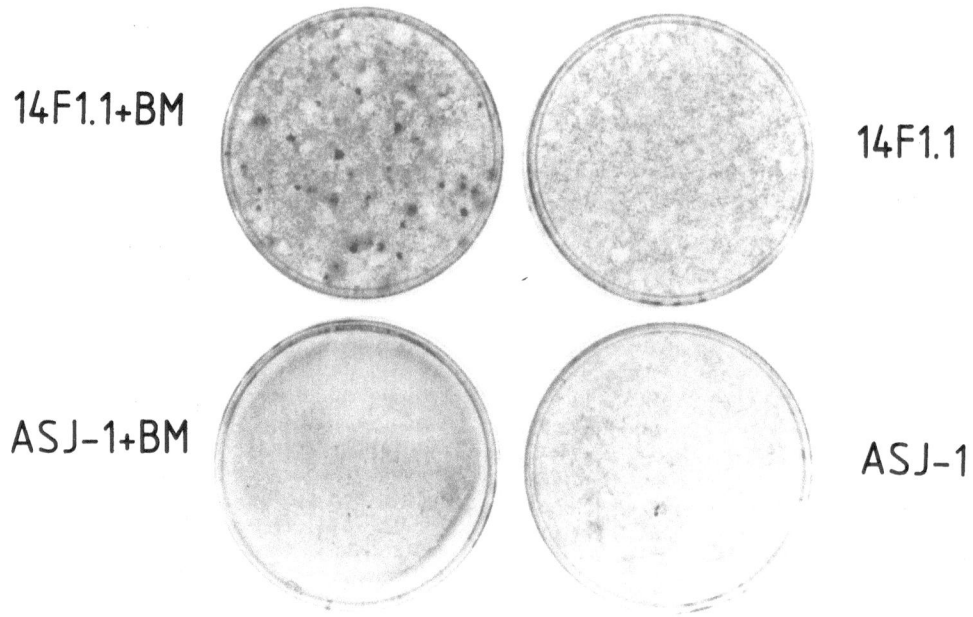

14F1.1+BM

14F1.1

ASJ-1+BM

ASJ-1

Fig. 3: Stroma-dependent colony formation. A photograph of fixed and stained cell cultures. Bone marrow cells (BM) were embedded in semisolid medium and seeded onto confluent cell layers of 14F1.1 cell line (top) and ASJ-1 cell strain (bottom). Formation of hemopoietic colonies occurred only on the stromal cells (top left). Cells were stained with May-Grunwald and Giemsa.

colony stimulating factors. The colonies were stroma-specific and did not appear in control cocultures of bone marrow with ASJ-1 foreskin fibroblast cells (Fig. 3). Colonies were microscopically counted at day 7 of culture, after staining with May-Grunwald and Giemsa. The colonies were composed of cells with blast-like morphology (33.9%), granulocytes (13.5%), macrophages (52.6%) or megakaryoid cells (infrequent) as illustrated in Fig. 4A, B, C & D, respectively. The incidence of each type of colony was constant over many repeated experiments (Table 1).

The incidence of GM-CFC in individual colonies was then determined. One hundred colonies were microscopically picked and independently assayed for the presence of myeloid progenitor cells. The colonies showed high variability in their GM-CFC content. Above 60% of the colonies contained more then one GM-CFC while about 30% contained more then twenty GM-CFC (Fig. 5). The highest number obtained was 426 GM-CFC per a single colony.

DISCUSSION

The data presented in this paper, considered with previous results [6, 7, 8, 12, 14], clearly show that the 14F1.1 endothelial-adipose cell line derived from mouse bone marrow is able to support long-term hemopoiesis in vitro. It has been shown that this cell line does not require the assistance of other cells from bone marrow stroma [14]. We further show here that the molecules allowing stem cell renewal were not secreted to the growth medium in significant amounts. Furthermore 14F1.1-ECM supported only the short term survival of mature myeloid cells but did not induce stem cell renewal. It seems therefore that the main function of the extracellular matrix is to give mechanical support to the different participants in the process. The attachment of hemopoietic cells to stromal elements was found to be mediated by fibronectin [10]. Erythroid precursor cells were shown to express fibronectin receptors on their membranes [16]. Furthermore, extracellular matrix components were found to adsorb colony stimulating factors [17].

Myeloid progenitor cells and B220 positive cells were recovered from long-term bone marrow cocultures as long as cobblestone-like areas were observed in the culture. This correlation might imply that the maintenance of long-term hemopoiesis requires close cell to cell associations between the stroma and hemopoietic stem cells. Indeed embedding bone marrow cells in methylcellulose semisolid medium allowed the formation of distinct hemopoietic colonies on the stroma. The main function of the methylcellulose was to prevent the migration of cells from one place to the other. This allowed the formation of dense colonies, each of which was derived from a single precursor cell. This semisolid medium coculture system has quite a few appealing features. The phenomenon is specific; the colonies did not appear when foreskin fibroblast cells were used instead of 14F1.1 cells. It is rapid; small colonies were microscopically recognizable following 48-h incubation, macroscopic colonies were observed within a week. It is simple; the analysis was carried out by staining the cells and differentially counting the colonies microscopically. The process is quantitative and reproducible; the incidence of each type of colony was constant over many repeated experiments. Furthermore quite a few colonies had high GM-CFC content; this might be correlated with stem cell

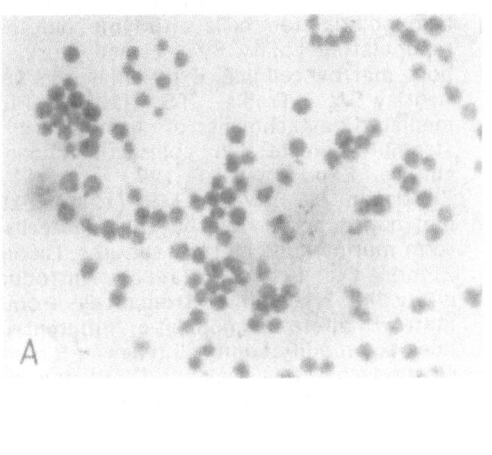
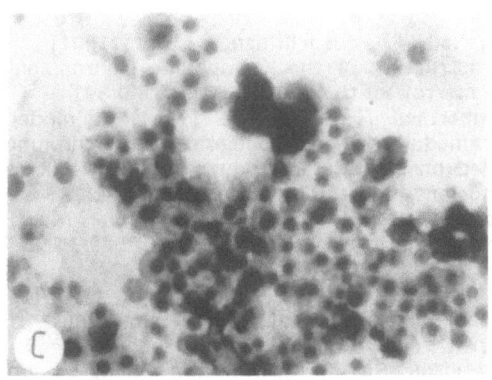
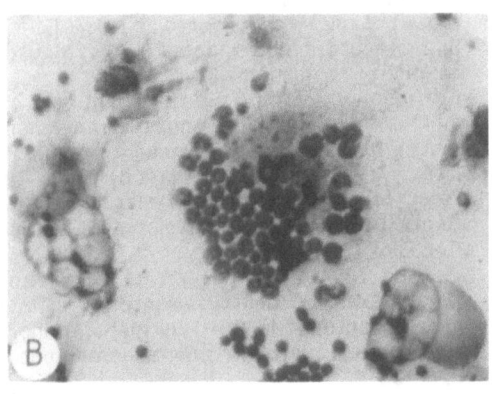
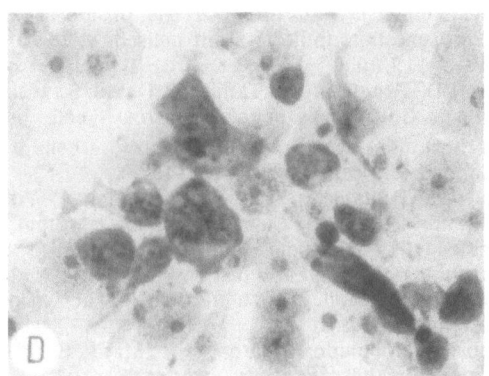

Fig. 4: Micrographs of the four types of stroma dependent hemopoietic colonies: (A) Blast-like cells; (B) granulocytes; (C) macrophages; (D) megakaryoid cells. Disperse colonies were especially chosen in order to show cell morphology. Arrow indicates fat globules. Magnification (x300).

phenotype and blast-like morphological appearance. All these suggest that stroma-dependent hemopoietic colony formation can be used as a short term, quantitative, simple assay for the study of the mechanisms that control stem cell renewal and hemopoiesis.

SUMMARY

A stromal cell line designated 14F1.1 supported long-term myelopoiesis and pre-B lymphopoiesis *in vitro*. Extracellular matrix and conditioned media from this cell line could not replace the intact cells in this function. The longevity of hemopoiesis in the system could be correlated to the formation and maintenance of cobblestone-like areas. As long as the latter were observed, GM-CFC and pre-B cells could be recovered from the cultures. Embedding nonadherent bone marrow cells in methylcellulose-semisolid medium and seeding them onto confluent 14F1.1 cell layers resulted, within one week, in the formation of distinct macroscopic colonies in the stromal cell layer. Microscopic examination revealed four types of colonies: blast-like (33.9%), granulocyte (13.5%), macrophage (52.6%) and megakaryoid (infrequent). The phenomenon was stroma-specific and reproducible. The GM-CFC content of nearly 30% of the colonies was above 20 per colony. The largest number was 426 per single colony. The data presented in this paper suggests that stroma-dependent hemopoietic colony

formation can be used as a rapid and simple method for the study of mechanisms controlling hemopoiesis.

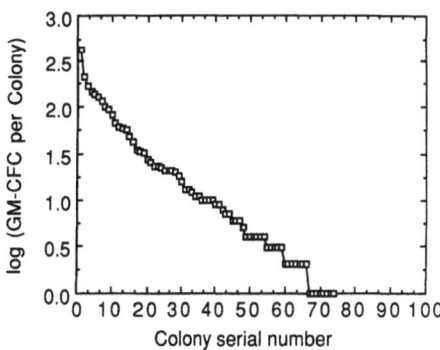

Fig. 5: The incidence of GM-CFC in individual stroma dependent hemopoietic colonies. One hundred colonies were microscopically picked and independently assayed for the presence of myeloid progenitors.

ACKNOWLEDGMENTS

This study was supported in part by a grant from the Belle S. and Irving E. Meller Center for the Biology of Aging at the Weizmann Institute of Science.

REFERENCES

1. Marshall A, Lichtman MD (1981) The ultrastructure of the hemopoietic environment of the marrow: a review. Exp Hematol 9:391

2. Lambertsen RH, Weiss L (1984) A model of intramedullary haematopoietic microenvironments based on stereologic study of the distribution of endocloned marrow colonies. Blood 63:287

3. Bradley TR, Metcalf D (1966) The growth of mouse bone marrow cells in vitro. Aust J Exp Biol Med Sci 44:287

4. Sakakeeny MA, Greenberger JS (1982) Granulopoiesis longevity in continuous bone marrow cultures and factor-dependent cell line generation. Significant variation among 28 inbred mouse strains and outbred stocks. J Nat Can Inst 68:305

5. Kodama H, Amagai Y, Koyama H, Kosai S (1982) A new preadipose cell line derived from newborn mouse calvaria can promote the proliferation of pluripotent hemopoietic stem cells in vitro. J Cell Physiol 112:89

6. Zipori D, Friedman A, Tamir M, David S, Malik Z (1984) Cultured mouse marrow cell lines: interactions between fibroblastoid cells and monocytes. J cell Physiol 118:143

7. Zipori D, Duksin D, Tamir M, Argaman A, Toledo J, Malik Z (1985) Cultured mouse stromal cell lines. II. Distinct subtypes differing in morphology, collagen types, myelopoietic factors, and leukemic cell growth modulating activities. J Cell Physiol 122:81

8. Zipori D, Toledo J, Von der Mark K (1985) Phenotypic heterogeneity amongst stromal cell lines from mouse bone marrow disclosed in their extracellular matrix composition and interactions with normal and leukemic cells. Blood 66:447

9. Li CI, Johnson GR (1987) In vitro maintenance of hemopoietic stem cells with lymphoid and myeloid repopulating ability by a cloned murine adherent bone marrow cell line. Exp Hematol 15:989

10. Bentley SA, Tralka TS (1983) Fibronectin-mediated attachment of hemopoietic cells to stromal elements in continuous bone marrow culture. Exp Hematol 11:129

11. Zukerman KS, Wicha MS (1983) Extracellular matrix production by the adherent cells of long-term murine bone marrow cultures. Blood 61:540

12. Zipori D, Lee F (1988) Introduction of interleukin-3 gene into stromal cells from the bone marrow alters hemopoietic differentiation but does not modify stem cell renewal. Blood 171:586

13. Gospodarowicz D, Delgado D, Vlodavsky I (1980) Permissive effect of the extracellular matrix on cell proliferation in vitro. Proc Natl Acad Sci USA 77:4094

14. Otsuka T, Arai K, Kalai M, Gluck U, Zipori D (1988) Long term survival of CFU-GM from human cord blood induced by a mouse stromal cell line. submitted.

15. Coffman RL (1985) Surface antigen expression and immunoglobulin gene rearrangement during mouse pre-B cell development. Immunol Rev 69:5

16. Patel VP, Lodish HF (1986) The fibronectin receptor on mammalian erythroid precursor cells: Characterization and developmental regulation J Cell Biol 102:449

17. Gordon MY, Riley GP, Watt SM, Greaves MF (1987) Compartmentalization of a haematopoietic growth factor (GM-CSF) by glycosaminoglycans in the bone marrow microenvironment. Nature 326:403

6   Matrix Glycoproteins May Regulate the Local Concentrations of Different Hemopoietic Growth Factors

M.Y. Gordon, A.D. Bearpark, D. Clarke, and L.E. Healy

ABSTRACT

We compared the binding properties of granulocyte-macrophage colony-stimulating factor (GM-CSF), erythropoietin (epo) and interleukin-3 (IL-3) to test the hypothesis that binding to extracellular matrix (ECM) governs growth factor distribution in the marrow microenvironment. Since mitogens for cells of mesodermal and neuroectodermal origin bind to heparin [1], heparin-like molecules (e.g. heparan sulphate) are candidate growth factor-binding structures in the ECM associated with marrow-derived stromal cells. In contrast to the heparin-binding fibroblast growth factor (FGF), GM-CSF did not bind to heparin-sepharose beads. However, most recombinant (r), human (h), gibbon (g) and murine (m) IL-3 bound to the beads; intermediate levels of binding were shown by epo and native mIL-3. Like GM-CSF [2], epo and h/gIL-3 did not bind to intact marrow-derived stromal layers but more than 85% of mIL-3 (native and recombinant) bound under these conditions. Murine IL-3 and GM-CSF [2] bind to marrow-derived ECM but epo does not. The different binding properties of haemopoietic growth factors vis-a-vis matrix glycoproteins suggests that their distributions and concentrations may be determined by the local composition of the ECM.

INTRODUCTION

The extracellular matrix (ECM) produced by bone marrow-derived stromal cells is a highly ordered, complex structure consisting of fibronectin, collagen and the glycosaminoglycans (GAGs) which are linked to protein to form proteoglycans [3,4]. It provides specific anchorage sites for haemopoietic cells of different lineages and at different stages of maturation [5-8] and can bind haemopoietic growth factors [2,9]. Hence, the ECM of marrow stromal cells may provide a framework for the interaction of haemopoietic growth factors and progenitor cells [4,10]. This model predicts that haemopoietic growth factors, as well as their target cells, differ in their binding properties so that local regulation of growth factor concentrations can be accomplished.

We tested haemopoietic growth factor-binding to heparin-sepharose beads because a class of heparin-binding growth factors has been identified [see 1 for review]; because heparan sulphate (an analogue of heparin) is one of the GAGs present in the ECM produced by marrow stromal cells in vitro [11,12] and because GM-CSF binds to GAGs [2,9]. Also, we tested the binding of epo and IL-3 to intact stroma and to ECM which have been used previously to investigate the binding properties of GM-CSF [2]. Finally, we used enzymatically degraded stroma to investigate the binding of GM-CSF to the GAGs chondroitin sulphate and hyaluronic acid which are represented in the ECM of marrow stroma [11,12, unpublished observations].

Our results show that haemopoietic growth factors differ in their binding properties and suggest a role for the ECM in orchestrating interactions between haemopoietic cells and haemopoietic growth factors.

MATERIALS AND METHODS

Growth factors:
    The growth factors used in this study were fibroblast growth factor (FGF) from bovine brain (Boehringer Mannheim); granulocyte-macrophage colony-stimulating factor (GM-CSF) in 5637 conditioned medium (CM) [13]; erythropoietin (epo) from human urine (Terry Fox Laboratories); recombinant (r), murine (m), human (h) (Genzyme) and gibbon (g) (Genetics Institute) interleukin-3 (IL-3) and native mIL-3 in WEHI-3B CM [14].

Target cells:
   Human marrow cells were obtained from
informed, consenting normal donors of
marrow for transplantation. The mononuclear cells were separated using Lympho-
prep (Nyegaard, Oslo), washed 3 times in
α-medium (GIBCO), resuspended in α-
medium supplemented with 15% fetal calf
serum (FCS; GIBCO) and depleted of
adherent cells by incubation in tissue
culture flasks at 37°C overnight. Murine
marrow cells were flushed from the femurs
of 12 week old male CBAca mice into
sterile medium. The mIL-3-dependent cell
line, FDCP-2 [15] was a gift from Dr E
Spooncer and the NIH-3T3 cells were provided by Dr C Marshall.

Heparin-sepharose affinity chromatography:
   Individual growth factors were stirred
overnight with heparin-sepharose CL-6B
beads swollen in phosphate buffered saline
(PBS). The volume of growth factor and
weight of beads were always 1ml and 0.1g
so that the beads were in excess of binding requirements. Unbound growth factor
was retrieved by centrifuging the beads at
700rpm for 1 min and aspirating the supernatant. To recover bound growth factors,
the beads were washed in PBS and extracted
with 0.5-2M NaCl. The extracts were
collected by centrifugation and dialysed
against 3 changes of α-medium. Bound and
unbound fractions were sterilised by
filtration.

Growth factor-binding to stromal cell
cultures:
   Human marrow mononuclear cells (5x10^5/
ml) were cultured in α-medium supplemented
with 10% FCS, 10% horse serum and 2x10^{-6}M
methylprednisolone (MP; Upjohn, Crawley,
UK) and incubated at 37°C in 5% CO_2 in
air. They were fed weekly until confluent
stromal layers had formed and haemopoietic
cells had disappeared. Murine stromal
cultures were set up in the same way using
total nucleated femoral cells.
   Some stromal cultures were dehydrated
before use. Others were extracted with 2M
NaCl prior to dehydration. Glycosaminoglycans were degraded by treatment with
heparitinase, hyaluronidase and chondroitinase ABC (Sigma). The enzyme activities
were simultaneously checked by digestion
of standard GAG substrates.
   Each growth factor was incubated overnight with treated or untreated stromal
layers and sterilised by filtration.

Growth factor bioassays:
FGF: NIH-3T3 cells were seeded at 5x10^3/
200μl in 96-well microtitre plates in
Dulbecco's modified Eagle's medium (DME)
supplemented with 5% calf serum. Once the
cells were confluent, the medium was
replaced with DME containing 1% calf serum
and FGF fractions equivalent to 25ng/ml
of the unfractionated material. After 18
hours, the wells were pulsed with 1μCi of
^{3}H-thymidine/well for 4 hrs, washed 3

times with DME, harvested with trypsin and
deposited onto 25cm Whatman GF/C filters.
The DNA was precipitated sequentially with
10% trichloroacetic acid (TCA) and 5% TCA
in 95% ethanol in the cold. The radioactivity was measured in a liquid scintillation counter.

GM-CSF: To assay GM-CSF, 1x10^5 nonadherent mononuclear marrow (NAMM) cells
were suspended in 1ml of 0.3% agar in α-
medium supplemented with 15% FCS and 5% or
10% v/v 5637 CM. The cultures were incubated for 14 days in humidified 5% CO_2 in
air at 37°C before colonies of more than
50 cells were counted.

Epo: Erythroid burst-forming units (BFU-E)
were measured by plating 1x10^5 NAMM cells
in 30% FCS, 1% bovine serum albumin,
1x10^{-4}M β-mercaptoethanol, 0.8% methyl
cellulose in Iscove's modified Dulbecco's
medium and 1 or 2.5U epo to a volume of
1ml per plate. Erythroid bursts were
scored after 14 days incubation in humidified 5% CO_2 in air at 37°C.

Human and gibbon rIL-3: Gibbon IL-3 was
assayed at concentrations of 12.5 and 25U/
ml and human IL-3 at 25 and 50U/ml. The
bioassay conditions were identical to
those used for measuring GM-CSF.

Murine IL-3: Factor-dependent FDCP-2 cells
[15] were used to measure mIL-3. The
cells were washed, diluted to 10^4/ml in
α-medium containing 10% horse serum and

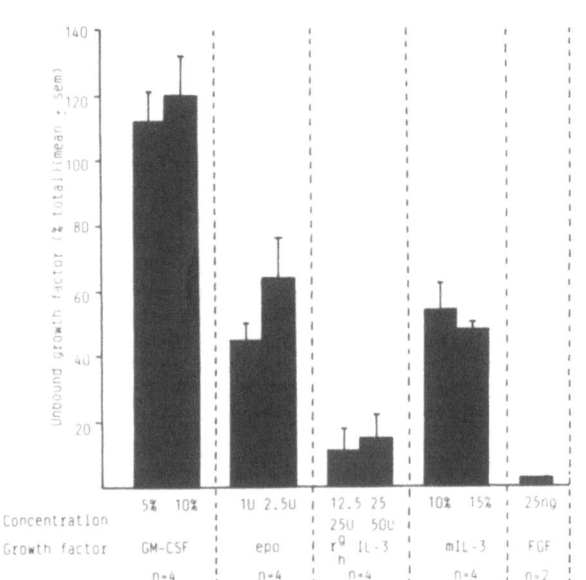

Fig 1. Measurement of growth factor-
binding to heparin-sepharose. The concentrations of growth factors (U/ml, ng/ml,
%v/v) are those used in the bioassays.
The results obtained using rg and rhIL-3
were indistinguishable and have been
pooled. n=the number of replicate
experiments. The vertical bars represent
the s tandard error of the interexperi-
mental mean.

10 or 15% WEH-3B CM. Aliquots of 100μl were placed in 20 replicate wells per group. After 48 hours incubation in 5% CO_2 in air at 37°C, the viability of the cells was determined by trypan blue dye-exclusion.

RESULTS

More than 95% of the growth-stimulating activity of FGF bound to heparin-sepharose beads but no GM-CSF was removed from 5637 CM by contact with heparin-sepharose (Fig 1).

However, the unbound fractions of epo and mIL-3 represented about 50% of the total activity whilst most of the rhIL-3 and rgIL-3 bound to the beads (Fig 1). Other experiments (data not given) showed that rmIL-3 binds to heparin-sepharose as well as native mIL-3 and that neither binds to sepharose CL-6B. Attempts to elute epo, rhIL-3 and rgIL-3 failed because these growth factors, in our hands, were inactivated by exposure to salt concentrations as low as 0.5M. However, native and recombinant mIL-3 were unaltered by 2M NaCl and mIL-3 eluted completely in 0.25-0.5M NaCl.

We have shown previously that GM-CSF does not bind to intact stromal layers but does bind to salt extracted cultures [2]. The data in Fig 2 shows that native mIL-3 binds to intact stromal layers (85% of rmIL-3 also binds - data not shown), However, epo, rhIL-3 and rgIL-3, like GM-CSF, do not bind (Fig 2).

The growth factors that did not bind to intact stroma were tested for binding to ECM. Epo did not bind under these conditions but mIL-3 bound efficiently and rh/gIL-3 to an intermediate extent (Fig 3).

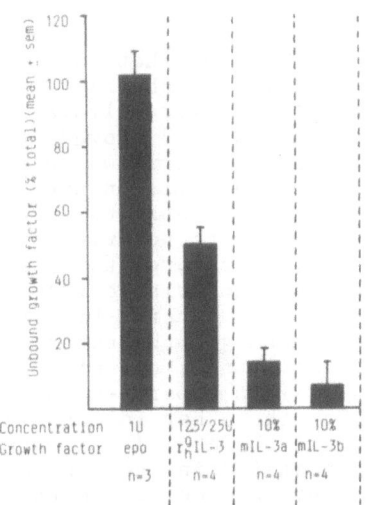

Fig 3. Measurement of growth factor-binding to ECM prepared by salt extraction of stromal layers. For notation, see Fig 1.

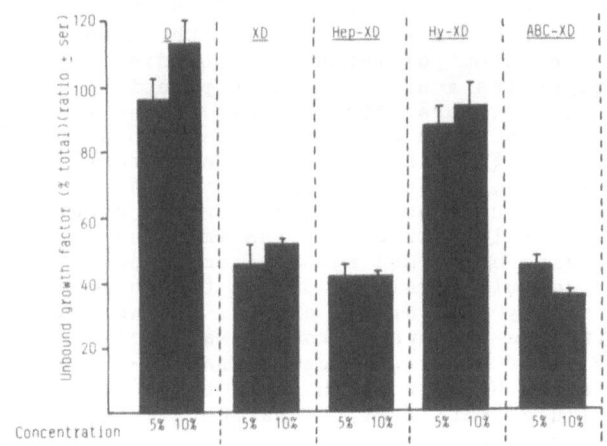

Fig 4. Measurement of GM-CSF-binding to enzymatically-treated ECM. D=dehydrated; XD=salt extracted, dehydrated; hep=heparitinase; hy=hyaluronidase; ABC=chondroitinase ABC. The vertical bars represent the standard error of the ratio between the numbers of GM-CFC stimulated by unfractionated GM-CSF and the numbers stimulated by the unbound fraction [16].

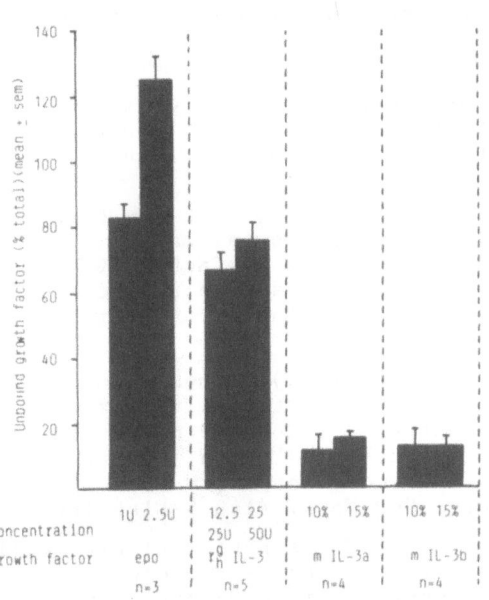

Fig 2. Measurement of growth factor-binding to intact bone marrow-derived stromal layers. For notation see Fig 1. The stromal layers were grown from human marrow cells with the exception of the experiments in mIL-3(a) which used mouse.

The results of a single experiment confirmed that GM-CSF binds to ECM prepared by salt extraction of stromal layers [2]. Treatment of the ECM with heparitinase or chondroitinase ABC did not affect binding but there was an apparent reduction in binding following treatment with hyaluronidase (Fig 4).

DISCUSSION

Our results have shown that haemopoietic growth factors differ in their affinities for components of the extracellular matrix. Both epo and IL-3 showed some degree of binding to heparin whereas GM-CSF did not bind at all. In the case of murine IL-3, we were able to demonstrate that the binding to heparin was of low affinity and, therefore, that it differs in this characteristic from the well-defined class of heparin-binding mitogens [1]. These growth factors can be divided into two classes, one of which (class I) elutes from heparin-sepharose in 1.0M NaCl whilst the other (class II) elutes in about 1.6M NaCl. It is noteworthy that a transforming growth factor derived from leukaemic cells and the bone-inductive protein, osteogenin, bind to heparin with low affinity [17, 18].

The results obtained using heparin-sepharose beads indicate that epo and IL-3, but not GM-CSF, might bind to heparan sulphate in the ECM of marrow stromal cultures. Although the results of the experiments testing growth factor-binding to intact stromal layers did not parallel exactly those obtained in the heparin-binding experiments, there were, again, clear differences in binding properties amongst the haemopoietic growth factors (Fig 2). Salt extraction has been shown to facilitate the binding of GM-CSF to the ECM in marrow stromal cultures, presumably by removing an endogenous growth factor from the matrix [2]. Salt extraction also improved the binding of rh and rgIL-3 from barely detectable levels to 50% of the total activity (cf. Figs 2 and 3). In agreement with the results of Roberts and colleagues [9], our findings suggest that IL-3 can bind to heparan sulphate in the extracellular matrix of the marrow microenvironment.

We have demonstrated that GM-CSF binds to glycosaminoglycans in the ECM of marrow stromal cultures [2] and Roberts et al [9] found that GM-CSF, like IL-3, binds to heparan sulphate. We were unable to demonstrate GM-CSF-binding to heparin-sepharose and our preliminary results using enzymatically degraded matrix suggest that hyaluronic acid rather than heparan sulphate is required for binding this particular growth factor. The evidence we have obtained about the binding properties of erythropoietin does not permit any conclusions about its putative binding site in the marrow. It is possible that epo binds to other matrix glycoproteins that we have not yet investigated. In particular, TGFβ has been shown to associate with fibronectin [19].

The finding that haemopoietic growth factors differ in their affinities for extracellular matrix glycoproteins suggests that their distribution in the marrow microenvironment might be influenced by the local composition of the matrix. It is relevant that the lineage-related growth factors, erythropoietin and GM-CSF, exhibit widely divergent binding properties and that their respective target cells bind to different structures in the ECM of marrow stroma [5-7]. The evidence for the binding of non-haemopoietic growth factors to matrix glycoproteins suggests that binding and immobilisation might be a general mechanism for regulating the distribution of growth factors and presenting them to the appropriate target cells. It is relevant that a low binding affinity of haemopoietic growth factors is likely to be favourable for interaction with high affinity receptors on the responding cells.

In conclusion, the results of this study have provided further support for the idea that the extracellular matrix of the marrow microenvironment functions as a framework for the regulated sequestration of growth factors for presentation to immobilised haemopoietic progenitor cells [2,4,9,10].

ACKNOWLEDGEMENT

The work was supported by the Leukaemia Research Fund of Great Britain.

REFERENCES

1. Lobb RR, Harper JW, Fett JW (1986) Purification of the heparin-binding growth factors. Analyt Biochem 154: 1-14

2. Gordon MY, Riley GP, Watt SM, Greaves MF (1987) Compartmentalization of a haematopoietic growth factor (GM-CSF) by glycosaminoglycans in the bone marrow microenvironment. Nature 326:403-405

3. Zuckerman KS (1984) Composition and function of the extracellular matrix in the stroma of long-term bone marrow cell cultures. In: Wright DG, Greenberger JS (eds) Long-term bone marrow culture. New York: Alan R Liss Inc

4. Gordon MY (1988) Annotation: Extracellular matrix of the marrow microenvironment. Brit J Haematol 70:1-4

5. Patel VP, Lodish HF (1986) The fibronectin receptor on erythroid progenitor cells: characterization and developmental regulation. J Cell Biol 102: 449-456

6. Tsai S, Patel V, Beaumont E, Lodish,

HF, Nathan DG, Sieff CA (1987) Differential binding of erythroid and myeloid progenitors to fibronectin. Blood 69:1587-1594

7. Campbell AD, Long MW, Wicha MS (1987) Haemonectin, a bone marrow adhesion protein specific for cells of granulocytic lineage. Nature 329:744-746

8. Gordon MY, Hibbin JA, Dowding C, Gordon-Smith EC, Goldman JM (1985) Separation of human blast progenitors from granulocytic, erythroid, megakaryocytic and mixed colony-forming cells by "panning" on cultured marrow-derived stromal layers. Exp Hematol 13:937-940

9. Roberts R, Gallagher J, Spooncer E, Allen TD, Bloomfield F, Dexter TM (1988) Heparan sulphate-bound growth factors: a mechanism for stroma cell-mediated hemopoiesis. Nature 332: 376-378

10. Keating A, Gordon MY (1988) Hypothesis: Hierarchical organisation of haematopoietic microenvironments: role of proteoglycans. Leukemia. Submitted for publication.

11. Gallagher JT, Spooncer E, Dexter TM (1983) Role of extracellular matrix in haemopoiesis. I. Synthesis of glycosaminoglycans by mouse bone marrow cell cultures. J Cell Science 63:155-171

12. Wight TN, Kinsella MG, Keating A, Singer JW (1986) Proteoglycans in human long-term bone marrow cultures: biochemical and ultrastructural analyses. Blood 67:1333-1343

13. Myers CD, Katz FE, Joshi G, Millar JL (1984) A cell line secreting stimulating factors for CFU-GEMM culture. Blood 64:152-155

14. Bazill GW, Haynes M, Garland J, Dexter TM (1983) Characterisation and partial purification of a haemopoietic cell growth factor in WEHI 3 cell conditioned medium. Biochem J 210:747-759

15. Dexter TM, Garland J, Scott D, Scholnick EM, Metcalf D (1980) Growth of factor-dependent haemopoietic precursor cell lines. J Exp Med 152:1036-1047

16. Blackett NM (1974) Statistical accuracy to be expected from cell colony assays: with special reference to the spleen colony assay. Cell Tissue Kinetics 7:407-412

17. Zack J, Smith RG, Ozanne B (1987) Characterization of a leukaemia-derived transforming growth factor. Leukemia 1:737-745

18. Sampath TK, Muthukumaran N, Reddi AH (1987) Isolation of osteogenin, an extracellular matrix-associated bone-inductive protein by heparin affinity chromatography. Proc Natl Acad Sci USA 84:7109-7113

19. Fava RA, McClure DB (1987) Fibronectin-associated transforming growth factor. J Cell Physiol 131:184-189

7 Binding of Growth Factors to Heparan Sulphate: Implications for the Regulation of Hemopoiesis by Bone Marrow Stromal Cells

J.T. Gallagher and T.M. Dexter

INTRODUCTION

The haemopoietic system is composed of a variety of different types of cell that carry out separaate but often inter- dependent functions. The system is a dynamic one in which a fine balance is maintained between cell renewal and cell loss. With the notable exception of lymphocytes, mature haemopoietic cells have little or no capacity for cell division and the rate and extent of cell renewal is largely determined by the degree of amplification that occurs in lineage - restricted 'transit' cell popula- tions that are maturing in distinct regions of the bone marrow prior to re- lease into the general blood circulation. Transit cells are found in all haemo- poietic cell lineages. They originate from a series of primitive progenitor cells, (identifiable in clonogenic assays in the presence of appropriate growth factors(whose progeny are committed to differentiate along defined pathways. However, the developmental potential of progenitor cells is restricted. They may be unipotent (ie progeny mature along one developmental pathway eg G-CFC) or bipotent such as the granulocyte-macro- phage precursor GM-CSF (1,2). Progeni- tors are themselves derived from a more primitive type of cell, the so-called haemopoietic stem cell, commonly identi- fied for experimental purposes as the CFU-S, a cell which gives rise to mixed colonies in the spleens of irradiated mice (2). The stem cell has the unique properties of self-renewal and pluri- potentiality. The self-renewal capability is essential for stem cells to maintain a relatively constant population whilst their pluripotentiality enables differen- tiation to all classes of haemopoietic progenitor cells and ultimately to all types of cell found in peripheral blood.

Progenitor cells possibly have some ability to self-renew but this is limited, and they must be continuously replaced by differentiation of stem cells. In general it is likely that the frequency of occurrence of the different cell types is inversely correlated with their develop- mental potential, with the mature cells being present in greatest abundance and stem cells being the minority population. Thus a hierarchical model has been formu- lated to describe 'the structure' of the haemopoietic system in terms of cellular diversification from the stem cell pool, lineage restriction and controlled growth and maturation to functional end cells (2,3). The model may be represented in simplified form as follows:

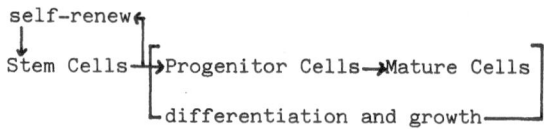

Notable features of the model are the inferred unidirectionality (ie differen- tiation is irreversible) and the large increases in cell number that occur as the progenitor cells develop into morphologic- ally and functionally distinct blood cells. Feedback mechanisms operate at different levels within the hierarchy to control proliferation in the different cell lineages.

From the above considerations, it is apparent that in order to understand the highly-complex processes of self-renewal and diversification, we need to identify the mechanisms that regulate mitosis and differentiation in the stem cell and pro- genitor cell compartments. Prevailing evidence, widely-discussed and documented in a number of recent reviews and original articles (4,5) suggests that discrete microenvironments in bone marrow exert a major regulatory influence on primitive haemopoietic cells. Humoral influences, characterised as freely-soluble molecules (eg. erythropoietin) are generally believed to influence the degree of amplification of committed transit cells rather than the actual committment process. However, recent data from our own laboratories suggest that some, perhaps all, of the haemopoietic growth factors may also have a role to play in "committment" of stem cells, and that the microenvironment of the bone marrow may be structured so as to present high local concentrations of particular growth stimulating molecules (6).

THE EXTRACELLULAR MATRIX (ECM)

An ECM is present in all organs and tissues. It is largely composed of a three-dimensional fibrous network of structural proteins (eg. collagens, fibronectin, laminin) and proteoglycans (PGs): an 'amorphous' soluble phase, in which PGs are also prominent components, exists within the fibre network. The ECM is a highly-ordered structure responsible for the shape and physical stability of tissues and it also provides sites for cell attachment and thus determines cellular organisation. However, the ECM is much more than simply a supportive structure. There is a wealth of literature indicating that the ECM is a vital component in the control of cell development in embryonic and adult tissues and in maintaining the stability of the differentiated phenotype (7). How can the matrix exert different developmental influences on different types of cell? The question is not easy to answer. It seems likely that despite some general similarities in composition, there are important quantitative and qualitative differences in ECMs prepared from various sources. The differences may be quite subtle in character; for example there can be variations in the expression of the collagen multigene family (8) and different isoforms of fibronectin may be produced by splicing of the fibronectin gene (9).

Wide variations also exist within the PG family. These macromolecules are composed of sulphated polysaccharide chains (the glycosaminoglycans or GAGs) linked covalently to protein. The structural diversity within the PG family is enormous due to variations in composition, number, size and sulphation patterns of the GAG chains and to the existence of many different types of protein core. Detailed descriptions of PG structure have been given in several recent reviews and will not be described here (10,11). However, we will provide some basic information on a particular group of PGs,

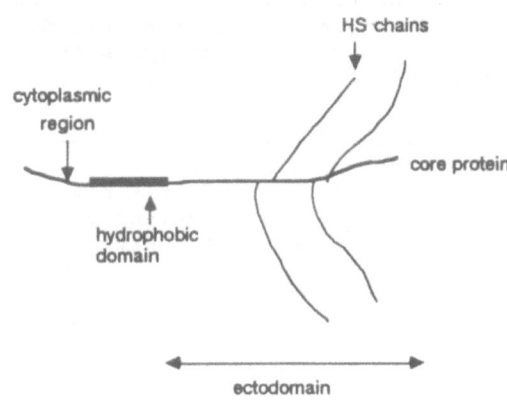

Fig. 1 - MODEL STRUCTURE OF A PLASMA MEMBRANE ASSOCIATED HSPG: HSPG is shown as a monomer composed of a core protein containing typically four HS side chains. In intact cells the hydrophobic amino acid sequence in the protein would span the lipid bilayer, with the polysaccharide chains located on the external surface of the cell where they are able to interact with other ECM components, specific cell surface receptors or growth factors.

called the heparan sulphate PGs (HSPGs), because they have some unique structural features, a strategic localisation in the pericellular domain and they might play a critical role in the organisation of the ECM and in the formation of 'mitogenic domains'.

STRUCTURE OF HEPARAN SULPHATE PROTEOGLYCANS

i) Macromolecular Organisation:

HSPGs are ubiquitous components of mammalain organs and tissues. They are principally located at cell surfaces or in the ECM but they have been detected in intracellular secretory granules and even in the cell nucleus (12,13). Several types of core protein have been identified in HSPGs. The PGs associated with plasma membranes are mostly hydrophobic molecules and probably contain a lipophilic amino acid sequence that enables the core protein to span the lipid bilayer (Fig.1; 14,15). Inositol phospholipids seem to be involved in retaining HSPGs on liver cell membranes though it is not clear whether the phospholipid is covalently attached to the HSPG or a component of a membrane receptor for the PG (16). Other HSPGs seem to be strictly extracellular constituents, the best described of these being the basement membrane PG that is composed of a large core protein and 3-4 HS side chains (17).

Overall the HSPGs comprise a diverse group of macromolecules and this diversity is most apparent from analyses that have been carried on the HS chain.

ii) The HS Polysaccharide Chain (Fig.2)

Like the majority of GAGs, HS is a linear polysaccharide composed of a disaccharide repeat unit which is covalently-linked to serine residues in the core protein of the PG (Fig.2a). In HS the disaccharide is formed from glucosamine and uronic acid. The glucosamine residue is either N-acetylated or N-sulphated to produce two basic types of disaccharide repeat (12,13):

N-acetylated:GlcNAcα1,4GlcUA

N-sulphated:GlcNSO$_3$$\alpha$1,4GlcUA or IdUA (Fig 2b)

These two disaccharides are present in approximately equal quantities in heparan sulphate chains, irrespective of the molecular weight of the polysaccharide, which can vary from about 20Kd-80Kd (ie 50-200 disaccharide units). The uronic acid is either glucuronic acid (GlcUA) or iduronic acid (IdUA) but the latter is only found at C-1 of GlcNSO$_3^-$ units (see Fig.2b). An intriguing feature of the HS chains is the predominantly segregated manner in which the disaccharides are arranged: extended sequences of N-sulphated and N-acetylated units co-exist in the same molecule (18). In the N-sulphated domains variable concentrations of additional sulphate residues (ester or 'O' sulphates) are found accentuating the high charge density in these regions. In contrast the N-acetylated domains contain low concentrations of sulphate groups. Ester sulphates are found most commonly at C-2 of IdUA and C-6 of GlcNSO$_3^-$. Other minor but biologically significant sulphate substitutions also occur. For example an O-sulphate at C-3 of GlcNSO$_3^-$ is important for the anticoagulant properties of some heparan sulphate species (13) and sulphation at C-2 of GlcUA is found in an HS fraction that is translocated from the cell surface to the nucleus in cultured liver cells (19). The key point is that considerable variations in the organisation and concentration of the sulphate residues may be found in heparan sulphate from different cell types. This stuctural polymorphism seems to arise in a regulated manner (12, 18) and accounts for the wide range of biological functions that have been attributed to HS. Properties of special relevance to our present discussion are growth factor binding, modulation of cellular interactions and the organisation of the ECM (for reviews - 10,12,13).

Fig 2a)

ser—Xyl—Gal—Gal—GlcUA—(GlcN.R—HexUA)$_n$

gly

```
         ┌──────────────── GAG chain ────────────────┐
ser ┌Xyl—Gal—Gal—GlcUA┐┌(GlcN.R—HexUA)₎ₙ           ┐
gly └──── Linkage Region ────┘└ Disaccharide
                                  Repeat
```

Core Protein   R = NAc or NSO$_3^-$

Fig 2b)

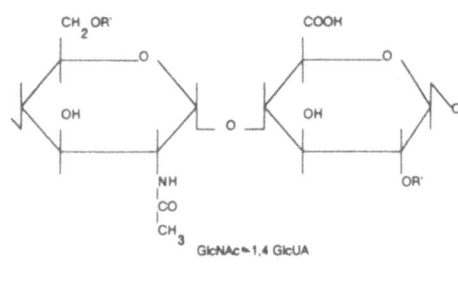

GlcNAcα1,4 GlcUA

ii) N-sulphated   R'= H or SO$_3^-$

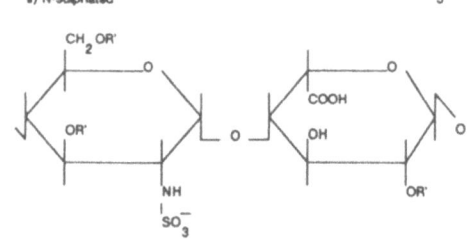

GlcNSO$_3$$\alpha$1,4 IdUA

Fig. 2 STRUCTURE OF THE HEPARN SULPHATE (HS) CHAIN:

a) Linkage to protein: HS chains are largely composed of a disaccharide repeat region that is connected to protein via a tetrasaccharide linkage sequence at the reducing end of the molecule. A xylose (xyl) residue forms an O-glycosidic linkage with serine in the protein core of the PG. Serines substituted with HS chains are normally adjacent to glycine residues. The disaccharide repeat unit consists of glucosamine (GlcN) and hexuronic acid (HexUA), the GlcN being either N-acetylated (Nac) or N-sulphated (NSO$_3^-$). The HexUA component may be glucuronic acid (GlcUA) or iduronic acid (IdUA).

b) Disaccharides repeat units in HS. N-acetylated (i) and N-sulphated (ii) disaccharides are joined in linear sequence in HS. The polymer is composed of approximately equal concentrations of these two disaccharides which are largely segregated from each other to create N-acetylated and N-sulphated domains. GlcUa is always present in the N-acetylated disaccharides whereas IdUA, most commonly, (as shown) or GlcUA may be found in the N-sulphated constituents. In panel (b) the letter 'R' indentifies potential sites of ester (O) sulphation. In reality, O-sulphates occur mainly in the N-sulphated domains. Variations in the concentration and distribution of O-sulphate groups are the major cause of the microheterogeneity observed in the HS family (12,18), and the synthesis of distinct sulphation isomers leads to the formation of specific protein binding sequences in the polymer (13).

HSPG IN BONE MARROW

HSPG is the major PG associated with the cell surface and ECM of mouse bone marrow cell cultures, (20,21). It is produced by the stromal cells and not by the haemopoietic cells (20). HSPG is a dynamic component of the pericellular domain ($t_{1/2}$, 5 hours) but turnover in marrow stroma is not accompanied by release into the growth medium. The PG is retained in the cell layer and is presumed to be internalised by endocytosis, and then degraded. This confinement of the PG to the cell surface is compatible with an important role in generating specialised microenvironments. Significantly, ultrastructural studies on embryonic chick bone marrow have identified HSPG in areas of active granulopoiesis and the PG seems to be located on the plasma membranes of fibroblastic stromal cells (22).

HSPG is also produced by cultures of human bone marrow (23) and by various stromal cell lines capable of supporting haemopoiesis (24) but in these culture systems it is generally less abundant than chondroitin sulphate PG. However, despite the relatively low concentration of HSPG in human marrow stroma we have recently found (in collaboration with Dr Myrtle Gordon) that this PG is enriched in mild trypsin extracts of human stromal cells, implying a specialised function at the cell surface or in the cell-associated ECM. Under the conditions used in our experiments trypsin did not release intracellular PGs.

It is perhaps worthwhile recalling that haemopoiesis in vitro and in vivo

requires direct contact between stromal cells of bone marrow and the haemopoietic cells (4). During this process specific and complex cellular interactions must take place. However, it is significant that bone marrow cultures do not release measurable quantities of growth factors into the culture medium. The mitogens responsible for initiating and sustaining haemopoietic activity seem to be retained on the stromal cell surface or in the ECM. This restricted distribution parallels that of HSPG. The question then arises of whether the PG is in some way involved in determining the distribution and metabolism of growth factors in bone marrow. In this context it is notable that Gordon et al (25) have demonstrated that GM-CSF would bind to a GAG extract prepared from human bone marrow cultures. Furthermore, there have been a number of reports indicating that fibroblast growth factor (FGF) is bound to HS in basement membranes.

INTERACTION OF FGFs WITH HEPARIN AND HEPARAN SULPHATE

Studies on this topic are of special relevance to our work on haemopoietic cell growth factors. FGFs are a family of structurally-related molecules that occur in both acidic and basic forms and which stimulate the growth of fibroblasts, smooth muscle cells and endothelial cells (26,27,28). FGFs also influence cell migration and morphology and they have been implicated in wound healing. It seems that basic FGF is the active component in preparations of tumour angiogenesis factor (29). FGFs bind with relatively high affinity to heparin-Sepharose columns and this property has been widely exploited to facilitate their purification from cell cultures or tissue extracts (27,30,31). Recently, the heparin binding site has been localised to a discrete region of the FGF molecule (32).

The physiological relevance of the interaction between heparin and FGF is uncertain because heparin is produced only by serosal mast cells (13). However, heparin is chemically-related to HS (13,18), and FGF found in tissues could be complexed to this polysaccharide. This seems to be the case in basement membranes produced by endothelial cells in culture and in vivo (31, 33). FGF-like activity can be released from such membranes by treatment with heparan sulphate-degrading enzymes or hypertonic salt (32). HS may not simply be an 'immobilising' agent for FGF. Heparin potentiates the mitogenic activity of basic FGF (26) and protects it from proteolytic enzymes and from denaturation in acidic conditions (34). Thus, in principle HS-FGF complexes in vivo could be very stable, biologically potent aggregates. Therefore we decided to examine the reactivity of HS with haemopoietic cell growth factors.

GROWTH FACTOR DEPENDENT CELL LINES

In the Paterson Institute cell lines have been developed that require the addition of growth factors to survive and proliferate in suspension culture. In our studies we used the IL3-dependent, multipotential cell line FDCP-mix A4 and the myeloid progenitor cell line FDCP-mix A7, which responds to both IL3 and GM-CSF (35). Growth and differentiation of the FDCP-mix cell lines also takes place in the absence of added growth factors if the cells are seeded onto irradiated mouse bone marrow stroma or onto embryo-derived 3T3-monolayers (36). Cell development requires intimate contact between the FDCP-mix cells and the stromal cells because, as discussed above, growth factors are not released in a soluble form by stromal cells in culture. It seems clear therefore that the mitogenic molecules must be located in the ECM or on the plasma membranes of the inductive stromal cells. However, it must be stressed that the mitogens produced by marrow stroma have not been indentified,
   although it is likely that they will be similar to IL3 and GM-CSF, at least in functional terms, and perhaps also in relation to their molecular structures. The heparan sulphate binding properties of IL3 and GM-CSF were examined in several different ways.

HAEMOPOIESIS INDUCED BY GROWTH FACTOR BINDING TO THE HEPARAN SULPHATE COMPONENT IN MATRIGEL

In our initial experiments we used a commercial preparation called Matrigel (purchased from Collaborative Research Ltd) which is an extract of basement membranes produced by the EHS sarcoma. Matrigel contains laminin, collagen type IV and other structural components of basement membranes, together with PGs, including HSPGs. Matrigel was layered onto the surface of a plastic culture dish and incubated (37°C) with solutions of single growth factors in physiological saline. Following incubation, the saline solutions were removed and the Matrigel was washed thoroughly to remove all traces of unbound growth factors. FDCP-mix A4 and FDCP-mix A7 cells were then seeded onto the appropriate Matrigel surfaces (pretreated with IL3 and GM-CSF respectivly) in the absence of any soluble growth factors.

In a typical experiment 2×10^5 cells in 1ml of culture medium were inoculated onto 35mm diameter plastic petri dishes coated with Matrigel and growth factor. Between 30% and 50% of the cells bound to the Matrigel surface over a 2 hour incubation period. Media containing non-adherent cells were removed and replaced with fresh media. The cultures were incubated at 37°C for 6-8 days. Small aliquots of media were removed at intervals to monitor release of cells from the Matrigels. The results were quite striking. Both FDCP mix A4 and FDCP-mix A7 cells increased steadily in number and by day 4 about 8×10^5 cells were present in the media whilst flourishing layers of adherent cells, which were not counted, were observed on the Matrigel surfaces. Cell numbers were maintained at this level for a further 48 hours before undergoing a gradual decline.

Throughout these experiments we could not detect any growth factor activities in the culture medium itself, so the IL3 and GM-CSF were acting as mitogens whilst retained on the Matrigel substrata. Both recombinant and naturally-occurring growth factors were active under these conditions. Since the recombinant growth factors were bacterially-derived this indicates that glycosylation of IL3 and GM-CSF is not required for the binding of these molecules to Matrigel. There was little or no cell binding or growth on Matrigel that had not been pre-treated with growth factors. Significantly, the adsorption of growth factors to the Matrigel, as measured by the proliferation of the growth factor-dependent cell lines, was markedly reduced (approx. 5% of control values) when the Matrigel substrata were first treated with the specific poly-saccharide lyases, heparitinase and heparinase, which degrade HS. These results strongly implicated HS as the component responsible for the growth factor binding properties of Matrigel. Treatment of Matrigel with another glycosaminoglycan-degrading enzyme, chondroitinase ABC; which breaks down chondroitin and dermatan sulphate and hyaluronic acid, did not influence its ability to bind IL3 or GM-CSF (37).

GROWTH FACTOR BINDING TO HEPARAN SULPHATE FROM INDUCTIVE STROMAL CELLS

Having established the principle that haemopoietic cell growth factors would bind to HS in a complex biomatrix, it was then necessary to see if heparan sulphate chains isolated from inductive stromal cells would display similar properties. HSPGs were extracted from 3T3 cells and the HS chains were released from the core protein by treatment with dilute NaOH. The free polysaccharides were then co-valently linked to a phospholipid (38) and this hydrophobic complex could be readily adsorbed to the surfaces of plastic petri dishes, providing a means of immobilising HS in the absence of core protein or other matrix macromolecules. In a parallel study, the purified intact

HSPG from bone marrow cultures was adsorbed onto a matrix substratum formed
by drying down a solution of fibronectin and type I collagen (1mg/ml of
each protein). Using similar cell growth
assays to those described for the
Matrigel experiments, we were able to
clearly demonstrate that IL3 and GM-CSF
would bind to thse HS-containing substrata and support the extensive proliferation of the appropriate growth-
factor dependent cell lines (37).
Binding of growth factors was largely
eliminated by treatment of the substrata
with heparinases but not by treatment
with chcndroitinase ABC. These results
indicate that free HS chains and HS
chains attached to protein (ie. HSPG)
display binding activity for IL3 and GM-
CSF. Activity of HS-chains was established in the absence of other ECM components.
Full quantitative data on the Matrigel
experiments and the studies with lipid-
derivatised HS are given in ref.37

SUMMARY AND PERSPECTIVE

Haemopoietic microenvironments in the
bone marrow must provide stimuli that
initiate cell division and differentiation in stem cells and their derived
progenitors. The foregoing data demonstrate that HSPG in plasma membranes and
the ECM of the marrow stroma express the
necessary binding properties to facilitate the loclisation of mitogens for
immature haemopoietic cells. The agonist
function of the bound mitogens is retained
and might even be potentiated. Haemopoiesis is focal in nature but it is
likely that HSPGs are widely distributed
in bone marrow. Mitogens could become
localised by secretion from specific
cell collectives followed by rapid binding to the ubiquitous HSPGs in the pericellular domain. It is also possible
that the HSPGs could play an important
role in regulating the intracellular
transport of growth factor along the
secretory pathway. Growth regulation in
marrow stroma may be depicted as a paracrine mechanism in which complexes of HS
and mitogens on stromal cell surfaces
interact with haemopietic stem cells and
progenitor cells (Fig.3). In order to
extend this model to normal haemopoiesis
it will be necessary to identify the
actual mitogenic molecules in bone
marrow and examine their interactions with
HS and other sulphated polysaccharides.
The structural characteristics of HS that
are responsible for mitogen binding also
need to be elucidated. It is conceivable
that the appropriate sugar sequences that
enable such interactions to take place
are found in the majority of HS species
but minor variations in sulphation pattern
might modify the mitogenic potency and
stability of the bound molecules. Determination of the structure-activity
relationships of the heparan sulphates

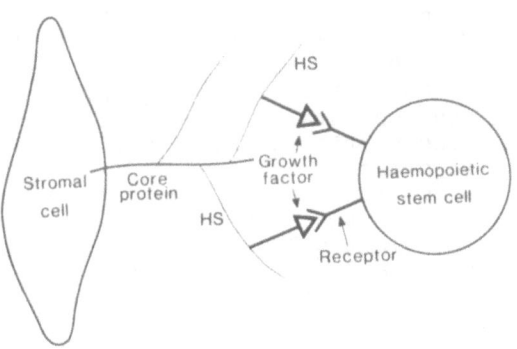

Fig.3: HEPARAN SULPHATE-MEDIATED PRESEN-
TATION OF GROWTH FACTORS BY BONE MARROW
STROMAL CELLS:A MODEL FOR PARACRINE
GROWTH STIMULATION: In this model growth
factors produced by stromal cells are
expressed on the cell surface by binding
to HS chains of trans-membrane HSPGs.
Mitosis in haemopoietic stem cells is
initiated by an interaction between
specific plasma membrane receptors and
the HS-bound growth factors. This
'paracrine' mechanism provides a means
for restricting growth factors to distinct
microenvironments in the bone marrow.

----

could lead to their use as modulators of
the activities of growth factors in
clinical use.

REFERENCES

1. Metcalf D (1984) The haemopoietic
   colony stimulating factors.
   Amsterdam: Elsevier
2. Dexter TM (1987) Stem cells in normal
   growth and disease. Br Med J 295:
   1192-1194
3. Dexter TM, Spooncer E, Schofield R,
   Lord PI and Simmons P (1984)
   Haemopoietic stem cells and the
   problem of self-renewal. 10:315-339
4. Dexter TM, Simmons P, Pomell RA,
   Spooncer E and Schofield R (1984)
   The regulation of haemopoietic cell
   development by the stromal cell
   environment and diffusible regulating
   molecules. In: Aplastic Anaemia:
   Stem Cell Biology and Advances in
   Treatment pp 13-33. Alan R Liss Inc:
   New York

5. Wolf NS and Treutin JJ (1968) Haemopoietic colony studies V Effect of haemopoietic organ stroma on differentiation of pluripotent stem cells. J Exp Med 127: 205-214

6. Heyworth CM, Ponting IO and Dexter TM (1988) The response of haemopoietic cells to growth factors: developmental implications of synergistic interactions. J Cell Sci in press

7. Grobstein C (1975) Developmental role of the extracellular matrix. In: Slautin H and Grenlich RD (Eds) Extracellular matrix influences on gene expression, pp 9-16. New York, Academic Press

8. Martin GR, Timpl R, Muller PK and Kuhn K (1985) The genetically distinct collagens. TIBS 10: 285-287

9. Dufour S, Duband JL, Kornblihtt AR and Thierry JP (1988) The role of fibronectins in embryonic cell migrations. TIG 4: 198-203

10. Poole AR (1986) Proteoglycans in health and disease, structures and functions. Biochem J 236: 1-14

11. Lindahl U and Höök M (1978) Glycosaminoglycans and their binding to biological macromolecules. Ann Rev Biochem 47: 385-418

12. Gallagher JT, Lyon M and Steward WP (1986) Structure and function of heparan sulphate proteoglycans. Biochem J 236: 313-325

13. Lindahl U and Kjelten L (1987) Biosynthesis of heparin and heparan sulphate. In: Wight TM and Mecham RP (eds) pp 59-104 Biology of Extracellular Matrix. Vol II Biology of Proteoglycans. New York, Academic Press.

14. Repraeger A, Jalkanen M and Bernfield M (1987) Integral membrane proteoglycans as matrix receptors: roles in cytoskeleton and matrix assembly at the epithelial surface. In: Wight TM and Mecham RP (eds) pp129-154. Biology of Extracellular Matrix. Vol II Biology of Proteoglycans. New York, Academic Press.

15. Woods A, Couchman JR and Höök M (1985) Heparan sulphate proteoglycans of rat embryo fibroblasts. A hydrophobic form that may link the cytoskeleton and matrix components. J Biol Chem 260: 10872-10879

16. Ishihara M, Fedarko NS and Conrad HE (1987) Involvement of phosphatidylinositol and insulin in the coordinate regulation of proteoheparan sulphate metabolism and hepatocyte growth. J Biol Chem 262: 4708-4716

17. Hassel JR, Leyshan WC, Ledbetter SR Tysee B, Suzuki S, Kato M, Kimata K and Kleinman HK (1985) Isolation of two forms of basement membrane proteoglycans J Biol Chem 260: 8098-8105

18. Gallagher JT and Walker A (1985) Molecular distinctions between heparan sulphate and heparin. Analysis of sulphation patterns indicates that heparan sulphate and heparin are separate families of N-sulphated polysaccharides. Biochem J 230: 665-674

19. Ishihara M, Fedarko NS and Conrad HE (1986) Transport of heparan sulphate into the nuclei of hepatocytes. J Biol Chem 261: 13575-13580

20. Gallagher JT, Spooncer E and Dexter TM (1983) Role of the cellular matrix in haemopoiesis. Synthesis of glycosaminoglycans by mouse bone marrow cell cultures. J Cell Sci 63: 155-171

21. Spooncer E, Gallagher JT, Krisza F and Dexter TM (1983) Regulation of haemopoiesis in long term bone marrow cultures. IV Glycosaminoglycan synthesis and the stimulation of haemopoiesis by β-D-xylosides. J Cell Biol 96: 510-514

22. Sorrell JM, Voci M and Weiss L (1987) Ultrastructural localisation of heparan sulphate and chondroitin sulphates associated with granulopoiesis in embryonic chick bone marrow. Amer J Anat 179: 186-197

23. Wight TN, Kinsella MG, Keating A and Singer JW (1986) Proteoglycans in human long term bone marrow cultures. Biochemical and ultrastructural analyses. Blood 67: 1333-1338

24. Kirby SL and Bentley SA (1987) Proteoglycan synthesis in two bone marrow stromal cell lines. Blood 70: 1777-1783

25. Gordon MV, Riley GP, Watt SM and Greaves MF (1987) Compartmentalisation of an haemopoietic growth factor (GM-CSF) by glycosaminoglycans in the bone marrow microenvironment. Nature 326: 403-405

26. Schrieber AB et al (1985) A unique family of endothelial polypeptide mitogens: the antigenic and receptor cross-reactivity of bovine endothelial cell growth factor and eye-derived growth factor II. J Cell Biol 101: 1623-1626

27. Lobb RR and Fett JW (1984) Purification of two distinct growth factors from bovine neural tissue by heparin affinity chromatography. Biochemistry 23: 6295-6298

28. Baird A, Esch F, Bohlen P, Ling N and Gospodarowicz D (1985) Isolation and partial characterisation of an endothelial cell growth factor from the bovine kidney: Homology with basic fibroblast growth factor. Regal Pept 12: 201-213

29. Klagsbrun M, Sasse J, Sullivan R and Smith JA (1986) Human tumour cells synthesize an endothelial cell growth factor that is structurally related to basic fibroblast growth factor. Proc Natl Acad Sci USA 83: 2448-2452

30. Maciag T, Mehlman T, Friesel R and Schrieber AB (1984) Heparin binds endothelial cell growth factor, the principal endothelial cell mitogen in bovine brain. Science 225: 932-935

31. Folkman J, Klagsbrun M, Sasse J, Wadzinski M, Ingber D and Vlodavsky I (1988) A heparin-binding angiogenic protein, basic fibroblast growth factor is stored within basement membrane. Amer J Path 130: 393-400

32. Baird A, Schubert D, Ling N and Guillemin R (1988) Receptor and heparin-binding domains of basic fibroblast growth factor. Proc Natl Acad Sci USA 85: 2324-2328

33. Baird A and Ling N (1987) Fibroblast growth factors are present in the extracellular matrix produced by endothelial cells in vitro: Implications for a role of heparinase-like enzymes in the neovascular response. Biochem Biophys Res Commun 142: 428-435

34. Gospodarowicz D and Cheng J (1986) Heparin protects basic and acidic FGF from inactivation. J Cell Physiol 128: 475-484

35. Spooncer E, Heyworth CM, Dunn A and Dexter TM (1986) Self-renewal and differentiation of interleukin-3 dependent multipotent stem cells are modulated by stromal cells and serum factors. Differentiation 31: 111-118

36. Roberts RA, Spooncer E, Parkinson EK, Lord BI, Allen TD and Dexter TM (1987) Metabolically inactive 3T3 cells can substitute for marrow stromal cells to promote the proliferation and development of multipotent haemopoietic stem cells. J Cell Physiol 132: 203-214

37. Roberts RA, Gallagher JT, Spooncer E Allen TD, Bloomfield F and Dexter TM (1988) Heparan sulphate bound growth factors: a mechanism for stromal cell mediated haempoiesis. Nature 332: 376-378

38. Tang PW, Gooi HC, Hardy M, Lee YC and Feizi T (1985) Novel approach to the study of the antigenicities and receptor functions of carbohydrate chains of glycoproteins. Biochem Biophys Res Commun 132: 474-480

Part III. Granulopoietic Regulators

Chairperson: D.H. Pluznik

A.R. Dunn, R.A. Cuthbertson, and R.A. Lang

INTRODUCTION

Studies of genes encoding products thought to play a role in growth and differentation have for the most part been carried out using in vitro systems. Often these genes have been introduced into cultured cells which consequently assume new characteristics and in some cases, in contrast to the parental cell, when introduced into the syngeneic animal develop into tumors. While these studies have been of undeniable value in identifying various oncogenes the relevence of tumors initiated in this way to those arising naturally but perhaps involving, in part, the same oncogene has been a matter of some concern.

In the past 10 to 12 years a major technological breakthrough has been the development of techniques that allow genes to be introduced into the germline of laboratory animals (1,2,3). Often the presence of the newly introduced gene results in profound affects on the founder animals and when the transgene is inherited the same phenotype is perpetuated in offspring. In this way colonies of affected animals become available for detailed studies over an indefinate period of time. Accordingly, transgenic technology provides a model of oncogenesis that is conceptually acceptable in our pursuit to understand the molecular basis of multistage malignant disease.

Experiments carried out in a large number of laboratories have revealed that the genes that can contribute to multistage carcinogenesis are both numerous and diverse in their nature. In spite of their diversity many of them are related in that, in one way or another, their products form part of signal transduction pathways originating from normal growth factor receptors. Thus, it is now firmly established that oncogenes can correspond to growth factors themselves, growth factor receptors or some of the genes whose products form part of the biochemical circuitry that link receptors to the cellular genome (for recent reviews see 4,5). Alterations in growth may occur if the concentrations of the products of oncogenes is increased, (either by being coupled to strong promotors or sometimes by changes in gene copy number), or by subtle alterations in the genetic composition of specific genes resulting in qualitative changes in cellular proteins. A long held belief is that one contributing factor to the malignant process might be that a cell normally requiring an interaction with a specific growth factor, produced by an opposing cell, might be released from this constraint if the same cell were to produce and respond to this factor (6,7). In this scheme, autocrine stimulation would liberate the cell from any territorial constraints imposed by the requirment for soluble mitogenic growth factors. While this notion provides a rational explanation for the derivation of autonomous cells within solid tissues its relevence to haemopoietic cells is less obvious. Nonetheless, experiments carried out in a number of laboratories including our own have demonstrated that non leukemic cells that depend on specific haemopoitic growth factors, such as granulocyte macrophage colony stimulating factor (GM-CSF), can be converted into autonomous and fully leukaemic cells by the introduction of a gene encoding GM-CSF (8). While this observation provides a satisfying explanation for oncogenic transformation of haemopoitic tissues in vitro its relevence to naturally arising leukaemia is less obvious. In this context we have undertaken a study involving the introduction of the murine GM-CSF gene under the control of a retroviral promotor into the germline of laboratory mice. At the outset we reasoned that if autocrine stimulation might represent one of the events in multistage leukaemogenesis then animals whose cells produced GM-CSF might be more susceptible to developing neoplastic changes in haemopoietic tissues than control animals. Moreover, since the CSF's are biologically potent molecules in vitro we reasoned that this approach may illuminate the role that GM-CSF might play in the regulation of normal haemopoiesis in vivo.

RESULTS

GM-CSF transgenic mice display a number of pathological abnormalities.

Since our purpose was to examine the consequences of abberant GM-CSF expression in the whole animal the coding sequence of the GM-CSF gene (9) was positioned downstream of a retroviral promotor. To aid identification of transcripts encoded by the transgene, and to distinguish them from endogenous GM-CSF mRNAs, a fragment of bacteriophage ϕX174 DNA was inserted into the 3' untranslated region of the construct (Fig 1) which was subsequently microinjected into the pronuclei of fertilised C57/SJL F1 mouse eggs. Following transfer of eggs into the fallopian tubes of pseudo-pregnant mice, progeny animals were generated which were subsequently screened for the presence of the transgene by Southern blotting of tail DNAs. In this way two founder animals were identified which on the basis of mapping with specific restriction endonucleases were certified as containing two copies of the transgene arranged in

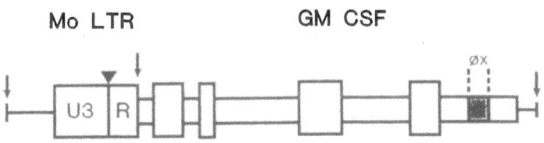

Figure 1. GM-CSF transgene. Fragments derived from the MoMLV LTR, the GM-CSF gene and ϕX174 are indicated by slashed lines. The site initiation of transcription within the LTR is at the border of U3 and R and is indicated by a triangle.

a tail-to-tail configuration. A breeding program was initiated and progeny mice descended from the founder animals were screened for the presence of the transgene. The pattern of inheritence amongst progeny animals led us to conclude that the transgene was integrated into one X chromosome of the founder female and that the germ tissue of the founder male was a mosaic of normal and transgenic cells. The different chromosomal integration sites within the independently derived founder animals strongly suggests that the observed abnormalities are a consequence of abberant expression of the transgene rather than fortuitous disruption or activation of a proximal cellular gene.

A striking feature of mice that inherited the GM-CSF transgene was a bilateral opacity of the eye. Histological examination of the eyes of affected animals revealed the presence of abnormal accumulations of macrophages with damage to the retina and cornea and in some cases associated cataractous changes to the lens (8). Since the occular abnormalities were evident as soon as mice opened their eyes 14 days after birth the syndrome clearly had its origin in development. By studying transgenic animals from early neonatal life we have established that retinal development follows a normal pattern until 5 days post partum at which time abnormal cells are present in the aqueous humour and vitreous and subsequently continue to accumulate (10). On morphological grounds coupled with evidence for expression of specific cell surface markers (Mac 1 and F4/80) the abnormal cells are probably macrophages which are thought to play a pivotal role in modelling occular tissues (11).

In contrast to the eyes of littermate control animals the eyes of GM-CSF transgenic mice contained biologically active GM-CSF. Moreover, analysis of pooled occular tissue by Northern blotting using a φX174 radiolabeled probe revealed the presence of transcripts corresponding to the GM-CSF transgene (12). This gross analysis has recently been refined by examining sections of eyes by hybridization histochemisty which has allowed us to establish that the occular macrophages represent a major site of transcription of the transgene (10, 13, Cuthbertson and Lang, manuscript in preparation).

In addition to occular abnormalities transgenic mice also display high levels of GM-CSF in serum and urine and have abnormal accumulations of macrophages in the peritoneal and pleural cavities. With increasing age the mice become lethargic, suffer weight loss and assume a hunched posture often involving some degree of paralysis associated with the hind limbs. At this advanced stage mice developed a pronounced tremor when handled and when this was evident mice usually succumbed in 2 to 3 days. Examination of the mice at autopsy revealed the presence of highly abnormal focal accumulations of macrophages in striated muscle. Where these accumulations of macrophages existed there was obvious atrophy and destruction of surrounding muscle cells.

To determine whether these macrophages express the transgene we have examined histological sections of striated muscle of affected animals by hybridization histochemistry using single stranded RNA probes corresponding to the φX174 DNA sequences contained within the 3' untranslated region of the GM-CSF transgene. Two probes have been utilised that differ only in the polarity of the φX174 sequences within the vector. As shown in Fig 2, when the RNA probe has the same transcriptional polarity as the transcript of the transgene no grains above background are observed. By contrast, when the φX174 sequences within the probe are complementary to transgene mRNAs significant

numbers of grains are found to be associated with focal accumulations of macrophages within striated muscle. Thus these macrophages, like those within eyes of transgenic animals (10), express the GM-CSF transgene.

Auto-stimulated Macrophages express genes encoding a number of cytokines.

At the outset it seemed unlikely that the gross tissue destruction observed primarily at the sites where macrophages had accumulated could be explained solely on the basis of unusual numbers of otherwise normal macrophages. Rather, it seemed that the damage might have occured due to the activated state of the macrophages leading either to a heightened capacity for phagocytosis or, alternatively, by the indirect effects of one or more cytokines known to be the products of activated macrophages. That the basis for tissue damage was the activated state is supported by the morphological appearance of macrophages in the eyes, peritoneal cavity and histological sections of affected tissues as well as a demonstrated capacity of peritoneal cells to phagocytose red blood cells in vitro.

To explore the possibility that the basis for tissue damage in transgenic mice was the production of one or more cytokines we have undertaken a survey of tissues of transgenic animals by hybridization histochemistry using probes corresponding to murine IL-1, basic FGF and TNF. Frozen sections of muscle were established from the thighs of GM-CSF transgenic mice sacrificed when the animals displayed the characteristic symptoms of advanced pathological disease. Sections were challenged with single-stranded radiolabeled probes that were either synonomous or complementary to particular monokine mRNAs. Following hybridization, processing and autoradiography sections were examined microscopically for the presence of autoradiographic grains. Fig 3 shows a typical example of such an analysis using single stranded radiolabeled probes corresponding to murine TNF. Significant labelling was observed when the probe was complementary to TNF mRNAs. By contrast, no labelling above background was evident in similar sections of transgenic tissue using a

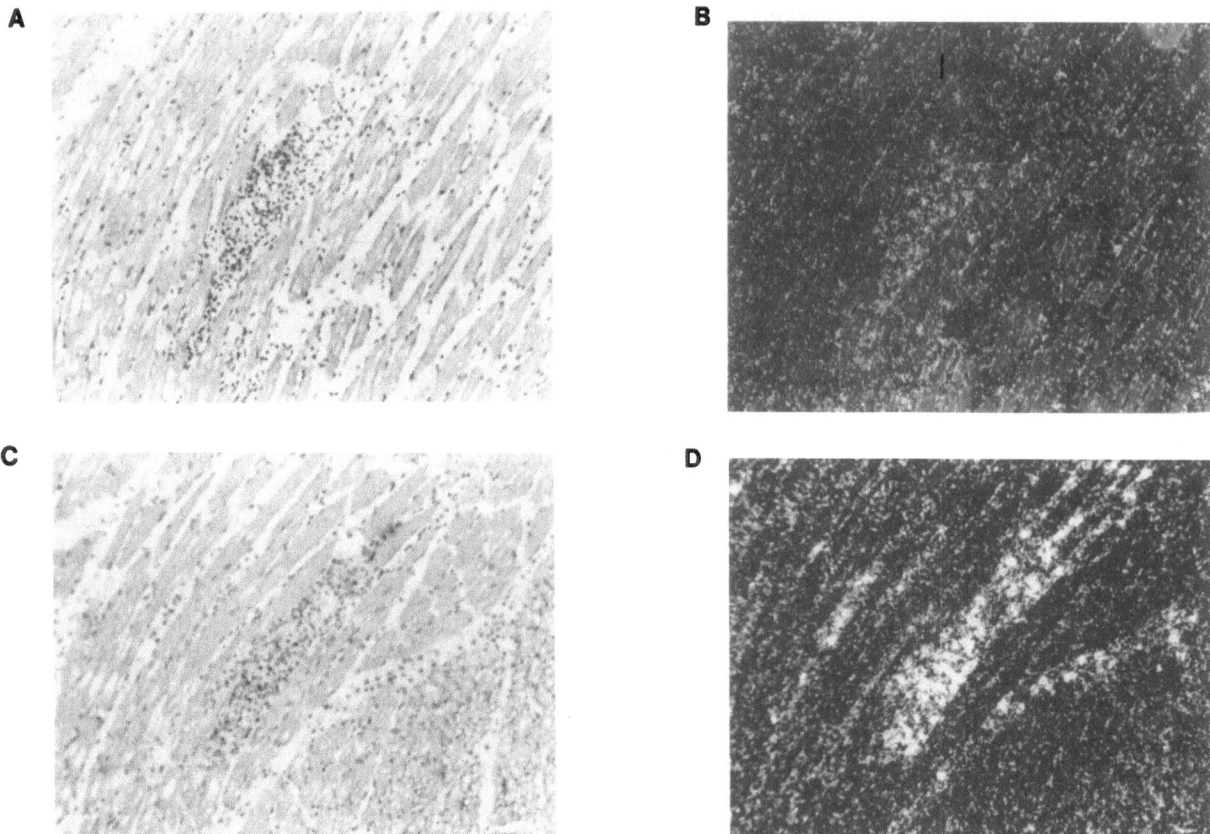

Figure 2. Expression of the transgene in macrophages within striated muscle. Hybridization histochemical analysis of frozen sections of striated muscle from an 18 day old female mouse from the female line. Hybridization was performed by the method of Penschow et al, (17) and the conditions detailed by Lang et al, (12). Single stranded RNA probes were generated by in vitro transcription of a 310 base pair fragment of φX174 DNA cloned into pGEM 3 in both transcriptional orientations using SP6 polymerase and α[32P]UTP. Upper panels show autoradiographs of sections treated with a probe synonomous to transgene mRNAs and photographed using light(a) and dark (b) field illumination. Lower panels show autoradiographs of sections exposed to a probe complementary to transgene mRNAs in light (c) and dark (d) field illumination (x125).

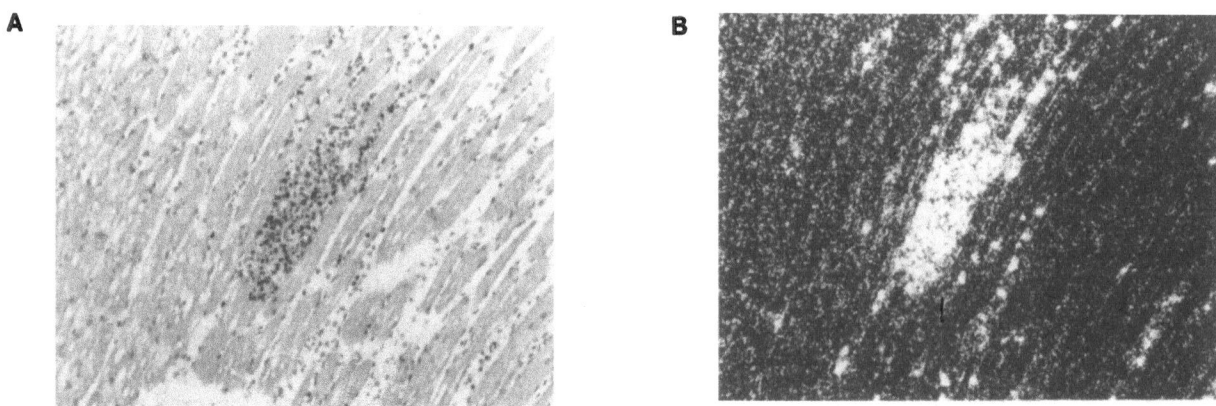

Figure 3. Expression of TNF in macrophages within striated muscle. Hybridisation histochemical analysis of sections from the same animal as Fig 2, but using a radiolabelled probe complementary to murine TNF. The field is shown in light (a) and dark (b) field illumination (x125).

radiolabelled TNF probe of the same transcriptional polarity as TNF mRNA.

DISCUSSION

The gross pathological abnormalities displayed by the transgenic mice generated in the present study occur, directly or indirectly, as a consequence of the unscheduled expression of GM-CSF. Some of the abnormalities were unexpected while others were predicted on the basis of the spectrum of activities displayed by GM-CSF on haemopoietic cells in vitro.

To a large extent the outcome of our study depended on the range of tissues that were permissive for the retroviral promotor used to direct expression of the transgene. In the event, our data indicate that the macrophages themselves are the principal sites of expression. The consequence of this is to create an unusual situation where cells that display receptors for GM-CSF are also the cells that produce it. This is of relevence in considering one of the primary roles attributed to CSFs, namely their role as survival factors. Thus, in normal health the distribution of macrophages within the body might be the regulated by the availability of a particular CSF. In transgenic mice where the macrophages constitutively produce their own GM-CSF this level of regulation is lost and this may account for the appearance of macrophages at abnormal sites. By the same token we can account for the large numbers of macrophages within transgenic animals on the basis of an inability to withdraw CSF. Thus, macrophages accumulate rather than turnover. The conclusion that the large numbers of macrophages evident in transgenic mice is based on enhanced survival is supported by the cellularity and composition of the bone marrow of transgenic animals that is similar to that of littermate mice. Accordingly, the increase in macrophages cannot be attributed to enhanced numbers of progenitor cells.

In addition to the fact that transgenic mice contain accumulations of macrophages it is also interesting that, based on a number of criteria, they appear to be functionally activated. This is in line with another well established role for CSFs in serving to functionally activate terminally differentiated haemopoietic cells. The presence of large numbers of activated macrophages in the transgenic mice is reminiscent of an infection or acute inflammatory response. In the normal course of an inflammatory response macrophages and other inflammatory cells enter the injured area and participate in a variety of physiological and pathological events including phagocytosis, the release of growth factors essential for the recruitment of other cells, the formation of new matrix, angiogenesis and the modulation of the immune response.

The way in which these various processes are regulated and coordinated is poorly understood. It is difficult therefore to equate the activated state of macrophages expressing the GM-CSF transgene with those activated during infection, inflammation or wound repair. Whatever the parallels might be it is clear from an ongoing analysis that the macrophages occupying sites within normal tissues of transgenic mice are expressing a number of monokines normally associated with activated macrophages. The demonstration, in the present study, that macrophages comprising focal lesions within striated muscle express TNF is particularly interesting in light of the muscle wasting and damage to normal tissue observed in GM-CSF transgenic animals. Not only is TNF a certified product of a macrophage cell line (U937) exposed to GM-CSF, as well as monocytes stimulated with endotoxin or phorbol myristate (14) it is also known to induce muscle wasting and necrosis (15, 16).

In ongoing studies in our laboratories we have also shown that macrophages within the eyes and muscle coordinately express IL-1α and basic FGF. While these, and other regulators, have specific properties they also display overlapping activities and thus it is not possible to attribute a specific pathological feature of the GM-CSF transgenic mice to the influence of a specific regulator. Whether one or more of these regulators plays a causal role in the abnormalities observed in the transgenic mice will ultimately require the long term infusion of specific neutralising antibodies to potentially important regulators in an attempt to alleviate or prevent some of the pathology we have observed.

Finally, what relevence do the GM-CSF mice have to a better understanding of myeloid leukemia? Despite this question being, in part,the basis of establishing these animals it is clear that they have not provided a good in vivo model for neoplastic haemopoiesis. To a large extent this conclusion is related to the fact that the retroviral promotor utilised in our study shows no evidence of being expressed in haemopoietic cells that are sufficiently immature to serve as the founder cells for a leukemic clone. Equally importantly the high levels of GM-CSF associated with transgenic mice lead to a progressively debilitating disease with few mice surviving for more than one quarter of the normal lifespan of a laboratory mouse.

Clearly if these interests are to be pursued it will be necessary to identify a promotor/enhancer which functions in primitive haemopoietic cells but not in progeny derived from them. This is a challenging objective but one that needs to be pursued if we are to establish an in vivo model for exploring the role, if any, that autocrine stimulation plays in the multistage nature of myeloid leukemia.

ACKNOWLEDGMENTS.

We are grateful to Don Metcalf, Tony Burgess and John Coghlan for their continued support and contributions to this work. R.A.C. is an NH&MRC post doctoral fellow and R.A.L. is the recipient of a Commonwealth Postgraduate Research Award. The work was supported in part by grants from the National Health and Medical Research Council, the Ian Potter Foundation and the Howard Florey Biomedical Foundation.

REFERENCES.

1. Brinster, R.L., Chen, H.V., Trumbauer, M.E., Yagle, M.K., Palmiter, R.D. (1985) Factors effecting the efficiency of introducing foreign DNA into mice by injecting eggs. Proc. Natl. Acad. Sci. USA 82, 4438-4442.

2. Palmiter, R.D., and Brinster, R.L. (1986) Germ line tranformation of mice. Ann. Rev. Genet. 20, 465-499

3. Jaenisch, R. Transgenic animals. (1988) Science 240, 14681474.

4. Heldin, C-H., Betsholtz, C., Claesson-Welsh, L., and Westermark, B. (1988) Subversion of growth regulatory pathways in malignant transformation. Biochimica et Biophysica Acta 907, 219-244.

5. Bishop, J.M. (1988) The molecular genetics of cancer: 1988. Leukemia 2, 199-208.

6. Todaro, G., and DeLarco, J.E. (1977) MSA and EGF receptors on sarcoma virus transformed cells and human fibrosarcoma cells in culture. Nature 267, 526-528.

7. Sporn, M.B., and Roberts, A.B. (1985) Autocrine growth factors and cancer. Nature 313, 745-747.

8. Lang, R.A., Metcalf, D., Gough, N.M., Dunn, A.R., and Gonda T.J. (1985) Expression of a hemopoietic growth factor cDNA in a factor-dependent cell line results in autonomous growth and tumorigenicity. Cell 43, 531-542.

9. Stanley, E., Metcalf, D., Sobieszczuk, P., Gough, N.M. and Dunn, A.R. (1985) The structure and expression of the murine gene encoding granulocyte-macrophage colony stimulating factor: evidence for utilisation of alternative promoters. EMBO J. 4, 3569-2573.

10. Cuthbertson, R.A., Lang, R.A., Dunn, A.R., Penschow, J.D., Lyons, I., Klintworth, G.K., Metcalf, D. and Coghlan, J.P. (1988) Developmental ocular disease is mediated by macrophages in GM-CSF transgenic mice. In: UCLA Symposium on Molecular and Cellular Biology, New Genes, Vol. 88 (J. Piatagorsky, P. Zelenka and T. Shinohara, eds.), Alan Liss, In., New York. In Press.

11. Balasz, E.A., L.Z. Toth, and V. Ozanics. (1980) Cytological studies of the developing vitreous as related to the hyaloid vessel system. Albrecht v. Graefes Arch klin exp Ophthal 213, 71-85.

12. Lang, R.A., D. Metcalf, R.A. Cuthbertson, I. Lyons, E. Stanley, A. Kelso, G. Kannourakis, J. Williamson, G.K. Klintworth, T.J. Gonda, and A.R. Dunn. (1987) Transgenic mice expressing a hemopoietic growth factor gene (GM-CSF) develop accumulations of macrophages, blindness and a fatal syndrome of tissue damage. Cell 51:675-86.

13. Dunn, A.R., Lang, R.A., Cuthbertson, R.A., and Metcalf D. (1988) The aberrant expresssion of a murine haemopoietic growth factor (GM-CSF) in transgenic mice. In: UCLA Symposium on Molecular and Cellular Biology, Vol. 88 (R. Ross, A. Burgess and T. Hunter, eds.), Alan Liss, Inc., New York. In Press.

14. Cannistra, S.A., Rambaldi, A., Spriggs, D.R., Herrmann, F., Kufe, D., and Griffin, J.D. (1987) Human granulocyte-macrophage colony stimulating factor induces expression of the tumor necrosis factor gene by the U937 cell line and by normal human monocytes. J. Clin Invest. 79, 1720-1728.

15. Carswell, E.A., Old, L.J., Kassel, R.L., Green, S., Fiore, N. and Williamson, B. (1975) An endotoxin-induced serum factor that causes necrosis of tumours. Proc. Natl. Acad. Sci. USA 72, 3666.

16. Oliff, A., Defeo-Jones, D., Boyer, M., Martinez, D., Kiefer, D., Vuocolo, G., Wolfe, A. and Socher, S. (1987) Tumors secreting human TNF/cachectin induce cachexia in mice. Cell 50, 555-563.

17. Penschow, J.D., J. Haralambidis, P. Aldred, G.W. Tregear, and J.P. Coghlan. (1986) Location of gene expression in the CNS using hybridization histochemistry. Methods in Enzymology. 124:534-548.

9 Granulocyte Colony-Stimulating Factor

K.McD. Taylor, G. Spitzer, and L. Souza

The independent (1,2) establishment of *in vitro* culture systems for hemopoietic progenitor cells has allowed the identification of various hormones which regulate blood cell production and differentiation. These glycoproteins, known as colony-stimulating factors (CSFs) were originally purified from the conditioned media of a number of diverse sources, including placenta, tumor cell lines, and peripheral blood mononuclear cells. The factors were initially named according to the predominant colony type their addition produced in cultures of bone marrow cells (3). Currently in the human system, several CSF's have been defined which act in a hierarchical fashion and which possess varying lineage specificity. Two of the factors have a relatively restricted proliferative action; macrophage-CSF (M-CSF) stimulates the formation of mainly monocyte/macrophages (4), whilst granulocyte-CSF (G-CSF) induces neutrophilic granulocytes and their precursor cells (5). The multi-lineage CSF's, granulocyte-macrophage CSF(GM-CSF) and multi-CSF (Interleukin 3 or IL-3), have a broader range of actions. GM-CSF stimulates neutrophil and monocyte-macrophage colonies, while at higher concentrations, eosinophil, megakaryocyte and some erythroid precursors are formed. IL-3 yields erythroid and myeloid colonies including neutrophils, eosinophils, basophils, monocyte-macrophages and megakaryocytes. Thus G-CSF and M-CSF cause proliferation of late-committed progenitor cells, while GM-CSF and IL-3 act on earlier progenitors capable of multiplicity of differentiation.

A number of additional hemopoietic growth factors have been defined. Erythropoietin (EPO) regulates terminal red cell development (6) and Interleukin-2 (IL-2) is a growth and regulatory factor for mature lymphoid cells (7). Hemopoietin-1, now shown to be identical to Interleukin-1 (IL-1) (8), appears to synergize with M-CSF to promote proliferation and differentiation of cells thus making them IL-3 responsive. It may also have a maintenance or survival-enhancing effect on primitive stem cells (9). Interleukin-4 (IL-4) acts on mast cells as well as T and B cells, and probably is a synergistic factor for proliferation of myeloid progenitors (10), while Interleukin-5 (IL-5) is a cytokine that stimulates the proliferation, differentiation and function of eosinophilic granulocytes (11,12). Interleukin-6 (IL-6) may be important in the proliferation and differentiation of normal B cells (13).

Intriguingly, there is clustering of the genes for GM-CSF, IL-3, IL-5, M-CSF and the M-CSF receptor (the c-fms proto-oncogene) on the long arm of chromosome 5 (14-19), IL-3 and GM-CSF being arranged in tandem and separated only by 9 kilobases (kb) of DNA (20). The frequent association of partial or whole deletions of chromosome 5 in myelodysplastic syndromes or myeloid leukaemia and the possible dysregulation that ensues, is currently the subject of intense research. Although located on a different chromosome, namely 17, the G-CSF gene is near the translocation breakpoint in acute promyelocytic leukemia (21).

Recombinant DNA technology has allowed cloning and expression of CSF's in bacteria, yeasts or mammalian cell lines yielding sufficient quantities of purified growth factors for large scale *in vivo* testing and clinical applications. G-CSF, GM-CSF, M-CSF, IL-2 and EPO have entered clinical trials and others will shortly follow. This review will focus on the preclinical and early clinical data on G-CSF.

Characterization and Preclinical Data

G-CSF is known to be produced by endothelial cells (22), fibroblasts (23) and macrophages (24), cell types intimately linked with resistance to infection and generation of an appropriate inflammatory response. In keeping with this, tissue concentrations of G-CSF are higher than those of the circulation.(25) G-CSF was purified and characterized from a human bladder carcinoma cell line 5637 (26) and then molecularly cloned (27) and expressed in *E coli*. Human G-CSF has also been isolated from the human squamous cell carcinoma line CHU-2 and the gene cloned and expressed in monkey COS cells (28). This latter G-CSF contains an additional three codons resulting in a longer molecule which appears less active in supporting granulocyte colony formation. It remains unresolved whether this is an aberrant product of a tumor cell line or whether different biological activities of G-CSF arise from altered versions of the same gene product (29). G-CSF profoundly influences multiple differentiated functions of mature neutrophils (30-32). These include: *(a)* promotion of superoxide anion production in the presence of chemotactic peptides such as f-Met-Leu-Phe (fMLP); *(b)* enhanced specific binding of fMLP either by increasing receptor number or affinity to fMLP; *(c)* enhanced antibody-dependent cellular cytotoxicity of tumor target cells (e.g. human leukemia and lymphoma cell lines); *(d)* heightened arachidonic acid release in response to ionophore and multiple chemoattractant agents; *(e)* improved survival of polymorphs (PMN) in the presence of G-CSF; and *(f)* modulation of PMN cytotoxicity in response to other mediators (e.g. GM-CSF, tumor necrosis factor and gamma-interferon).

A number of effects on leukemic cell lines and fresh leukemic cells have been demonstrated (33). *In vitro* differentiation of the murine myelomonocytic leukemia cells WEH1-3B(D+) and human promyelocytic leukemia cells HL60 have been seen. Similarly fresh cells from patients with M2 and M3 AML have been shown to differentiate, while leukemic megakaryoblasts have shown increased membrane expression of platelet glycoprotein IIb/IIIa in response to G-CSF. Proliferation rather than differentiation has been seen with KG1 human leukemia cells, some murine IL-3 dependant cell lines and fresh cells from patients with chronic granulocytic leukemia.

Preclinical studies confirmed the promise of G-CSF in promoting neutrophilia. Injection of G-CSF in hamsters and non-human primates (Cynomolgus monkeys) produced a dose-dependant increase in neutrophils (34,35). Analysis of the neutrophils in the latter study showed them to be functional with enhanced ability to kill and phagocytize bacteria. In monkeys treated with cyclophosphamide to marrow ablation, greatly accelerated recovery was seen with G-CSF therapy, while in mice myelosuppression after various cytotoxic drugs was prevented or minimized (36). No evidence of depletion of bone marrow reserves has been seen in animals given sustained G-CSF or when treated with repeated cycles of chemotherapy (37).

Human Studies

Studies in humans followed, initial objectives being to see if G-CSF could safely prevent or shorten the period of neutropenia after intensive chemotherapy. Gabrilove and colleagues, in a phase I-II study (38), evaluated the ability of G-CSF to prevent or shorten days of neutropenia in patients with transitional cell carcinoma of the urothelium receiving MVAC chemotherapy. Given as a short infusion prior to any chemotherapy G-CSF led to dose-dependant neutrophil increases. When given after chemotherapy, significant reductions in duration of neutropenia and need for antibiotic therapy for neutropenic fever were noted. In addition maintenance of chemotherapy schedule (D14 cytotoxics) was possible in all patients and incidence of mucositis was significantly reduced. G-CSF was well-tolerated, mild to moderate bone pain usually of a transient nature being the only notable finding. Leucocyte alkaline phosphatase was noted to increase, a not surprising finding as synthesis is thought to be regulated by G-CSF (39).

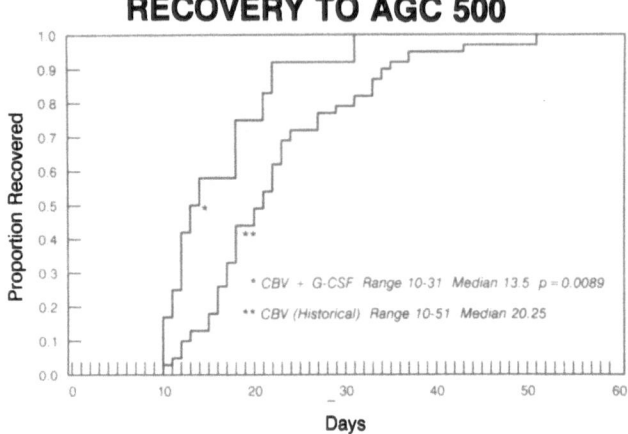

RECOVERY TO AGC 500

* CBV + G-CSF Range 10-31 Median 13.5 p = 0.0089

** CBV (Historical) Range 10-51 Median 20.25

Using a continuous infusion mode of administration, patients with advanced small cell carcinoma of the lung receiving combination chemotherapy were treated with G-CSF on alternate cycles (40). The regimen significantly protected against infection, no infective episodes being seen on cycles with G-CSF. Neutrophils produced were functionally normal in terms of motility and bactericidal activity (41). In a further study (42) in patients with a variety of malignancies receiving melphalan G-CSF administration reduced the period of chemotherapy induced neutropenia. Heartened by the minimal toxicity and reduction of neutropenia in these studies, investigators looked at other clinical settings. High-dose chemotherapy with cyclophosphamide, BCNU, and VP-16 (CBV) and autologous bone marrow rescue has been shown to be effective therapy for relapsed/refractory Hodgkin's disease (43,44). G-CSF was added to this regimen (45) to see if it could hasten granulocyte recovery thus reducing morbidity/mortality. Using a 30-minute G-CSF infusion, a carefully designed dose reduction/cessation schedule was designed to prevent (a) the theoretical risks of a restricted lineage growth factor in the transplant setting (i.e. diversion of hematopoietic recovery to granulopoiesis alone); and (b) the unknown toxicity of excessive neutrophilia in patients who had received high dose chemotherapy. Comparison was made with an historical control cohort who had optimal marrow collection. Significantly hastened recovery to absolute granulocyte count (AGC) 100, 500 (Fig. 1) and 1000 was seen whilst platelet recovery remained unchanged. Toxicity was minimal with mild to moderate bone pain being observed in some patients.

Figure 1. Hematologic recovery to absolute granulocyte count (AGC) $500 \times 10^3/mm^3$ from day of marrow infusion in 12 Hodgkin's disease patients treated with CBV + G-CSF (*) compared with an historical cohort of 39 patients who received CBV (**) alone.

G-CSF is also being evaluated in settings other than chemotherapy-induced neutropenia. Cyclic hematopoiesis, an inherited disease of man and gray collie dogs characterized by regular oscillations of peripheral blood cells may be due to a regulatory defect at the pluripotent stem cell level. In the collie model, G-CSF administration was able to eliminate predicted neutropenic episodes (46). Human studies are currently in progress.

In vitro studies (47) have suggested a differentiative, rather than proliferative role for G-CSF in myelodysplastic syndromes (MDS), disorders of ineffective hematopoiesis where dysmaturational defects and resultant cytopenias occur. In addition, studies where neutrophils of MDS patients with myelodysplastic syndrome (48) were incubated with G-CSF have shown improved function, suggesting that the defect in MDS neutrophils may be repairable.

Using a low-dose subcutaneous schedule, four of five patients thus far treated by Greenberg's group with G-CSF have exhibited a striking rise in AGC. Myeloid differentiation and reduction in marrow blasts were also noted (49). Progression to acute leukemia has not been seen.

Kostmann's syndrome, a rare usually autosomal recessive disorder characterized by severe neutropenia and high morbidity/mortality in early life, has been another target for G-CSF therapy (50). *In vitro* CFU-GM assays of bone marrow from such patients demonstrated that G-CSF promoted growth and differentiation of neutrophil colony formation. Administration of G-CSF to 5 patients has resulted in beneficial neutrophil increase and improvement of pre-existing infections.

such as azido-deoxythymidine or high-dose Cotrimoxasole therapy of Pneumocystis Carinii pneumonia) to be given more optimally.

A trial is underway in neutropenic fever in settings other than acute myeloid leukaemia, while other areas which will need to be investigated include neonatal sepsis with neutropenia where granulocyte transfusions have been useful (54-56), post-operative and chronic infections. Obviously caution will have to be adopted in acute myeloid leukaemia until the advantages of differentiation induction (57) versus proliferation (58) can be addressed more precisely in particular patients.

Regulation of granulocyte production by G-CSF and similar factors should herald new therapeutic endeavors in management of cancer, disorders of neutrophil production and infectious disease.

Toxicity and Future Prospects

To date, administration of G-CSF to humans has not resulted in undue toxicity. Theoretical fears of depletion of marrow reserves of stem and progenitor cells or diversion of recovery lineage have been unfounded. Capillary leak syndrome seen with IL-2 (51,52) and high-dose GM-CSF (53) therapy has not occurred, and perhaps this reflects the restricted specificity of G-CSF compared to those lymphokines. However, whether long term usage (e.g. accompanying repeated cycles of chemotherapy) will produce indirect toxic effects, perhaps by mediation of or interplay with other cytokines, remains uncertain. In addition, the direct effects of G-CSF on tumours, whether beneficial or detrimental remains to be seen. G-CSF however, may allow higher doses of chemotherapeutic agents previously limited by myelosuppressive toxicity. Many other clinical settings beckon for study of G-CSF. In acquired immunodeficiency syndrome, where marrow function may be suboptimal, increased neutrophil counts may theoretically reduce infection or allow drugs associated with high risks of neutropenia (anti-retroviral agents

REFERENCES

1. Bradley TR, Metcalf D (1966) The growth of mouse bone marrow cells in vitro. Aust J Exp Biol Med Sci 44:287

2. Pluznik DH, Sachs L (1966) The induction of clones of normal mast cells by a substance from conditioned media. Exp Cell Res 43:553-563

3. Metcalf D (1984) The Hemopoietic Colony Stimulating Factor. Amsterdam: Elsevier

4. Stanley ER, Heard PM (1977) Factors regulating macrophage production and growth. Purification and some properties of the colony stimulating factor from medium conditioned by mouse L cells. J Biol Chem 252:4305-12

5. Nicola NA, Vadas M (1984) Hemopoietic colony-stimulating factors. Immunol Today 5:76

6. Fisher J (1983) Control of Erythropoietin production. Proc Soc Exp Biol Med 173:289-305

7. Muraguchi A, Kehnl JH, Longo DL, Wolkman DJ, Smith KA, Fauci AJ (1985) Interleukin 2 receptors on human B cells. Implications for the role of interleukin 2 in human B cell function. J Exp Med 161:187-97

8. Moore MAS, Warren DJ, Souza LM (1987) Synergistic interaction between Interleukin 1 and CSF's in hematopoiesis. In: Gale RP, Golde DW (eds) Recent Advances in Leukemia and Lymphoma, UCLA Symposia on Molecular and Cellular Biology, vol. 61, New York: Alan R. Liss

9. Zsebo KM, Wypychi J, Yuschenkoff VN, Luff, Hunt P, Dukes PP, Langley RE (1988) Effects on Hematopoietin 1 and Interleukin 1: Activities on Early Hematopoietic cells of the bone marrow. Blood 71:962

10. Broxmeyer HE, Cooper S, Gillis S, William DE (1987) Synergistic effects of purified recombinant human B cell growth factor-1/ Interleukin 4 on colony formation in vitro by human hematopoietic cells. Blood 70 (suppl 1):525

11. Campbell HD, Tucker WQS, Hort Y, Martinson ME, Mayo G, Clutterbuck EJ, Sanderson CS, Young IG (1987) Molecular cloning and expression of the gene encoding human eosinophil differentiation factor (Interleukin-5). Proc Natl Acad Sci USA:84:6629

12. Lopez AF, Sanderson CJ, Gamble JR, Campbell HD, Young IG, Vadas MA (1988) Recombinant human interleukin-5 (IL-5) is a selective activator of human eosinophil function. J Exp Med 167:219-24

13. Freeman G, Freedman A, Robinowe S, Segil J, Nadler LM (1987) Expression of IL-6 (BCDF) and IL-4 (BSF-1) genes in normal activated and neoplastic B cells. Blood 70(suppl):882

14. Sherr CJ, Rettenmier CW, Sacca R, Roussel MF, Look AT, Stanley ER (1985) The c-fms proto-oncogene product is related to the receptor for the mononuclear/phagocyte growth factor CSF-1. Cell 41:665-676

15. Le Beau MM, Westbrook CA, Diaz MO Larson RA, Rowley JD, Gasson JC, Golde DW, Sherr CJ (1986) Evidence for the involvement of GM-CSF and FMS in the deletion (5q) in myeloid disorders. Science 231:984-987

16. Huebner K, Isobe M, Croce CM, Golde DW, Kaufmann SE, Gasson JC (1985) The human gene encoding GM-CSF is at 5q21-q32 the chromosome region deleted in the 5q- anomaly. Science 230:1282-1285

17. Le Beau MM, Epstein ND, O'Brien SJ, Nienhuis AW, Yang YC, Clark SC, Rowley JD (1987) The interleukin 3 gene is located on human chromosome 5 and is deleted in myeloid leukemia with a deletion of 5q. Proc Natl Acad Sci (USA) 84:5913-5917

18. Pettenati MJ, Le Beau MM, Lemons RS, Shima EA, Kawasaki ES, Larson RA, Sherr CJ, Diaz MO, Rowley JD (1987) Assignment of CSF1 to 5q33.1 evidence for clustering of genes regulating hematopoiesis and for their involvement in the deletion of the long arm of chromosome 5 in myeloid disorders. Proc Natl Acad Sci (USA): 84:2970-2974

19. Sutherland GR, Baker E, Callen DF, Campbell HD, Young ID, Sanderson CS, Garson OM, Lopez AF, Vadas M (1988) Interleukin 5 is at 5q31 and is deleted in the 5q- syndrome. Blood 71:1150-52

20. Yang Y-Chung, Kovacic S, Kriz R, Woll S, Clark SC, Wellens TE, Nienhuis A, Epstein N (1988) The human genes for GM-CSF and IL-3 are closely linked in tandem on chromosome 5. Blood 71:958-961

21. Simmers RN, Webber LM, Shannon MF, Garson OM, Wong G, Vadas MA, Sutherland GR (1987) Localization of the G-CSF gene on chromosome 17 proximal to the breakpoint in the

t(15,17) in acute promyelocytic leukemia. Blood 70:330-332

22. Broudy VC, Kaushansky K, Harlan JM, Adamson JW (1987) Interleukin 1 stimulates human endothelial cells to produce Granulocyte Macrophage colony-stimulating factor and granulocyte colony-stimulating factor. J Immunol 139:464-468

23. Kaushansky K, Lin N, Adamson JW (1987) Interleukin 1 stimulates fibroblasts to synthesize granulocyte-macrophage and granulocyte colony-stimulating factor. J Clin Invest 81:92-97

24. Kaushansky K, Miller JE, Morris DR, Hammond WP (1986) Hematopoietic growth factors (HGF) derived from human peripheral blood mononuclear cells (MNC): Evidence for dependant production by purified lymphocytes (Lymphs) and monocytes (monos). Blood 68(suppl):549

25. Metcalf D (1987) The role of the colony stimulating factors in resistance to acute infection. Immunol Cell Biol 65:35-43

26. Welte K, Platzer E, Lu L, Gabrilove JC, Levi E, Mertelsmann R, Moore MAS (1985) Purification and biochemical characterization of human pluripotent hematopoietic colony stimulating factor. Proc Natl Acad Sci (USA) 82:1526

27. Souza LM, Boone TC, Gabrilove JL, Lai PH, Zsebo KN, Murdock DC, Chazin VR, Bruszewski J, Lu L, Chen KK, Burendt J, Platzer E, Moore MAS, Mentelsmann R, Welfe K (1986) Recombinant human granulocyte colony stimulating factor: Effects on normal and leukemic myeloid cells. Science 232:61

28. Nagata S, Tsuchiya M, Asano S, Kazino Y, Yamazaki T, Yamamoto O, Hirata Y, Kubota N, Oheda M, Nomura H, Ono M (1986) Molecular cloning and expression of cDNA for human granulocyte colony stimulating factor. Nature 319:415

29. Clark SC, Kamen R (1987) The human hematopoietic colony stimulating factors. Science 236:1229-36

30. Platzer E, Welfe K, Gabrilove J, Lu E, Harris P, Mentelsmann R, Moore MAS (1986) Biological activities of human pluripotent hemopoietic colony stimulating factor on normal and leukemic cells. J Exp Med 162: 1788-1801

31. Platzer E, Ocz S, Welfe K, Gabrilove JL, Mentelsmann R, Moore MAS, Kalder JR (1986) Human pluripotent hemopoietic colony stimulating factor: Activities on human and marine cells. Immuno Biol 172:185

32. Avalos BR, Hedzat C, Baldwin GC, Golde DW, Gasson JC, DiPersio JF (1987) Biological activities of human G-CSF and characterization of the human G-CSF receptor. Blood 70(suppl):517

33. Welte K, Bonilla MA, Gabrilove JL, Gillio AP, Potter GK, Moore MAS, O'Reilly RJ, Boone TC, Souza LM (1987) Recombinant human granulocyte-colony stimulating factor: in vitro and in vivo effects on myelopoiesis. Blood Cells 13:17-30

34. Cohen AM, Zsebo KM, Inove H, Hines D, Boone TL, Chazin VR, Tsai L, Ritch T, Souza L (1987) In vivo stimulation of granulopoiesis by recombinant human granulocyte colony-stimulating factor. Proc Natl Acad Sci (USA) 84:2484-2488

35. Welte K, Bonilla MA, Gillio AP, Boone TC, Kotter GK, Gabrilove JL, Moore MAS, O'Reilly RJ, Souza LM (1987) Recombinant human granulocyte colony-stimulating factor. Effects on hematopoiesis in normal and cyclophosphamide-treated primates. J Exp Med 165:941

36. Shimamura M, Kobayashi Y, Yuo A, Urabe A, Okabe T, Komatsu Y, Itoh S, Takaku F (1987) Effect of human recombinant granulocyte colony-stimulating factor on

hematopoietic injury in mice induced by 5 fluorouracil. Blood 69:353-355

37. Bonilla MA, Gillio AP, Potter GK, O'Reilly RJ, Souza LM, Welfe K (1987) Effects of recombinant human G-CSF and GM-CSF on cytopenias associated with repeated cycles of chemotherapy in primates. Blood 70(suppl):377

38. Gabrilove JL, Jakubowski A, Scher H, Sternberg C, Wong G, Grous J, Yagoda A, Fain K, Moore MAS, Clarkson B, Oettgen HF, Alton K, Welfe K Souza L (1988) Effect of granulocyte colony-stimulating factor on neutropenia and associated morbidity due to chemotherapy for transitional cell carcinoma of the urothelium. New Engl J Med 318:1414-1422

39. Chikkappa G, Wong G, Santella D, Pasquale D (1987) Neutrophil Alkaline Phosphatase (NAP) synthesis is regulated by Granulocyte colony-stimulating factor (G-CSF). Blood 70(suppl):195

40. Bronchud MH, Scarffe JH, Thatcher N, Crowther Dm, Souza LM, Alton NK, Testa NG, Dexter TM (1987) Phase 1/11 study of recombinant human granulocyte colony-stimulating factor in patients receiving intensive chemotherapy for small cell lung cancer. Br J Cancer 56:809-813

41. Bronchud M, Potter MR, Morgenstern G, Blasco MJ, Scarffe JH, Thatcher, Crowther D, Souza LM, Alton NK, Testa NG, Dexter TM (1988, in press) In vitro and in vivo analysis of the effects of recombinant human granulocyte colony-stimulating factor in patients. Br J Cancer 58

42. Morstyn G, Campbell L, Souza LM, Alton NK, Keech J, Green M, Sheridan W, Metcalf D, Fox R (1988) Effect of granulocyte colony stimulating factor on neutropenia induced by cytotoxic chemotherapy. Lancet 667-671

43. Spitzer G, Dicke KA, Litam J, Verma DS, Zander A, Lanzotti V, Valdivieso M, McCredie K, Samuels ML (1980) High-dose combination chemotherapy with autologous bone marrow transplantation in adult solid tumors. Cancer 45:3075-85

44. Jagannath S, Dicke KA, Armitage JO, Cabanillas F, Horwitz LJ, Vellekoop L, Zander AR, Spitzer G (1986) High dose cyclophosphamide, carmustine and etoposide and autologous bone marrow transplantation for relapsed Hodgkin's disease. Ann Intern Med 104:163

45. Taylor K McD, Spitzer G, Dicke KA, Cabanillas F, Horwitz LJ, Vincent M, Jagannath S, Souza L (1988, in preparation) Recombinant human granulocyte-colony stimulating factor hastens granulocyte recovery in Hodgkin's disease after high dose chemotherapy and autologous bone marrow transplantation (ABMT)

46. Lothrop GD, Jones JB, Warren DJ, Moore MAS Souza LM (1987) Correction of cyclic hematopoiesis with recombinant human granulocyte colony-stimulating factor. Blood 70(suppl):407

47. Nagler A, Ginzton N, Bungs C, Donlon T, Greenberg PL (1987) In vitro effects of recombinant human (rh) granulocyte (G-CSF) and granulocyte-monocyte (GM-CSF) colony stimulating factors on hemopoiesis in the myelodysplastic syndromes. Blood 70(suppl):417

48. Yuo A, Kitagawa S, Okabe T, Urabe A, Komatsu Y, Hoh S Takaku F (1987) Recombinant human granulocyte colony-stimulating factor repairs the abnormalities of neutrophils in patients with myelodysplastic syndromes and chronic myelogenous leukemia. Blood 70:404-411

49. Negrin RS, Haeuber DH, Nagler A, Souza LM, Greenberg PL (1988) Treatment of myelodysplastic syndromes with recombinant human granulocyte colony stimulating factor. Exp Hematol 16(suppl):240

50. Bonilla MA, Gillio AP, Ruggiero M,
    Kernan NA, Brochstein JA, Fumagalli
    L, Bordignon C, Vincent M, Welte K,
    Souza L, O'Reilly RJ (1988)
    Correction of neutropenia in
    patients with congenital
    agranulocytosis with recombinant
    human granulocyte colony
    stimulating factor in vivo. Exp
    Hematol 16(suppl):243

51. Lotze MT, Matory YL, Rayner AA,
    Ettinghauser SE (1986) Clinical
    effects and toxicity of
    Interleukin-2 in patients with
    cancer. Cancer 58:2764-2772

52. Rosenstein M, Ettinghausen SE,
    Rosenberg SA (1986) Extravasation
    of intravascular fluid mediated by
    the systemic administration of
    recombinant Interleukin 2.
    J Immunol 137:1735-1742

53. Brandt SJ, Peters WP, Atwater SK,
    Kurtzberg J, Borowitz MJ, Jones
    RB, Shpall EJ, Bart RC, Gilbert CJ,
    Oettte DH (1988) Effect of
    recombinant human
    granulocyte-macrophage colony
    stimulating factor or
    Hematopoietic reconstitution after
    high dose chemotherapy and
    autologous bone marrow
    transplantation. New Engl J Med
    318:869-875

54. Wheeler JG, Chauveret AR, Johnson
    CA, Dillard R, Block SM, Boyle R,
    Abramson JS (1984) Neutrophil
    storage pool depletion in septic
    neutropenic neonates. Pediatr
    Infect Dis 3:407-409

55. Christensen RD, Rothstein G,
    Anstall HB and Bykee B (1982)
    Granulocyte transfusions in
    neonates with bacterial infection,
    neutropenia and depletion of mature
    marrow neutrophils. Paediatrics
    70:1-6

56. Cairns MS, Worchester C, Rucker R,
    Bennetts GA, Anlie R, Perkin R,
    Anas N, Hicks D (1987) Role of
    circulating complement and
    polymorphonuclear leukocyte
    transfusion in treatment and
    outcome in critically ill neonates
    with sepsis. J Pediatr 110:935-41

57. Nicola NA, Begley CG, Metcalf D
    (1985) Identification of the human
    analogue of a regulator that
    induces differentiation in murine
    leukaemic cells. Nature
    314:625-628

58. Begley CG, Metcalf D, Nicola NA
    (1987) Primary human myeloid
    leukaemia cells: comparative
    responsiveness to proliferative
    stimulation by GM-CSF or G-CSF and
    membrane expression of CSF
    receptors. Leukaemia 1:1-8

10 Regulation of Granulocyte-Macrophage Colony-Stimulating Factor in Murine Lymphocytes

M. Bickel, S.E. Mergenhagen, and D.H. Pluznik

ABSTRACT. The coordinated expression of the hematopoietic growth factors, granulocyte-macrophage colony-stimulating factor (GM-CSF) and interleukin 3 (IL3), as well as the lymphokines interleukin 2 (IL2) and interferon-γ (INF-γ) in activated T cells suggests common control mechanisms involved in their synthesis. However, cyclosporin A (CsA) which inhibits production of IL2 and IL3 by lymphocytes, has no effect on the production of GM-CSF. This observation led to the hypothesis that GM-CSF gene expression is under a different control mechanism. Since CsA is known to inhibit inducible lymphokine gene expression at the transcriptional level, it appears likely that GM-CSF regulation could differ from IL2 and IL3 regulation at this level. Thus different approaches were undertaken to demonstrate that the GM-CSF gene is under post-transcriptional control. Upon examining the GM-CSF mRNA steady-state levels, no differences were detected between GM-CSF and IL2 mRNA. However, when analyzing the transcriptional activity of the GM-CSF gene, we detected active transcription in unstimulated T-and B-cell lines. IL2, which is known to be transcriptionally controlled, was silent in unstimulated T-cells, but active in stimulated cells.

We conclude that the GM-CSF gene is under a post-transcriptional control mechanism and that this distinguishes its biosynthetic pathway from that of IL2 and IL3.

Key words: GM-CSF - IL3 - IL2-Cyclosporin A - gene regulation

INTRODUCTION

The murine hematopoietic growth factor, granulocyte-macrophage colony stimulating factor (GM-CSF) appears in various tissue culture supernatants after induction with mitogens or antigens. Biological activity of GM-CSF has been detected in culture supernatants of spleen cells, T-lymphocytes, macrophages, endothelial cells, fibroblasts and even in certain B cell lines (1,2,3,4,5). This induction of GM-CSF production is not specific, i.e. a variety of other lymphokines are coordinately expressed. In mitogen-stimulated T cells, interleukin 3 (IL3), interleukin 2 (IL2) and interferon-γ (INF-γ) are each simultaneously induced with the same mitogen.

However, a recent study showed that it is possible to selectively induce GM-CSF production and not IL3 in T cells stimulated by exogenous IL2 (6). In previous studies on T-cells, we suggested a separate control mechanism for GM-CSF based on the observation that the immunosuppressant drug cyclosporin A (CsA) was not capable of inhibiting GM-CSF production, but did inhibit the production of IL3 and IL2 (1).

In this paper we review our findings and present data which add to the understanding of the regulatory mechanism that is involved in the production of GM-CSF in B- and T-cell lines.

MATERIALS AND METHODS

Reagents. Concanavalin A (Con A) was purchased from Calbiochem (San Diego, CA), the phorbol ester, 12-*o*-tetradecanoyl-phorbol-13-acetate (TPA) from Consolidated Midland Co. (Brewster, NY), lipopolysaccharide (LPS) from *Salmonella abortus equi* from Difco (Detroit, MI), cyclosporin A from Sandoz (Basel, Switzerland). TPA was dissolved in dimethyl sulfoxide. Fetal

calf serum (FCS) was purchased from HyClone (Logan, UT), RPMI-1640 and NCTC 109
medium from M.A. Bioproducts (Walkersville, MD), L-glutamine, sodium pyruvate,
nonessential amino acids, penicillin,
streptomycin and HEPES buffer from GIBCO
(Grand Island, NY). 2-Mercaptoethanol was
obtained from Eastman Kodak (Rochester,
NY). Purified and recombinant murine GM-
CSF, IL3 and IL2 were purchased from Gen-
zyme (Boston, MA). Rabbit anti-mouse recombinant GM-CSF antibodies were kindly
provided by D. Mochizuki (Immunex, Seattle, WA) (7).

Cell cultures. Spleen cells from 6- to 8-
week-old female CBA/J mice (The Jackson
Laboratory) were cultured at 5 x 10^6/ml in
growth medium (RPMI 1640 medium supplemented with 1 mM L-glutamine, 1 mM sodium
pyruvate, nonessential amino acids at 0.1
mM each, penicillin at 100 units/ml, streptomycin at 100 μg/ml, 25 mM HEPES buffer,
50 μM 2-mercaptoethanol) and 10% FCS. The
EL-4 thymoma (8) and the 2B4 T-cell hybrid
(9) cell lines were maintained by twice
weekly passage in growth medium (RPMI
1640) supplemented with 5% FCS. EL-4 and
2B4 cells (2 x 10^6/ml) were stimulated in
serum-free growth medium with 20 nM TPA or
5 μg/ml Con A. The B cell lines, M12.4.1
(10) and TH2.2 (11). were maintained by
twice weekly passage in a culture medium
containing RPMI-1640 supplemented with 10%
NCTC 109 medium, 1 mM L-glutamine, 100
units/ml penicillin, 100 μg/ml streptomycin, 50 μM 2-mercaptoethanol and 5% FCS.
M12.4.1 and TH2.2 cells were cultured in
serum-free growth medium supplemented with
50 μg/ml LPS for 24 h, unless otherwise
indicated, after which the supernatants
were harvested and assayed for lymphokine
activity.

Determination of lymphokine activity. GM-CSF
activity was determined by its ability to
stimulate growth of the mast/basophil cell
line PT-18 (12) which is dependent for
growth on GM-CSF and/or IL3 (13). IL3
activity was assayed on the myeloid cell
line DA-1 (14) and IL2 on the CTLL cell
line (15). Threefold serial dilutions of
supernatants were added (100 μl) to 10^4
cells (100 μl) per well in 96-well plates
in triplicate. After a 40 h incubation,
the cultures were pulse labeled with ^{3}H-
thymidine [^{3}H-TdR, 1 μCi/well, (37 kBq/
well) Amersham, Arlington Heights, IL] for
an additional 4 h, harvested and ^{3}H-TdR
incorporation measured. Lymphokine activity was expressed as a stimulation index
(SI) which represents the ratio between
counts per minute (cpm) for stimulated
cells and cpm for cells incubated in
growth medium alone ($\leq$ 500 cpm). In order
to allow comparison between different
assays, standard curves were generated in
each experiment by assaying the same lot
of the respective purified growth factors
in 2-fold serial dilutions. Data represent
the mean SI ($\pm$ S.D.) of 3 experiments.

cDNA probes. The plasmid pJL4 containing 750
bp of cDNA specific for GM-CSF (pGM5'Δ19)
(16) and the plasmid pJL3 containing 600
bp of specific cDNA for IL3 (pMu5.1) (17)
were gifts of Drs. N.M. Gough and A.W.
Burgess (Ludwig Institute for Cancer Research, Melbourne, Australia). The plasmid
(pcDX) containing cDNA specific for IL2
(18) was a gift form Dr. K. Arai (DNAX
Research Institute of Molecular and Cellular Biology, Palo Alto, CA). The glycer-
aldehyde-3-phosphate dehydrogenase (GAPDH)
probe has been described by Piechaczyk et
al. (19). For Northern analysis, insert
cDNA were ^{32}P-labeled with the large fragment of DNA polymerase I using random
oligonucleotides as primers (20).

RNA isolation and molecular probing of mRNA. Total
RNA was extracted from cells by lysis in
6M guanidine thiocyanate buffer (21),
homogenization and separation by ultra-
centrifugation on cesium trifluoro-acetate
(Pharmacia, Uppsala, Sweden) gradients.
mRNA was isolated by passing total RNA
twice over oligo-d(T) affinity cellulose
columns (Collaborative Research, Bedford,
MA). Polyadenylated [Poly(A)$^+$] RNA was
electrophoresed on 1% agarose gels containing formaldehyde and transferred to
Gene Screen Plus transfer membranes (New
England Nuclear, Boston, MA). The blots
were prehybridized and the membrane-bound
RNA was hybridized under stringent conditions as outlined in the Gene Screen Plus
manual. Membranes were exposed for different time intervals at -70°C to Kodak
XAR films.

Transcriptional analysis. Nuclear run-off
transcription assays were performed according to the method of Groudine et al.
(22). Nuclei were prepared by lysing 10^8
cells for 5 minutes on ice in buffer (10
mM Tris HCl pH 7.4, 3 mM CaCl$_2$, 2 mM MgCl$_2$)
containing 0.2% NP-40. Linearized plasmids containing the cDNA probes (10 μg/
slot) were blotted onto nitrocellulose
using a slot blot apparatus (Schleicher
and Schuell, Keene, NH). Following overnight prehybridization, the membrane-bound
DNAs were hybridized at 65°C with ^{32}P-la-
beled RNA that was generated from 2 x 10^7
nuclei *in vitro*. Membranes were dried,
baked at 80°C *in vacuo*, and autoradio-
graphed.

RESULTS AND DISCUSSION

Inducible GM-CSF production.
Table 1 summarizes a series of experiments that demonstrate the inducible
release of CSF from spleen cells, two T-
cell lines and two B-cell lines. CSF
activity was detected in the supernatants
of spleen cells isolated from CBA/J mice
24 h after induction with 5 μg/ml Con A.
In two T-cell lines of independent origin
(8,9), CSF activity was detected 24 h
after stimulation with either 5 μg/ml Con
A or 20 nM of the tumor promoting phorbol

ester, TPA. In the B-cell hybridoma TH2.2 and lymphoma M12.4.1, LPS at 50 μg/ml proved to be a potent inducer of CSF production when added to the cultures for 24 h. No CSF activity was detected in the supernatants of unstimulated cells. Dose and time dependency of CSF production by these cells have been reported elsewhere (5,22). With the exception of the B-cell supernatants, all supernatants of stimulated cells contained activities which supported the growth of both PT-18 and DA-1. B-cell supernatants contained activity detectable on PT-18 cells only. All induction of T- and B-cells were performed under non-proliferating growth conditions, i.e. in the absence of serum and at a cell density of 1 to 2 x 10^6/ml. These results clearly demonstrate that the production of all of these lymphokines is inducible in the absence of cell proliferation.

Table 1. CSF Activity in Supernatants of Induced Normal Spleen Cells, T-cell and B-cell Lines.

Cells	Treatment	CSF Activity*	
		PT-18	DA-1
Spleen cells	untreated	3.2	1.1
	5 μg/ml Con A	25.0	23.2
EL-4 thymoma	untreated	1.8	1.2
	5 μg/ml Con A	33.0	18.6
	20 nM TPA	40.1	26.8
2B4 T cell hybrid	untreated	2.1	1.8
	5 μg/ml Con A	30.0	20.4
	20 nM TPA	39.5	32.3
M12.4.1 B lymphoma	untreated	2.1	1.8
	50 μg/ml LPS	36.2	3.5
TH2.2 B hybridoma	untreated	1.5	1.9
	50 μg/ml LPS	38.4	2.4

* CSF activity is expressed as stimulation index. PT-18 cells are GM-CSF and/or IL3 dependent, DA-1 cells are IL3 dependent.

Identification of GM-CSF activity in T and B cells and resistance to CsA.
Previous studies demonstrated a differential susceptibility of GM-CSF, IL3 and IL2 to the immunosuppressant drug CsA (1,23). When supernatants from Con A- or TPA-stimulated EL-4 or 2B4 cells that had been exposed to 0.5 μg/ml CsA were assayed on PT-18 cells, the growth promoting activity was reduced, but not abolished. Since IL3 is known to be inhibited by CsA (23,24) this reduction is likely due to the inhibition of IL3.

To assess whether GM-CSF contributed growth supporting activity to PT-18, we attempted to block growth on PT-18 with antibodies directed against rGM-CSF (7). Table 2 shows the results of experiments performed to identify the GM-CSF activities produced by either T-cell or B-cell lines upon stimulation with either Con A, TPA or LPS as well as the CsA resistant activity produced by stimulated T cells. These experiments demonstrate that GM-CSF is produced by all of the stimulated cell lines and appears in supernatants of stimulated T-cells even when they were treated with CsA.

Table 2. Neutralization of GM-CSF Activity with Antibodies Against Murine rGM-CSF.

Cells	Treatment	CSF Activity*
		PT-18
EL-4	Con A 5 μg/ml	38.5
	Con A + antibody	18.3
	Con A + CsA 0.5 μg/ml	15.8
	Con A + CsA + antibody	1.8
	TPA 20 nM	41.2
	TPA + antibody	27.1
	TPA + CsA 0.5 μg/ml	32.1
	TPA + CsA + antibody	2.5
2B4	Con A	31.2
	Con A + antibody	16.3
	Con A + CsA	14.0
	Con A + CsA + antibody	2.8
	TPA	38.7
	TPA + antibody	25.0
	TPA + CsA	22.8
	TPA + CsA + antibody	2.9
M12.4.1	LPS 50 μg/ml	34.0
	LPS + antibody	1.4
TH2.2	LPS	33.2
	LPS + antibody	0.8
control	GM-CSF 25 U/ml	26.5
	GM-CSF + antibody	0.4
	IL3 25 U/ml	26.6
	IL3 + antibody	25.8

* CSF activity is expressed as stimulation index. Final dilution of anti-rGM-CSF antibody was 1:400.

Northern analysis of GM-CSF mRNA.
Because CsA inhibits only inducible lymphokine expression at the transcription level (25, 26) as in the case of IL2 and IL3 and because GM-CSF activity is not inhibited by CsA, we postulated that the GM-CSF gene is transcribed constitutive-

ly. Northern analysis of EL-4 and 2B4 cells, as well as the B-cell lines M12.4.1 and TH2.2 revealed GM-CSF transcripts in stimulated cells only (Fig. 1). However, the absence of suspected transcripts in unstimulated cells does not necessarily mean that the gene is not transcribed, but may rather reflect a rapid turn-over or a very low amount of the mRNA, not detectable by Northern analysis.

Cycloheximide, an inhibitor of protein synthesis, is known to stabilize steady-state levels of mRNAs (27, 28). When un-stimulated EL-4 cells were treated with 10 μg/ml cycloheximide, a low level of GM-CSF transcripts was detected (data not shown). This observation points to the possible involvement of a regulatory protein in the control mechanism of GM-CSF that is rapidly turned over. Such a protein might interfere with AU-rich sequences present in the 3' untranslated region of mRNA. These regions, found in a broad spectrum of growth factor and on-cogene mRNAs were recently reported to affect the half-life of their RNAs (29).

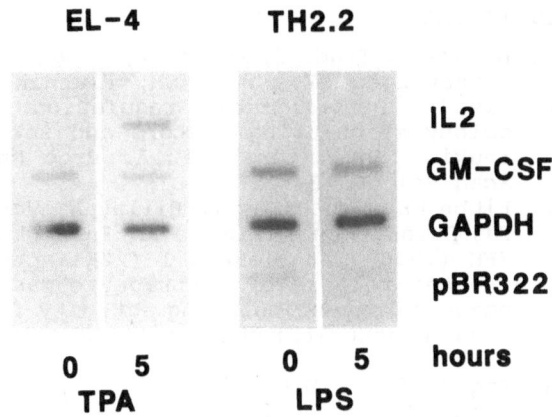

Fig. 2. Nuclear transcription analysis of T-cell and B-cell lines. Nuclei were iso-lated 5 h after stimulation with TPA (20 nM) or LPS (50 μg/ml) or in the absence of a stimulant. Nascent RNA chains were elon-gated *in vitro* in the presence of ^{32}P-labeled UTP. TCA-precipitable RNA was hybridized to GM-CSF, IL2 and GAPDH cDNAs and pBR322 immobilized on nitrocellulose filters.

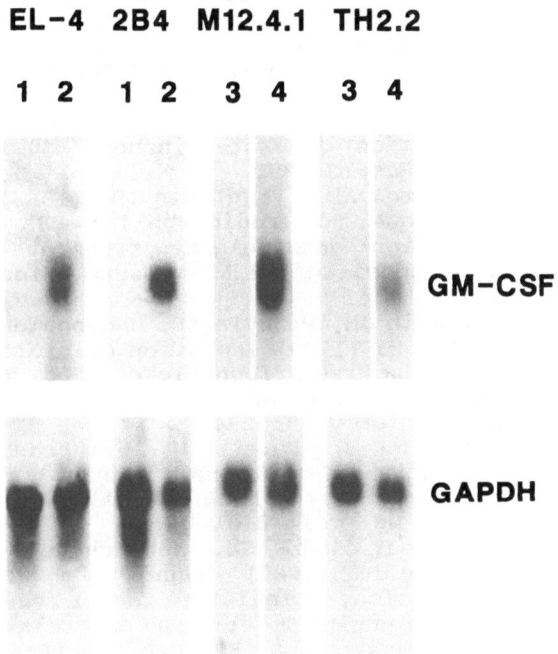

Fig. 1. Northern analysis of poly(A+) enriched RNA extracted from T-cell and B-cell lines. *Lane 1*, mRNA from unstimulated T-cells; *lane 2*, mRNA from TPA (20 nM) stimulated T-cells; *lane 3*, mRNA from unstimulated B-cells; *lane 4*, mRNA from LPS (50 μg/ml) stimulated B-cells. Cells were cultured for 8 h in the presence or absence of TPA, LPS RNA extracted, en-riched for poly[A+] and 10 μg/lane elec-trophoresed. Membrane-bound RNAs were first hybridized with the GM-CSF cDNA probe and subsequently with the control cDNA probe for GAPDH.

Transcription analysis of the GM-CSF gene.

More direct evidence excluding the pos-sibility of a transcriptional control regulating GM-CSF biosynthesis was obtained by performing nuclear run-off experiments. The results obtained from the experiments with cycloheximide mentioned above imply that GM-CSF mRNA is present in unstimulated cells at very low levels. Stimulation of the cells with TPA resulted in an increase in steady-state levels of this RNA. Nuclear transcription analysis was performed in vitro with isolated nuclei from EL-4 and M12.4.1 cells that had been treated with TPA and LPS, respectively, or left untreated. As seen in Fig. 2, the GM-CSF gene is transcribed in both unstimulated and stimulated cells, and the transcriptional activity does not appear to be signifi-cantly affected by the addition of the respective stimulants. The IL2 gene, known for its transcriptional activation and therefore used as an internal control, was silent in unstimulated EL-4 cells and expressed in stimulated cells.

The different experimental approaches used in the present study points to a post-transcriptional control mechanism of GM-CSF biosynthesis. This mechanism of lymphokine regulation may not be that uncommon, as a post-transcriptional con-trol mechanism similar to that found for GM-CSF has recently been proposed for the production of murine tumor necrosis factor α by macrophages (30).

ACKNOWLEDGEMENTS

We are indebted to Dr. Gerald M. Feldman for editorial assistance and Ms. Lynda L. Weedon for technical assistance.

REFERENCES

1. Bickel M, Tsuda H, Amstad P, Evequoz V, Mergenhagen SE., Wahl SM, Pluznik DH (1987) Differential regulation of colony stimulating factors and interleukin 2 by cyclosporin A. Proc Natl Acad Sci USA 84:3274

2. Fibbe WE, van Damme J, Billau A, Voogt PJ, Duinkerken N, Kluck PMC, Falkenburg JFH (1986) Interleukin 1 (22K factor) induces release of granulocyte-macrophage colony-stimulating activity from human mononuclear phagocytes. Blood 68: 1316

3. Bagby GC Jr, Dinarello CA, Wallace P, Wagner C, Hefeneider S, McCall E (1986) Interleukin 1 stimulates granulocyte-macrophage colony-stimulating activity release by vascular endothelial cells. J Clin Invest 78:1316

4. Zucali JR, Dinarello CA, Oblon DJ, Gross MA, Anderson L, Weiner RS (1986) Interleukin 1 stimulates fibroblasts to produce granulocyte-macrophage colony-stimulating activity and prostaglandin E_2. J Clin Invest 77:1857

5. Bickel M, Amstad P, Tsuda H, Sulis C, Asofsky R, Mergenhagen SE, Pluznik DH (1987) Induction of granulocyte-macrophage colony-stimulating factor by lipopolysaccharide and anti-immunoglobulin M stimulated B cell lines. J Immunol 139:2984

6. Kelso A, Metcalf D, Gough NM (1986) Independent regulation of granulocyte-macrophage colony stimulating factor and multi-lineage colony stimulating factor production in T lymphocyte clones. J Immunol 136:1718

7. Mochizuki DY, Eisenmann JR, Conlon PJ, Park LS, Urdal DL (1986) Development and characterization of antiserum to murine granulocyte-macrophage colony stimulating factor. J Immunol 136:3706

8. Farrar JJ, Fuller-Farrar J, Simon PL, Hilfiker ML, Stadler BM, Farrar WL (1980) Thymoma production of T cell growth factor (interleukin 2). J Immunol 125:2555

9. Samelson LE, Schwartz RH (1983) T cell clone-specific alloantisera that inhibit or stimulate antigen-induced T cell activation. J Immunol 131:2645

10. Hamano T, Kim KJ, Leiserson WM, Asofsky R (1979) Establishment of B cell hybridomas with B cell surface antigens. J Immunol 129:1403

11. Kim KJ, Kanellopoulos-Langevin C, Mervin RM, Sachs DH, Asofsky R (1979) Establishment and characterization of BALB/c lymphoma lines with B cell properties. J Immunol 122:549

12. Pluznik DH, Tare NS, Zata MM, Goldstein AL (1982) A mast/basophil cell line dependent on colony stimulating factor. Exp Hematol [Suppl 12] 10:211

13. Pluznik DH (1985) CSF dependent proliferation is triggered in the early G_1 phase of the target cell cycle. In: Sorg C, Schimpl A (eds) Cellular and Molecular Biology of lymphokines. New York: Academic Press. p.497

14. Ythier AA, Abbud-Filho M, Williams JM, Loertscher R, Schuster MW, Nowill A, Hansen JA, Maltezos D, Strom TB (1985) Interleukin 2- dependent release of interleukin 3 activity by T4[+] human T cell clones. Proc Natl Acad Sci USA 82:7020

15. Baker PE, Gillis S, Smith KA, (1979) Monoclonal cytolytic T-cell lines. J Exp Med 149:273

16. Gough NM, Metcalf D, Gough J, Grail D, Dunn AR (1985) Structure and expression of the mRNA for murine granulocyte-macrophage colony stimulating factor. EMBO J 4:645

17. Hapel AJ, Fung MC, Johnson RM, Young IG, Johnson G, Metcalf D (1985) Biologic properties of molecularly cloned and expressed murine interleukin-3. Blood 65:1453

18. Yokota T, Arai N, Lee F, Rennick D, Mosmann T, Arai K (1985) Use of a cDNA expression vector for isolation of mouse interleukin 2 cDNA clones: expression of T-cell growth-factor activity after transfection of monkey cells. Proc Natl Acad Sci USA 82:68

19. Piechaczyk M, Blanchard JM, Marty L, Dani Ch, Panabieres F, El Sabouty S, Fort Ph, Jeanteur Ph (1984) Post-transcriptional regulation of glyceraldehyde-3-phosphate-dehydrogenase gene expression in rat tissues. Nucleic Acids Res 12:6951

20. Feinberg AP, Vogelstin B (1983) A technique for radiolabeling DNA restriction endonuclease fragments to high specific activity. Anal Biochem 132:6

21. Chirgwin JM, Przbyla AE, MacDonald RJ, Rutter WJ (1979) Isolation of biologically active ribonucleic acid from sources enriched in ribonuclease. Biochemistry 18:5294

22. Groudine M, Peretz M, Weintraub H (1981) Transcriptional regulation of hemoglobin switching in chicken embryos. Mol Cell Biol 1:281

23. Bickel M, Wahl SM, Mergenhagen SE, Pluznik DH (1988) Granulocyte-macrophage colony-stimulating factor regulation in murine T cells and its relation to cyclosporin A. Exp Hematol 16:691

24. Palacios R, (1985) Cyclosporin A inhibits antigen- and lectin-induced but not constitutive production of interleukin 3. Eur J Immunol 15:204

25. Krönke M, Leonard WJ, Depper JM, Arya SK, Wong-Staal F, Gallo RC, Waldmann TA, Greene WC (1984) Cyclosporin A inhibits T cell growth factor gene expression at the level of mRNA transcription. Proc Natl Acad Sci USA 81:5214

26. Granelli-Piperno A, Andrus L, Steinman RM (1986) Lymphokine and nonlymphokine mRNA levels in stimulated human T cells. J Exp Med 163:922

27. Friedman AP, Manly SP, McMahon M, Kerr IM, Stark GR (1984) Transcriptional and posttranscriptional regulation of

interferon-induced gene expression in human cells. Cell 83:745

28. Reed JC, Alpers JD, Nowell PC (1987) Expression of *c-myc* proto-oncogen in normal human lymphocytes. Regulation by transcriptional and posttranscriptional mechanisms. J Clin Invest 80:101

29. Shaw G, Kamen R (1986) A conserved AU sequence from the 3′ untranslated region of GM-CSF mRNA mediates selective mRNA degradation. Cell 46:659

30. Beutler B, Krochin N, Milsark IW, Luedke C, Cerami A (1986) Control of cachectin (tumor necrosis factor) synthesis: mechanisms of endotoxin resistance. Science 232:977

Part IV. Genetic Manipulation of Hematopoiesis

Chairperson: M.J. Evinger-Hodges

11 Gene Transfer by Electroporation

A. Keating and F. Toneguzzo

INTRODUCTION

The development of efficient means of gene transfer into hematopoietic progenitor cells has permitted investigators to analyze stem cell relationships [1] and has increased the feasibility of gene therapy protocols [2]. Recent studies suggest that electroporation (electric-field-mediated DNA transfer) may be a particularly attractive technique for the latter purpose.

During electroporation, target cells are exposed to a brief electric impulse which results in the passive transfer into the cell of DNA present in the medium. The DNA is believed to enter via pores in the cell membrane created by local areas of depolarization produced by the external electric field [3].

This technique has been the subject of intense scrutiny over the past two years and several physical parameters have emerged as important in improving efficiency of gene transfer [4]. In addition, investigators need no longer rely on custom-built devices since numerous commercial instruments are now available (eg. BTX, Biorad Gene Pulser, BRL Cell Porator, Baekon 2000, Promega, Hoefer ProGenetor). The devices are designed to deliver one of two types of electric impulses: an exponential decay pulse or the square wave pulse. Preliminary data with human hematopoietic cells suggest that superior results are obtained with the exponential decay-type pulse (unpublished observations).

We have extensively characterized DNA transfer into a variety of lymphoid cell lines by electroporation [5]. We subsequently applied our methodology to the transfer of DNA into primary cultures of human hematopoietic cells and showed transient gene expression in the normal nucleated marrow cell population [6]. We further demonstrated that genes encoding antibiotic resistance transferred by electroporation were expressed in the progeny of committed hematopoietic progenitors [6]. The presence of the transferred DNA was confirmed by hybridization analysis. Up to 3% of CFU-GM expressed the resistant phenotype.

Having established that genes transferred by electroporation can be stably integrated and expressed in committed hematopoietic progenitors, the next important goal in developing gene therapy protocols is to demonstrate successful transfection of multipotent hematopoietic stem cells - the cells necessary to reconstitute hematopoiesis in marrow transplant recipients after marrow-ablative therapy. These cells are not measurable by colony formation in semi-solid media but may be indirectly assayed in the two-stage long-term marrow culture system [7]. The latter method permits the differentiation of pre-colony forming cells to committed progenitor cells but is cumbersome and time-consuming. Moreover, a potential drawback of gene transfer into stem cells is reflected by studies with embryonal cells in which reduced transcription is attributed to cellular factors operating on sequences within the enhancer region of the introduced gene [8]. Consequently, before experiments involving human hematopoietic stem cells can be conducted with any expectation of success, further studies are required to determine the most active regulatory sequences in equivalent cell populations.

In order to address the above issues we have examined the influence of different regulatory sequences on the transient expression of DNA introduced by electroporation into K562 cells, a multipotent leukemia cell line [9] and into passaged marrow stromal cells, a non-transformed highly proliferative population sharing the phenotypic expression of the predominant cell type in the adherent layer of human long-term marrow cultures [10].

MATERIALS AND METHODS

Cells:

Passaged bone marrow stromal cells were generated from the adherent layers of human long-term bone marrow cultures derived from normal consenting marrow transplant donors.

Long-term marrow cultures were generated as previously described [11]. When the layers were confluent and the cultures actively hemato-poietic (2-4 weeks), the adherent cells were passaged at least 10 times to yield a phenotypically homogenous population [10] in McCoy's 5A medium supplemented with 12.5% fetal bovine serum, 12.5% horse serum, 1% each of glutamine, sodium pyruvate, sodium bicarbonate and vitamins, 0.8% essential amino acids, 0.4% non-essential amino acids (all from Gibco) and 10^{-7} M hydrocortisone (Sigma).

K562, a multipotent myeloid leukemia cell line (ATCC, Rockville, MD) was cultured and maintained in RPMI with 10% fetal bovine serum.

Plasmids:

The plasmid pSV2CAT is an expression vector in which the simian virus 40 (SV40) early transcriptional promotor-enhancer region regulates expression of the chloramphenicol acetyltransferase (CAT) gene [12]. In pRSVCAT, the CAT gene is transcribed from the Long Terminal Repeat (LTR) element of Rous sarcoma virus (RSV). In pHßApr-1-CAT (kindly provided by A. Pawson) expression of the CAT gene is regulated by the human ß-actin promotor [13]. pLPV-5'-CAT (kindly provided by L. Mosthaf) was constructed by replacing the SV40 enhancer element in pSV2CAT with the lymphocytotropic papovavirus (LPV) enhancer element [14]. pUCRNmCMVX/HCAT and pUCRNtKCAT [15] were generously donated by J. de Villiers. In pUCRNmCMVX/HCAT, the regulatory sequence is the murine cytomegalovirus (MCMV) immediate early promotor and in pUCRNtKCAT, the herpes simplex (HSV) thymidine kinase (TK) promotor.

Table 1:

Plasmid	Regulatory Sequence	Regulatory Sequence Abbreviation
pSV2CAT	SV40 early promotor enhancer	SV40
pRSVCAT	Rous sarcoma virus LTR	RSV
pHßApr-1-CAT	human ß-actin promotor	ß-actin
pLPV-5'-CAT	lymphocytotropic papovavirus enhancer and SV40 early promotor	LPV
pUCRNmCMVX/H CAT (J. de Villiers)	murine cyto-megalovirus immediate early promotor	CMV
pUCRNtKCAT (J. de Villiers)	herpes simplex thymidine kinase promotor	HSV-TK

Electroporation:

Electroporation was performed according to our previously published method [4] except that supercoiled in place of linear plasmid DNA [prepared as described earlier [16] and purified through 2 cesium chloride gradients], was used at a concentration of 100 µg/ml for gene transfer. K562 cells underwent electroporation at a concentration of $2-6 \times 10^7$/ml, while stromal cells were used at a concentration of $1-2 \times 10^7$/ml in sterile calcium and magnesium-free phosphate-buffered saline (PBS).

Following electroporation, cells were incubated on ice for 10 minutes and then transferred to the appropriate medium at 37°C, in an atmosphere containing 5% CO2 for 48-72 hours.

CAT Assay:

Following 48-72 hour incubation, cells were harvested, washed three times in PBS and extracts prepared as described earlier [12]. Protein determinations were performed and 50-200 µg of cellular protein were assayed for the presence of the CAT activity using either the thin layer chromatography (TLC) method [12] or the recently published diffusion assay method which relies on the differential diffusion of the reaction product, radiolabelled acetyl-chloramphenicol, into non-aqueous scintillation fluid [17].

RESULTS

In this study the effect of different regulatory sequences on the expression of the electroporated chloramphenicol acetyltransferase (CAT) gene was monitored in a transient expression assay for CAT enzyme activity. The plasmids containing the CAT gene and the different regulatory sequences are listed in Table 1.

We found that after electroporation, cell viability for K562 and passaged marrow stromal cells was approximately 50 and 75% respectively, in contrast to values of 90% and greater in the case of human CFU-GM [6].

K562:

CAT activity was assayed in K562 cells transfected with the plasmids listed in Table 1 and with pSVOCAT as a negative control. Purified CAT protein (0.01 U) was included as a positive control.

The highest CAT activity was obtained with the RSV LTR but the immediate early CMV promotor and the herpes simplex thymidine kinase promotor were also active. Relative CAT activity is shown in Table 2 with activity from the SV40 early promotor/enhancer designated as unity.

Table 2: CAT Activity in Electroporated K562 Cells

Regulatory Sequence	Relative CAT Activity
LPV	0.4
RSV	6.5
SV40	1
CMV	2.4
HSV-TK	2.3
ß-ACTIN	0.6

Table 3: Relative CAT activity in passaged
marrow stromal cells.

Regulatory Sequence	Relative CAT Activity
LPV	0.04
RSV	0.1
SV40	1.0
CMV	15.0
HSV-TK	0.1
ß-ACTIN	0.04

Human Marrow Stromal Cells:

In comparison to other promotors, the
immediate early CMV promotor is highly active in
marrow stromal cells, as shown in Table 3. In
contrast, low activity was detected in cells
transfected with plasmids containing the other
regulatory sequences, except for the SV40 early
promotor which gave moderate levels.

DISCUSSION

Recent studies indicate that electroporation
is a relatively efficient means of transferring
genes into hematopoietic cells [4-6]. Further
work is in progress in several laboratories to
optimize gene transfer efficiency with this
technique. To date, a number of physical
parameters have been shown to be important in
this regard and include electric field strength,
pulse shape and duration, cell concentration,
DNA conformation and temperature after electroporation [4]. For optimal results, these parameters may differ with different cell types.

Characterization of DNA transfer by electroporation has disclosed several advantages of
this method. For example, unlike the calcium
phosphate coprecipitation method, in electroporation the transferred DNA appears to bypass
lysosomes and is consequently less prone to
mutation or degradation [18]. Also unlike the
former method, the DNA can be introduced in low
copy number [19,5]. In DNA-mediated gene
transfer by electroporation, manipulation into
specialized vectors is unnecessary. Moreover,
gene transfer by electroporation is not limited
to the size constraints on DNA imposed by
retrovirally-mediated transfer - DNA sequences
of over 100 kb can be transferred [20]. A
further advantage of electroporation lies in the
ability to co-transfect unlinked sequences
efficiently [4].

We have examined the effect of different
regulatory sequences in a transient expression
assay as a first step toward the investigation
of lineage-specific expression of genes transferred to hematopoietic stem cells. We used
K562 as a convenient model for the human hematopoietic multipotent stem cell and human passaged
marrow stromal cells as representatives of the
hematopoietic microenvironmental population. We
found that the transferred reporter (CAT) gene
was expressed at high levels by several regulatory sequences including RSV, CMV, HSV-TK and
SV40. However, the RSV LTR gave the highest
activity. Of interest was our observation that
the cellular promotor for human ß-actin gave low
levels of activity.

In contrast, results with marrow stromal
cells indicate that compared to the other

regulatory sequences, the CMV immediate early
promotor generated by far the highest level of
activity.

Our results suggest that of the regulatory
sequences tested, the most active promotor in
multipotent progenitor cells may be the RSV
LTR. Further studies are in progress to examine
the effect of this regulatory sequence in the
progeny of pre-colony forming cells.

Somewhat unexpectedly we have found that the
CMV immediate early promotor gave the highest
level of activity in marrow stromal cells. In
other studies, we have readily generated
transformed stromal cell lines with plasmids
containing the SV40 large T antigen gene,
suggesting that stromal cells can be transfected
efficiently by electroporation.

Taken together our data indicate that in
addition to hematopoietic precursors, marrow
stromal cells may also be a suitable target
population for the development of gene delivery
protocols.

ACKNOWLEDGEMENTS

This work was supported by the Medical
Research Council of Canada and the National
Cancer Institute of Canada (NCIC). A.K. is a
Research Scholar of the NCIC.

REFERENCES

1. Keller G, Paige C, Gilboa E, Wagner EF
   (1985) Expression of a foreign gene in
   myeloid and lymphoid cells derived from
   multipotent hematopoietic precursors.
   Nature 318:149.
2. Dick JE, Magli MC, Phillips RA, Bernstein A
   (1986) Genetic manipulation of hematopoietic
   stem cells. Trends in Genetics 2:165-170.
3. Neumann E, Schaefer-Ridder M, Wang Y,
   Hofschneider PH (1982) Gene transfer into
   mouse myeloma cells by electroporation in
   high electric fields EMBO J 7:841.
4. Toneguzzo F, Hayday A, Keating A (1986)
   Electric field mediated DNA transfer:
   transient and stable gene expression in
   human and mouse lymphoid cells. Mol Cell
   Biol 6(2):703-706.
5. Toneguzzo F, Keating A, Glynn S, McDonald K
   (1988) Electric field-mediated gene
   transfer: characterization of DNA transfer
   and patterns of integration in lymphoid
   cells. Nucleic Acids Res 16(12):5515-5533.
6. Toneguzzo F, Keating A (1986) Stable
   expression of selectable genes introduced
   into human hematopoietic stem cells by
   electric field mediated DNA transfer. Proc
   Natl Acad Sci USA 83:3496-3499.
7. Takahashi M, Keating A, Singer JW (1985) A
   functional defect in irradiated adherent
   layers from chronic myelogenous leukemia
   long-term marrow cultures. Exp Hematol
   13:926-931.
8. Gorman CM, Rigby PWJ, Lane DP (1985)
   Negative regulation of viral enhancers in
   undifferentiated embryonic stem-cells. Cell
   42:519-526.
9. Lozzio CB, Lozzio BB (1975) Human chronic
   myelogenous leukemia cell line with positive
   Philadelphia chromosome. Blood 45:321-334.

10. Keating A, Just-Mitchell K, Toor P, Klein M, Sodek J (1986) Passaged human marrow stromal cells: a unique cell population. Exp Hematol 14(6):426.

11. Keating A, Powell J, Takahashi M, Singer JW (1984) The generation of human long-term marrow cultures from marrow depleted of Ia (HLA-DR) positive cells. Blood 64(6):1159-1162.

12. Gorman CM, Moffat LF, Howard BH (1982) Recombinant genomes which express chlor-amphenicol acetyl transferase in mammalian cells. Mol Cell Biol 2:1044-1051.

13. Gunning T, Leavitt J, Muscat G, Ng S-Y, Kedes L (1987) A human ß-actin expression vector system directs high level accumu-lation of antisense transcripts. Proc Natl Acad Sci USA 84:4831-4851.

14. Mosthaf L, Pawlita M, Gruss P (1985) A viral enhancer element specifically active in human haematopoietic cells. Nature 315:597-600.

15. Xian-Jun F, Keating A, de Villiers J, Sherman M (submitted) Tissue specific activity of heterologous viral promotors in primary rat hepatocytes and Hep G2 cells.

16. Norgard MV (1981) Rapid and simple removal of contaminating RNA from plasmid DNA without the use of RNASE. Anal Biochem 113:34-42.

17. Neumann JR, Morency CA, Russian KO (1987) A novel rapid assay for chloramphenicol acetyl-transferase gene expression. Biotechniques 5:444-447.

18. Bertling W, Hunger-Bertling K, Cline MJ (1987) Intranuclear uptake and persistance of biologically active DNA after electro-poration of mammalian cells. J Biochem Biophys Methods 14:223-232.

19. Boggs SS, Gregg RG, Borenstein N, Smithies O (1986) Efficient transformation and frequent single-site, single copy insertion of DNA can be obtained in mouse erythroleukemia cells transformed by electroporation. Exp Hematol 14:988-994.

20. Jastreboff MM, Ito E, Bertino JR, Narayan R (1987) Use of electroporation for high-molecular-weight DNA-mediated gene transfer. Exp Cell Res 171(2):513-517.

12   Improved Retroviral Transfer of Genes into Canine Hematopoietic Progenitor Cells

F.G. Schuening, R. Storb, R.B. Stead, W.W. Kwok, R. Nash, and A.D. Miller

ABSTRACT

Canine hematopoietic progenitor cells were infected with amphotropic helper-free retroviral vectors containing either the neomycin phosphotransferase gene (NEO) or a mutant dihydrofolate reductase gene (DHFR*). Successful transfer and expression of both genes in hematopoietic progenitor cells was demonstrated by the ability of the viruses to confer resistance to either the aminoglycoside G418 or methotrexate (MTX). It was shown that the incidence of resistant granulocyte- macrophage colony-forming units (CFU-GM) after cocultivation for 24 hours with helper-free virus-producing cells was 10% (6% to 16%). Autologous marrow cocultivated for 24 hours with virus-producing packaging cells was then transplanted into six dogs after lethal total body irradiation. All dogs showed engraftment within two weeks and four dogs survived 5-7 months without adverse effects. One dog that had been given marrow infected with a DHFR* virus and that had received MTX as in vivo selection after marrow transplantation and survived, showed 0.1% and 0.03% MTX-resistant CFU-GM at weeks 3 and 5, respectively. In an attempt to increase the efficiency of gene transfer, which could improve our yield of cells with a transformed phenotype posttransplant, marrow was cocultivated for 24 hours with virus-producing packaging cells and then kept in long-term marrow culture which was fed with virus-containing super-natant. Cells were harvested after six days and cultured in CFU-GM assay with and without selective agent. The average rate of gene expression in CFU-GM was 41% (19% to 86%). Successful gene transfer into canine hematopoietic progenitor cells can be achieved using retroviral vectors and the efficiency of gene transfer is increased by combining cocultivation with long-term marrow culture. It is uncertain at this point whether canine pluripotent hematopoietic stem cells can be infected.

INTRODUCTION

Amphotropic retroviral vectors are effective in mediating the transfer of genes into mammalian cells of several different species. Helper-free retroviral vectors were designed such that viral replication occurs only in packaging cells [1,2] and not in target cells, thereby preventing spread of virus through host tissues. Retroviral vectors therefore appear to be suitable vehicles for gene transfer into mammalian cells.

Interest has focused on bone marrow as a target for gene transfer studies because of ease of access and the ability to transplant marrow after ex vivo manipulation of the marrow cells. Genetic diseases in man could potentially be treated with transfer of genetic material into marrow cells [3]. Successful gene transfer into hematopoietic progenitor cells using retroviral vectors has been demonstrated in different animals and in humans [4-8]. Infection rates were determined by growth of resistant granulocyte/macrophage colony-forming units (CFU-GM) were from 4% to 28%. The demonstration of drug-resistant CFU-GM does not necessarily indicate successful infection of pluripotent hematopoietic stem cells, but shows that infection is possible in the hematopoietic lineage of the dog.

Infection of pluripotent hematopoietic stem cells was first demonstrated in gene transfer studies in murine marrow transplants [4,5]. Marrow reconstitution occurred in these animals with a high percentage of progeny derived from infected stem cells. Expression of these genes was generally poor, and the frequency of cells expressing the gene of interest decreased over time. In other animal models, transplantation studies to determine the efficiency of infection of stem cells are few [6,9]. Very low infection rates have been demonstrated. The higher rates of infection of reconstituted marrow in mice with retroviral vectors may result from the relatively larger amount of marrow from congenic mice that can be preselected prior to transplant, as well as vectors which derive from murine retroviruses that may infect murine stem cells more readily.

The purpose of the current studies is to ascertain the feasibility of somatic gene transfer into canine hematopoietic precursor cells using retroviral vectors. The canine model provides a logical intermediate step in current investigations of gene transfer techniques, prior to application of these techniques to man, because the dog represents a well-established large, random-bred animal model for marrow transplantation. Different strategies may be needed to increase in vitro infection rates of nonmurine animals to efficiently reconstitute marrow with infected stem cells. These studies have implications for the therapy of human genetic diseases and for investigation of the behavior of pluripotent hematopoietic stem cells and their multilineage progeny after marrow transplantation using the unique proviral integration sites as clonal markers.

MATERIALS AND METHODS
Animals

Beagles, hounds, and mongrel dogs of both sexes, 6-14 months old, raised at the Fred Hutchinson Cancer Research Center or purchased from commercial kennels, were used as marrow donors. They were dewormed, vaccinated against distemper, leptospirosis, hepatitis, and parvovirus, and observed for disease for two months before use. Research was carried out according to the principles enunciated in the "Guide for Laboratory Animal Facilities and Care" prepared by the National Academy of Sciences, National Research Council.

Retroviral vectors and cell lines

The retroviral vectors SDHT [10,11] and pN2 [4] are replication-defective. SDHT contains a mutant dihydrofolate reductase (DHFR*) gene which confers resistance to methotrexate (MTX) and pN2 has the neomycin phosphotransferase gene (NEO) which confers resistance to G418. The SPN vector is similar to SDHT but has had the NEO gene substituted for the DHFR*. The NIH 3T3(TK⁻) [12] cell line was used as a target cell for virus titer assays. Packaging cell lines PA12 [13] and PA317 [2] enable replication and also package vectors so that they are competent to infect other cells. PA12-N2 has been noted to produce helper virus. Cell

lines were grown in Dulbecco's modified Eagle (DME) medium (Gibco, Grand Island NY) with 4.5 g/L glucose and 10% fetal bovine serum (FBS) (Gibco, Grand Island NY). Assays for viral titer and helper virus have been described [13]. Virus titers ranged from 10^6-10^7 CFU/ml for both viruses.

CFU-GM assay and assessment of infection rates

CFU-GM assays were carried out as described [8,14]. Mononuclear marrow cells were separated by Ficoll-Hypaque gradient (density 1·077) and then 7.5×10^4 cells were seeded in 2 ml of semisolid medium consisting of DME medium with 20 ug of L-asparagine per ml, 75 ug of DEAE-dextran per ml, 20% prescreened heat-inactivated human AB plasma, 5% phytohemagglutinin-stimulated lymphocyte-conditioned medium, and 0.3% agar with or without selective agent, either MTX at a concentration of 0.25 or 0.50 umol/L or G418 at a concentration of 1 or 2 mg/ml (concentration of powder, of which ~50% was active). Cultures were kept at 37°C in a humidified 10% CO_2 incubator. Colonies were counted after 14 days. To determine the rate of infection as a percentage, the number of resistant CFU-GM colonies grown in the presence of the selective agent was divided by the number of colonies grown in the same conditions without selective agent. Controls were established with uninfected marrow to confirm that CFU-GM could not grow in the presence of the selective agent. For colony morphology, agar cultures were overlaid for 6 minutes with 1 ml of 2.5% glutaraldehyde (Sigma Chemical Co., St. Louis MO) in 0.1 M phosphate buffer. The agar pellicles were transferred onto acid-clean glass slides. Slides were then stained with Gill's hematoxylin (Tago, Burlingame CA) for 5 min, washed, dried, mounted (Permount, Fisher Scientific Co., St. Louis MO), and assessed. Granulocytes and monocytes/macrophages were enumerated based on their morphology.

Long-term culture of canine marrow

The conditions for the long-term culture of canine marrow cells have been developed based on the method described for the murine long-term marrow culture [15]. Cultures were established with 6×10^7 mononuclear marrow cells (MNC) in 75 cm² canted-neck flasks (Corning, Corning NY) at 2×10^6 MNC/ml in RPMI Medium 1640 (M.A. Bioproducts, Walkersville MD) supplemented with 20% prescreened heat-inactivated horse serum (Flow Laboratories, Rockville MD), 10^{-7} M hydrocortisone-21-phosphate (Sigma, St. Louis MO), 1% nonessential amino acids, 1% pyruvate, 2% L-glutamine, and 1% penicillin-streptomycin. Cultures were maintained in a humidified atmosphere of 5% CO_2 in air at 37°C. Nonadherent cells and medium were removed from the flasks after 1 week and centrifuged at 400 g for 10 min. Half of the used medium (15 ml) together with 15 ml of fresh medium and 6×10^7 freshly aspirated autologous mononuclear marrow buffy coat cells from the same dog were returned into each flask (marrow boost).

Infection procedure of canine marrow cells

Bone marrow was diluted 1:3 with RPMI Medium 1640, and separated by Ficoll-Hypaque density gradient centrifugation (density 1·077). The low density cells at the interface were cocultivated for 24 hours at a ratio of 2:1 with the vector-producing packaging cells (~80% confluent) that had been irradiated with 45 Gy from two opposing ^{60}Co sources. Cells were kept at 37°C in a humidified 10% CO_2 incubator, and DME medium supplemented with 20% fetal bovine serum and 2 ug/ml Polybrene (Sigma, St. Louis MO) was used. The nonadherent marrow cells were removed from the culture dishes by gentle rinsing with DME medium.

Because of the large amount of canine marrow necessary for autologous marrow transplants, cocultivations with virus-producing packaging cells for marrow transplantation were done in roller bottles. Conditions for infection of bone marrow and procedures for transplantation have been described [9]. Briefly, bone marrow was aspirated from the humeri and femora of anesthetized dogs, diluted 1:3 with Waymouth's medium, and separated using Ficoll-Hypaque density-gradient centrifugation. Nucleated marrow cells were cocultivated with virus-producing packaging cells (PA12-SDHT or PA12-N2) at a ratio of 1:2 for 12-24 hours in 850 cm^2 roller bottles (Corning). Packaging cells were seeded in the roller bottles at $3-5\times10^7$ cells 48 hours prior to the addition of marrow. The roller bottles received 1500 cGy irradiation just prior to the addition of marrow cells, were supplemented with CO_2, and maintained at 37°C, rotating at 1-2 rpm on a Belco roller apparatus. Following cocultivation, marrow cells were removed carefully without dislodging the packaging cells, washed, resuspended in serum-free medium, and reinfused into the marrow donor. The dog had received 500 cGy midline tissue dose of total body irradiation from two opposing ^{60}Co sources at a rate of 10 cGy/minute approximately 2 hours prior to marrow infusion.

Marrow cells recovered after 24-hour cocultivation were incubated in long-term marrow cultures that had a stromal layer established one week before. When the long-term marrow cultures were given the marrow boost, nonadherent cells and medium were removed from the cultures and 6×10^7 marrow cells previously cocultivated on vector-producing packaging cells together with 15 ml of fresh medium and 15 ml of used long-term culture medium were returned into each flask. Cultures were fed on days 0, 3, and 5 of culture with 15 ml virus-containing fresh medium ($2-5\times10^6$ CFU/ml) and 15 ml used medium. Control long-term cultures were set up by recharging with marrow cells cocultivated for 24 hours on packaging cells which did not produce viral vectors (TK$^-$N2) or by recharging with marrow cells which were only Ficoll-separated without further cocultivation. Controls were fed three times with 15 ml virus-free fresh medium and 15 ml used medium. Control and virus-containing flasks were harvested after 6 days of culture, and cells were plated in the CFU-GM assay with and without selective agents.

RESULTS

Retroviral transfer of genes into canine CFU-GM after short-term virus incubation

To establish the CFU-GM assay as a tool for determining the percent of progenitor cells infected with the retroviral vector, the sensitivity of canine CFU-GM had to be determined to the selective agents. G418 completely inhibited CFU-GM colony growth at 1 mg/ml and MTX suppressed all growth at 0.25 uM. The infection rate obtained by cocultivation of marrow cells with virus-producing packaging cells was compared with the infection rate obtained following exposure of hematopoietic cells to virus-containing medium. Cocultivation resulted in three- to tenfold more efficient gene transfer and was therefore used in the following experiments. The efficiency of infection, as determined by the rate of drug resistance, ranged from 7% to 16% with PA12-SDHT, and was 6% with PA317-N2 (Table 1).

Autologous transplantation of marrow infected with retroviral vectors by short-term virus incubation

To assess the ability of the retroviral vectors to infect canine pluripotent hematopoietic stem cells, autologous transplants of infected marrow were done in six dogs. The vector-producing packaging cells used to infect marrow in five of six transplants were PA12-SDHT and in the other, PA12-N2. Infection rates of CFU-GM assayed immediately after cocultivation were 2.8% to 12.9%. Since PA12-SDHT confers MTX resistance, 0.5 mg/kg MTX was administered intravenously on days 1, 3, 5 posttransplant, and twice weekly thereafter to three of the dogs as an in vivo selection step. All dogs engrafted without delay, with peripheral polymorphonuclear blood counts >500/mm^3 occurring at an average of 14 days post-

Table 1.
Drug resistance of canine CFU-GM following retrovirus infection by 24-hour cocultivation

CFU-GM Colonies (no. drug-resistant/total no. colonies)			
PA12	PA12-SDHT	PA317	PA317-N2
0/682	86/551 (16%)		
0/853	59/670 (9%)	0/1560	119/1960 (6%)
0/474	62/480 (13%)		
0/958	55/486 (11%)		
0/439	17/238 (7%)		
0/3406	279/2425(12%)	0/1560	119/1960 (6%)

transplant. Two dogs died at three weeks after marrow transplantation from severe gastrointestinal toxicity due to MTX, and the remaining four survived >5-7 months with normal grafts. The one dog that survived the in vivo selection with MTX had received the autologous marrow with the higher rate of infection (12.9%).

Posttransplant marrow from the four dogs were studied for evidence of infection of stem cells with retroviral vectors. Resistant CFU-GM were present in only one dog; this dog received a marrow transplant which showed a high rate of infection immediately after cocultivation and he also received MTX postgrafting for in vivo selection (Table 2). The percentage of MTX-resistant CFU-GM in the reconstituted marrow was low and transient: 0.1% at week 3 and 0.03% at week 5 after marrow transplantation. At 7 and 10 weeks there was no growth of resistant CFU-GM. To detect proviral DNA which may not express the selectable marker to allow for the selection of resistant CFU-GM, Southern analysis was performed. No proviral DNA was detected at a level of sensitivity of about 1 copy per 50 cells. No adverse effects on engraftment or on survival were seen as a result of cocultivation of marrow with retroviral vectors.

Table 2.
MTX-resistant CFU-GM in dog #5 following autologous transplantation of marrow infected with DHFR*-virus

Week after marrow transplant	CFU-GM colonies (no. MTX-resistant[a]/ total no. colonies)
3	5/ 3480 (0.14%)
5	2/ 6900 (0.03%)
7	0/11960 (<0.008%)
10	0/ 276 (<0.36%)

[a] MTX was added to each tissue culture dish to 0.25 uM. This concentration was sufficient to inhibit growth of CFU-GM colonies completely in the uninfected controls.

Table 3.
MTX or G418 resistance of canine CFU-GM following virus infection by cocultivation and long-term marrow culture

CFU-GM colonies/10^5 MNC (no. resistant/total no. colonies)			
PA12-SDHT (MTX-resistant)		PA317-SPN (G418-resistant)	
619/1972	(31%)	71/82	(87%)
0/99	(0%)	63/73	(86%)
301/365	(82%)	12/61	(20%)
21/73	(29%)	7/37	(19%)
114/228	(50%)		
44/101	(44%)	153/253	(60%)
4/16	(25%)		
18/96	(19%)		
1121/2950	(38%)		

Long-term marrow culture enhances retrovirus-mediated gene transfer

In an attempt to increase the infection rate of canine hematopoietic progenitor cells, marrow was cocultured for 24 hours with vector-producing packaging cells PA12-SDHT as before, and then kept for six days in long-term culture, which was fed three times with virus-containing supernatant. Table 3 summarizes the results of eight independent experiments with this vector. No infection was seen in one experiment; in the others infection rates ranged from 19% to 82% (mean 38%). No drug-resistant CFU-GM were seen in uninfected controls. Morphology of drug-resistant CFU-GM was examined by light microscopy after fixing and staining the agar cultures as a whole. The majority (90%) of the drug-resistant colonies were monocyte-macrophage colonies. No fibroblast colonies were seen.

Results obtained with the DHFR* vector were confirmed by using a different viral vector, SPN, which contains the NEO gene. Table 3 summarizes the results of four other independent experiments with SPN. Infection rates ranged from 19% to 87% (mean 60%).

DISCUSSION

Amphotropic helper-free retroviral vectors with either the DHFR* gene or NEO gene inserts have been demonstrated to successfully transform canine CFU-GM to MTX and G418 resistance. The average rate of infection of canine CFU-GM (10%) was comparable to the infection rates of CFU-GM from other species such as mouse (10-20%) [4], monkey (7-28%) [6], and human cells (3-10%) [7]. Pluripotent hematopoietic stem cells, though, must be infected to achieve persistence of the transformed phenotype in grafts after marrow transplantation.

At this time the only way to study pluripotent hematopoietic stem cells in the dog is by confirming their ability to reconstitute marrow after lethal ablation of host marrow. Therefore, to study whether it was possible to infect canine pluripotent hematopoietic stem cells and obtain expression of the gene of interest carried by the retroviral vector, autologous marrow transplants were done after cocultivation of marrow with virus-producing packaging cells for 24 hours. Only one dog had evidence of drug-resistant hematopoietic progenitors posttransplant suggestive of successful infection, and this was at a low rate. These cells did not persist beyond seven weeks. Similar results of low rates of infection were also seen in a study on nonhuman primates [6] where in situ hybridization for vector-encoded mRNA in peripheral blood mononuclear cells was positive on 0.8% of cells counted.

The reason for infection rates which are low and transient in marrow cells obtained from reconstituted animals can be explained by several possible mechanisms. Infection may have been achieved in only a very small number of pluripotent stem cells with dilution of these cells by the noninfected stem cells in the regenerating marrow. Another alternative is that retroviral vectors may have infected only progenitor cells which have lost the capability for self-renewal and therefore have a limited life span. It is also possible that there is decrease in the level of gene expression secondary to cell regulatory events. If either of the first two possible mechanisms is significant, increasing the efficiency of infection could result in the successful long-term infection of pluripotent hematopoietic stem cells such that it could be detected in the graft after marrow transplantation.

Long-term marrow cultures have several possible advantages in attempts to increase the efficiency of infection. Prolonged exposure of marrow cells to retroviral vectors is achieved by feeding with virus-containing supernatant after an initial period of cocultivation with virus-producing packaging cells. Also, since the majority of pluripotent hematopoietic stem cells are in a G_0 phase and cell replication appears to be important for the integration of retroviral vectors, it is interesting to note that replication of progenitor cells in the adherent layer is triggered by feeding the long-term marrow culture [16,17]. Therefore, by using long-term marrow cultures, prolonged exposure to retroviral vectors while cells in culture are dividing can be achieved.

Infection rates of CFU-GM were improved with long-term marrow culture as compared to results obtained with cocultivation only. To determine if pluripotent hematopoietic stem cells have been infected in long-term marrow culture, the infected autologous marrow can be transplanted into lethally irradiated marrow donors. Reconstitution of the hematopoietic systems in mice [18] as well as in man [19] has been demonstrated from marrow growing in long-term culture. Cells from canine long-term marrow cultures that have been established for as long as six days can reconstitute lethally irradiated animals [unpublished data]. Further experiments will explore whether canine marrow cells cocultivated for 24 hours with virus-producing packaging cells, and then exposed to virus supernatant in long-term cultures for an additional six days, will lead to hematopoietic reconstitution after lethal TBI and to increased and lasting expression of retroviral vectors.

ACKNOWLEDGEMENTS

The authors wish to thank Drs. Eli Gilboa and David Trauber for supplying the plasmids pN2 and pSDHT, and Joey Meyer, Theodore Graham, Ray Colby, Greg Davis, and Robert Raff for excellent technical assistance.

REFERENCES

1. Mann R, Mulligan RC, Baltimore D (1983) Construction of a retrovirus packaging mutant and its use to produce helper-free defective retrovirus. Cell 33: 153-159

2. Miller AD, Buttimore C (1986) Redesign of retrovirus packaging cell lines to avoid recombination leading to helper virus production. Mol Cell Biol 6: 2895-2902

3. Lehn PM (1987) Gene therapy using bone marrow transplantation. Bone Marrow Transplant 1: 243-258

4. Keller G, Paige C, Gilboa E, Wagner EF (1985) Expression of a foreign gene in myeloid and lymphoid cells derived from multipotent haematopoietic precursors. Nature (London) 318: 149-154

5. Dick JE, Magli MC, Huszar D, Phillips RA, Bernstein A (1985) Introduction of a selectable gene into primitive stem cells capable of long-term reconstitution of the hemopoietic system of W/W^v mice. Cell 42: 71-79

6. Kantoff PW, Gillio AP, McLachlin JR, Bordignon C, Eglitis MA, Kernan NA, Moen RC, Kohn DB, Yu S, Karson E, Karlsson S, Zwiebel JA, Gilboa E, Blaese RM, Nienhuis A, O'Reilly RJ, French Anderson W (1987) Expression of human adenosine deaminase in nonhuman primates after retrovirus-mediated gene transfer. J Exp Med 166: 219-234

7. Hock RA, Miller AD (1986) Retrovirus-mediated transfer and expression of drug resistance genes in human haematopoietic progenitor cells. Nature (London) 320: 275-277

8. Kwok WW, Schuening F, Stead RB, Miller AD (1986) Retroviral transfer of genes into canine hemopoietic progenitor cells in culture: A model for human gene therapy. Proc Natl Acad Sci USA 83: 4552-4555

9. Stead RB, Kwok WW, Storb R, Miller AD (1988) Canine model for gene therapy: Inefficient gene expression in dogs reconstituted with autologous marrow infected with retroviral vectors. Blood 71: 742-747

10. Miller AD, Trauber DR, Buttimore C (1986) Factors involved in production of helper virus-free retrovirus vectors. Somatic Cell Mol Genet 12: 175-183

11. Wolff L, Ruscetti S (1985) Malignant transformation of erythroid cells in vivo by introduction of a nonreplicating retrovirus vector. Science 228: 1549-1552

12. Wei C, Gibson M, Spear PG, Scolnick EM (1981) Construction and isolation of a transmissible retrovirus containing the src gene of Harvey murine sarcoma virus and the thymidine kinase gene of herpes simplex virus type I. J Virol 39: 935-944

13. Miller AD, Law M-F, Verma IM (1985) Generation of helper-free amphotropic retroviruses that transduce a dominant-acting methotrexate-resistant dihydrofolate reductase gene. Mol Cell Biol 5: 431-437

14. Schuening F, Emde C, Schaefer UW (1983) Improved culture conditions for granulocyte-macrophage progenitor cells (abstr). Exp Hematol 11(Suppl 14): 205

15. Dexter TM, Allen TD, Lajtha LG (1977) Conditions controlling the proliferation of haemopoietic stem cells in vitro. J Cell Physiol 91: 335-344

16. Toksoez D, Dexter TM, Lord BI, Wright EG, Lajtha LG (1980) The regulation of hemopoiesis in long-term bone marrow cultures. II. Stimulation and inhibition of stem cell proliferation. Blood 55: 931-936

17. Cashman J, Eaves AC, Eaves CJ (1985) Regulated proliferation of primitive hematopoietic progenitor cells in long-term human marrow cultures. Blood 66: 1002-1005

18. Spooncer E, Dexter TM (1983) Transplantation of long-term cultured bone marrow cells. Transplantation 35: 624-627

19. Chang J, Morgenstern G, Deakin D, Testa NG, Coutinho L, Scarffe JH, Harrison C, Dexter TM (1986) Reconstitution of haemopoietic system with autologous marrow taken during relapse of acute myeloblastic leukaemia and grown in long-term culture. Lancet i: 294-295

13 The Use of Retroviral Vectors in Human Disorders

M. Scarpa and C.T. Caskey

With the application of recombinant DNA technology to medical practice, geneticists have gained powerful tools for the diagnosis and therapy of genetic disease at the molecular level.

It is now possible to isolate and compare equivalent genes from different individuals and identify mutations of the DNA responsible for various disorders. Therefore, corrections of genetic disorders via insertion of a normal gene into a human being to replace mutant gene function (gene therapy) may soon be possible.

The ethic of gene therapy in humans has been, and is, widely debated (1-7), but essentially all the investigators agree that it would be ethical to insert genetic material into human somatic cells for the sole purpose of correcting severe genetic defects - i.e. somatic cell gene therapy.

The development of human gene therapy requires the selection of a disease model system and of an efficient mammalian gene transfer system. (Table 1).

The selection of the disease model is tied to: a) the knowledge of the cause of the disease at the DNA level and b) to a tissue that can be easily accessible; that can be manipulated *in vitro*; and implanted into the patient who will be donor and receptor at the same time.

Since the main goal of gene therapy (as presently envisioned) is the addition of a normal gene to the cells, recessively inherited diseases, such as inborn errors of metabolism, are the most reasonable candidates.

CANDIDATES FOR GENE THERAPY

The first aim of gene transfer will be to apply sophisticated technology to improve the quality of life of patients suffering from highly debilitating or lethal diseases. It would also be desirable to consider diseases of such a frequency that their cure or improvement would result in a significant social impact. Cystic Fibrosis (CF) and Duchenne Muscular Dystrophy (DMD), for example are severe inherited diseases with the frequency of 1:2000-2500 and 1:3500-4000 newborns respectively. CF is the most common caucasian severe autosomal recessive disease. Clinically it is a generalized affection of the exocrine glands with production of thick mucous excretion resulting in chronic obstructive lung diseases, leading to res-piratory failure and death, exocrine pancreatic insufficiency, intestinal obstruction and male infertility. In late 1985, the CF locus has been mapped at chromosome 7 and two DNA markers, met on-cogene and the anonymous probe J3.11 (D7S8), were found to be tightly linked to CF locus. Actually these two probes are used for prenatal diagnosis (8,9). Unfortunately even if these two probes are linked to CF locus the gene responsible for the disease has not been isolated yet, therefore, it is not always possible to determine the carrier status of the disease.

The DMD is an X-linked disease which clinically involves progressive muscular weakness. Cardiac involvement also occurs in the majority of patients and together with muscular respiratory failure is one of the principle causes of death from DMD. Molecular genetics investigations have permitted the isolation of cloned probes, called pERT, which allow prenatal diagnosis and carrier detection (10). The full length cDNA sequence of 14 kb corresponding to the fetal skeletal muscle transcript has been isolated (11).

Since severe involvement of lung and muscle are the principle cause death in these two disorders, gene transfer has to be performed in these organs. To be able to apply gene transfer in CF and DMD, affections, such as metabolic diseases, that can be considered model systems to test safe protocols that, in the near future, could be widely applied, are necessary to study.

The pathophysiology of these diseases often involve organs that are distant from the tissue specific expression of the gene. The "gene therapist", then, has the possibility to target the gene replacement to an organ that is directly affected by the disease (e.g. α-1 antitrypsin, adenosine deaminase deficiency, ornithine transcarbamylase deficiency) or try to express the gene in more readily accessible organs when expression is not required in a specific tissue (e.g. β-galactosidase deficiency, arylsulfatase-B deficiency).

For example, the pathophysiology of α-1 an-

Table 1: Requirements for Gene Therapy

Disease model system
a) Knowledge of the exact causes of the disease at the DNA level.
b) Expression of the normal gene in tissues accessible for *in vitro* manipulation or at least improvement of the disease through the expression of the gene in an accessible tissue.
c) Possibility to improve the disease with low level expression of the normal gene replacing the mutant gene.

Mammalian gene transfer system
a) Capacity to introduce the gene into a variety of recipients, *in vitro* and *in vivo*.
b) High efficiency of transduction of genetic material into a minor component of an heterogeneous cell population (hematopoietic stem cells in bone marrow tissue).
c) Absence of undesirable effect such as activation of silent sequences eventually acting as oncogenes.

Table 2: Candidates for Gene Therapy: Cloned Genes

Hematopoietic
*Adenosine deaminase (immune deficiency)[d]
 *α1-Antitrypsin[a]
  Carbonic anhydrase II (osteopetrosis)
  Complement-C2, C3, C4, Factor B, and C9
  Fucosidase (fucosidosis)
  α-Galactosidase (Fabry)
*Globin α,B, (thalassemia, hemoglobinopathies)[g]
*Glucocerebrosidase (Gaucher)[c]
  B-Glucuronidase (Sly)
*Purine nucleoside phosphorylase (immune deficiency)[a]

Liver
  Arginase (argininemia)
  Argininosuccinic acid lyase (hyperammonemia)
*Argininosuccinic acid synthetase citrullinemia)[h]
  Carbamoylphosphate synthetase (hyperammonemia)
  Dihydrobiopterin reductase (hyperphenylalaninemia)
*Factor VIII, IX (hemophilia)[i]
*LDL receptor (familial hypercholesterolemia)[j]
*L-methylmalonyl-CoA mutase (methylmalonic acidemia)[l]
*Ornithine transcarbamylase (hypercholesterolemia)[b]
*Phenylalanine hydroxylase (PKU)[k]
  Propionyl CoA carboxylase, a and B subunits (propionic acidemia)
  Pyruvate carboxylase (congenital lactic acidosis

Muscle
  Duchenne muscular dystrophy (DMD)

Central Nervous System (CNS)
*Hypoxanthine-guanine phosphorybosyltransferase (HPRT)[f]

Targets for Gene Therapy: Uncloned Genes or Defect Unknown
Hematopoietic
 Defect known
  Arylsulfatase B (Maroteaux-Lamy)
  Aspartylglycosaminadase
  C3bi receptor
  B-Galactosidase (Morquio B)
  Galactosamine 6-sulfate sulfatase (Morquio A)
  α-L-Iduronidase (Hurler, Scheie)
  α-D-Mannosidase (mannosidosis)
  α-Neuraminadase (sialidosis)
  Iduronate sulfatase (Hunter)

 Defect unknown
  Ataxia-telangiectasia
  Bare lymphocyte syndrome
  Chediak-Higashi syndrome
  Chronic granulomatous disease
  Glanzman's thrombasthenia
  Hermansky-Pudlak syndrome
  Osteopetroses
  Severe combined immune deficiencies (autosomal recessive and X-linked)
  Wiskott-Aldrich syndrome
  X-linked agammaglobulinemia

Liver
 Defect known
  Acetoacetyl 3-ketothiolase
  Acyl-CoA dehydrogenases
  Amylo-1,6 glucosidase (GSD Type III)
  Amylo-1,4:1,6 transglucosidase (GSD Type IV)
  ATP:cobalamin adenosyltransferase (methylmalonic-acidemia cblB)
  Branched-chain ketoacid dehydrogenase (maple syrup urine disease)
  Cl inhibitor (hereditary angioedema)
  Cystathionine B-synthetase (homocystinuria)
  Dihydrobiopterin synthetase (hyperphenylalaninemia)
  Electron-transfer flavoprotein (glutaric aciduria Type II)
  Fructose-1 phosphate aldolase (hereditary fructose intolerance)
  Fructose 1,6 diphosphatase
  Fumarylacetoacetase (tyrosinemia Type I)
  Galactokinase
  Galactose 1-phosphate uridyl transferase (galactosemia)
  Glucose 6-phosphatase (GSD Type I)
  3-Hydroxy-3-methylglutaryl-CoA lyase
  Isovaleryl-CoA dehydrogenase (isovaleric acidemia)
  Lipoprotein lipase
  Pyruvate dehydrogenase complex
  Phosphorylase, hepatic (GSD Type VI)
  Phosphorylase B kinase (GSD Type VIII)
  Hyperlysinemia
  Hypervalinemia
  Hyperleycinemia-isoleucinemia
  Nonketotic hyperglycinemia
  Wilson's disease
  Byler's disease

Lung
  Cystic Fibrosis

* - under investigation for gene transfer
a-Ref.12; b-19; c-23; d-26-35; e-36; f-37-53;
g-71,72; h-80; i-81; j-82; k-82; l-84

titrypsin (α-1AT) deficiency is not completely understood. It appears that the hepatic accumulation of α-1AT is associated with cellular dysfunction and that the anti-elastase activity of the enzyme is involved as the major serum antiprotease in the protection of the parenchyma from damage during inflammation. The loss of the function results in liver cirrhosis and pulmonary emphysema.

A number of infants with liver cirrhosis have been successfully treated by liver transplantation.

The full length cDNA and genomic sequences of α-1AT has been isolated and characterized (12).

Recently, human cDNA for α-1AT has been successfully transferred and expressed in mouse fibroblasts via retroviral infection (13). Furthermore, mouse fibroblasts expressing human α-1 AT has been implanted into the peritoneal cavities of nude mice. Four weeks later, human α-1AT was detected in both sera and epithelial surface of the lungs (14) giving encouragement that similar results could be achieved in humans.

The ornithine transcarbamylase (OTC) deficiency is the most common urea cycle disorder in humans and is associated with a severe hyperammonemia evident within the first 3 days after birth (15). Mental retardation accompanies this disorder. OTC deficiency is a dominant X-linked disease (Xp21.1) and the homozygous males are more severely affected than heterozygote females. The frequency of the disease is about 1/30,000. For OTC deficiency, the delivery of the gene must be to the liver since the synthesis of citrulline requires carbamoylphosphate synthetase activity in the mitochondria. Cloned cDNA has been obtained from rat (16), mouse (17), and human (18), and all of them have been used for minigene constructions that express OTC enzyme in mammalian cells (19). OTC deficiency is remarkable for the availability of two murine models for the disease: the "sparse-fur" (20) and the "sparse-fur abnormal skin and hair" (SPF/ash) (21) which provide a system to develop tissue-specific gene transfer therapy. Recently in our laboratory the molecular defect responsible for the SPF phenotype has been identified (22).

Another interesting disease in which successful attempts of gene transfer has been done is the Type I Gaucher's disease. Type I Gaucher's disease is an autosomal recessive disorder characterized by a glycolipid storage caused by an inherited deficiency of the lysosomal enzyme glucocerebrosidase. The substrate, the glucosylceramide (glucocerebroside), is accumulated in the reticuloendothelial system. Hepatosplenomegaly, variable bone involvement, and the characteristic "Gaucher cell" in bone marrow aspiration and in visceral organs are associated with the disorder. The glucosylceremide is a portion of a larger glycosphingolipid which is generated during the degradation of glycosides of red and white blood cells glycolipid and endogenous membrane glychosphingolipids. Allogenic bone marrow transplantation has the potential to cure the disease but is associated with high morbidity and mortality. Nevertheless, type I Gaucher's disease is an important candidate for gene therapy because the phenotypic cellular defect is manifested only in the cells of the macrophage lineage which is accessible through bone marrow transplantation. The human glucocerebrosidase cDNA has been cloned

(23,24) and has been introduced into fibroblast and lymphoblast from a patient via retroviral vector. Normal level of glucocerebrosidase was obtained (25).

Although many other potential candidates for gene therapy could be described, we will focus our attention on three diseases that seem to be the best and most studied candidates for the first attempt of a gene therapy in human: the adenosine deaminase (ADA) deficiency, the hypoxanthine-guanine phosphoribosyltransferase (HPRT) deficiency, and the purine nucleoside phosphorylase (PNP) deficiency. The absence of each of these enzymes is responsible for a Severe Combined Immune Deficiency (SCID) syndrome, the Lesch-Nyhan syndrome, and a severe immune deficiency respectively.

ADA and PNP deficiencies are autosomally recessive inherited diseases and result from mutations in their structural genes. They both result in T and B cell disfunction with SCID. ADA deficiency is approximately 5-10 fold more frequent than PNP deficiency. In ADA deficiency the lack of T and B cells seems to be due to a cytotoxic accumulation of adenosine and deoxyadenosine not converted into inosine and deoxyinosine respectively. Accumulation of dATP, also, seems to play an important role. The clinical features of PNP deficiency are similar except for the persistence of antibodies production and B cells. ADA deficiency is a life-threatening SCID syndrome with death frequently occurring during the first year despite therapeutic effort of bone marrow transplantation and blood transfusions. The high level of ADA expression is normally found in T cells and in other tissues. The ADA enzyme is composed of a single peptide of 36,000 - 38,000 mw encoded by gene of 32 kb split in 12 exons and mapping to 20q13.2-qter (26). The cDNA is 1,533 bp long, the promoter activity is performed by a stretch of 135 bp immediately preceding the cap site. Analysis of ADA deficiency mutation is underway and indicates that the disease is heterogeneous at the molecular level (27). Both point mutations and larger rearrangements can be expected on the basis of the molecular analysis of the human mutations. Successful attempts to correct defect in fibroblast and cultured T and B cells from patient (28,29) has been performed. Furthermore, human ADA has been transferred and expressed into mice (30-34). Recently, long term low level expression of human ADA in mouse hematopoietic system was obtained in our laboratory. (35).

PNP is composed of a single subunit of 30,000-32,000 mw. A full length cDNA for human PNP has been cloned (36) but the rarity of the PNP deficiency is likely to retard its clinical trial for gene therapy.

The Lesch-Nyhan (L-N) syndrome is an X-linked inherited disease resulting from a deficiency of the enzyme hypoxanthine-guanine phosphoribosyltransferase (HPRT). The HPRT catalyzes the metabolic salvage of the purine bases hypoxanthine and guanine. The clinical manifestations of the disease are characterized by a severe mental retardation, self mutilating behavior, choreoathetosis and hyperuricemia which, if unattended, leads to early renal failure. The pathobiology accounting for the CNS disfunction of L-N syndrome is unclear. There is no evidence of accumulation of toxic metabolites in the CNS. Normally, HPRT is expressed at high level in the basal ganglia where, since low level purine biosynthesis is produced, HPRT may be required for the maintenance of the nucleotides pool. HPRT is a 218 amino acid protein with a mass of 24.470 dalton (37) and is present in all the tissues (housekeeping protein) at a relatively low level with the exception of CNS where the gene is expressed at high level. The single copy gene is localized on the X-chromosome at band Xq26-q27. The gene is 44 kb long and split into 9 exons (38,39). Genetic analysis of families with HPRT deficiencies reveal a number of point mutations responsible for single amino acid substitution in the peptide chain resulting in absent or markedly decreased HPRT activity (40-43). Actually, with the RNAse cleavage technique, point mutations and deletions undetectable by northern analysis without prior knowledge of the precise localization of the lesion (44-46). Despite the well characterized clinical phenotype associated with HPRT deficiency and an exhaustive search for potential animal models, no animal equivalent of the L-N syndrome has been found until 1987. After years of surgical and chemical attempts to reproduce HPRT deficiency in animal, recently two independent approaches have been taken to develop an authentic model for HPRT deficiency in mouse. The first approach involves the selection of HPRT⁻ embryo stem (ES) cells to produce HPRT⁺/HPRT⁻ animals (47,48). Another strategy employed the use of antisense of HPRT mRNA (RNA complementary to the endogenous HPRT message). Antisense recombinant constructions injected into fertilized mouse eggs have reduced the levels of HPRT to approximately 50% of the normal level in regions of the CNS that express the antisense transgene (49). Interesting, male mice with complete HPRT deficiency produced using the ES method are phenotypically normal, lacking the metabolic and neurological manifestations associated with L-N syndrome in human. Attempts to correct HPRT deficient cells in vitro have been made using retroviral vectors. Partial correction of the metabolic abnormalities such as elevated purine excretions and increased intercellular hypoxanthine have been obtained (50-53). cDNAs have been introduced into mice cultured bone marrow cells and animals have been reconstituted. HPRT provirus has been detected in mouse spleen and bone marrow cells as long as 133 days after transplantation.

GENE TRANSFER SYSTEM

The major gene transfer system proposed for the first attempt of a gene therapy in human is based on the use of the RNA viruses (retroviruses). There are a number of advantages in the use of retroviruses. Briefly, we can summarize four of them.

1) Efficiency of infection and expression.

Retroviruses are able to infect almost 100% of the cells submitted to infection and will express the foreign gene after integration. There are other methods to transfer genes into cells (calcium phosphate precipitation, electroporation, micro injection) but the efficiency of expression is very low, only one in 10^3 to 10^7 cells will stably express the foreign gene.

2) Simultaneous infection of as many cells as desired.

3) Integration of the DNA as a single copy.

Under appropriate conditions the DNA can be integrated as a single copy. Other techniques of gene transfer often result in integration of DNAs in multicopies. In those system, one or more copies of DNA have the possibility to associate head to tail and been transferred together into the cells.

4) Precise integration of the DNA with respect to the viral genome.

Although the integration of the DNA is random with respect to cellular genome, the structure of the cloned DNA is very well maintained with respect to the viral genome.

Another very important consideration is the safety of the retroviruses. No one of the retrovirus used as vector of foreign gene has ever harmed the whole cell, yet. Reconstituted mice harboring and expressing a human cDNA after six months have not manifested any harmful effect due to the retrovirus vector (John Belmont, personal communication). Evidently, a more thorough evaluation of the defective retrovirus vectors will be necessary before they can be considered perfectly safe.

STRUCTURE AND LIFE CYCLE OF RETROVIRUSES:

The retrovirus is composed of a RNA-protein core and a glycoprotein envelope. It is able to enter a cell where its RNA will act as a template for reverse transcriptase, a polymerase, which transcribes the RNA genetic information into a double stranded DNA copy. Double strand DNA, then, will integrate into the genomic DNA of the host cell (provirus) and will drive the expression of the genes.(See Fig.1) The provirus structure of Mo-MLV, for example, is composed of two LTRs at each end of the provirus, containing the regulating sequence for the transcription. In the left LTR the enhancer, the promoter, and the initiation signal (cap site), in the right LTR the termination site (a poly-A signal) for viral RNA synthesis are localized. Since the two LTRs are identical, the mechanism by which activation and suppression of the regulating sequences and of the terminal site occur is not clear. Each LTR is flanked by a short region necessary for reverse transcription. Between the two LTRs the packaging signal, the gag, pol and env coding sequences are contained. A splicing donor signal is present at the 5' of the packaging signal while a splicing acceptor is localized at the 5' of the env gene. The retrovirus containing all these components is called replication-competent retro-virus. The gag, pol and env protein encoding-genes are involved respectively in viral encapsidation and assembly, synthesis of viral DNA (reverse transcriptase) and integration, and viral budding and entrance. These are transacting genes. Retroviruses have also cis-acting sequences for reverse transcriptase (sequences involved in the viral DNA synthesis and primer binding site and polypurine tract sequences involved in priming DNA synthesis) for integration (attachment sites) and for encapsidation (Ψ site) (See Fig.2).

PACKAGING CELL LINES

For gene therapy purpose, ones wants to obtain a defective retrovirus. Such retrovirus that is

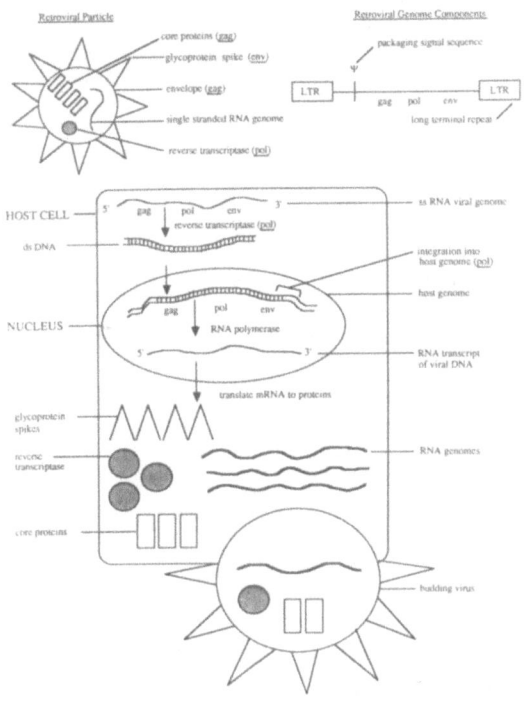

Fig. 1 Structure and life cycle of a retrovirus

However, there are several problems with these packaging cell lines, including limited host range and low titer of retrovirus. One of the major problems is the possibility of recombinational events between vector and retrovirus packaging system with the production of helper virus. Recently, to circumvent this problem, two different packaging cell lines have been produced: PA317 (59) and Gp+E-86 (60). PA317 contains two main deletions: the first one in the 3' LTR to remove the site for initiation of the second strand DNA synthesis and the second one in the 5' and of the 5' LTR. Consequently, a proper integration signal cannot be made from the remaining sequences. Despite the necessity of two recombinational events, production of helper virus has been observed (Belmont, personal communication). With various vectors, viral titer greater than 10^7/ml can be obtained in the PA317 cell line, Gp+E-86, for its parts, has been obtained by transfecting NIH3T3 cells with two different plasmids, one containing the gag and pol genes and the second the env gene. In addition, the two plasmids have deletions of the Ψ packaging sequence and the 3' LTR to further diminish the opportunity for recombination. In fact, to have the production of intact helper virus, three recombinational events would have to occur. The viral titer obtained with this cell line is comparable to other packaging cell lines (10^5, 10^6 with N2 vector) and sufficient to be used in gene transfer experiments.

RETROVIRAL VECTORS

Virus particle production, target-expression and *in vivo* expression are the major criteria to have a retroviral vector useful for gene therapy. For this purpose, a large number of retroviral vectors have been designed, some of which will be described here. We can summarize the different viral structures in three groups: 1) vectors with internal promoters, 2) double expression vectors, 3) self-inactivating vectors, 4) tissue-specific vectors.

1) Vectors with internal and LTR promoters.

The best example for this group is N2 (61). N2 is a Mo-MLV derived vector in which the selectable neomycin (Neo) resistance gene is flanked by two LTRs. The 5' LTR extends into the viral gag gene coding sequences, the Neo gene is so preceded by a functional codon as well as by 418 bp of coding sequences containing a total of 9 AUG start codons. The titer of N2 ranges from $1x10^6$ to $5x10^6$ CFU/ml despite the fact that the Neo and the gag initiation codons are out of frame. It appears that it is just the gag sequences that confers the higher titer to the virus since this DNA has unexpectedly introduced a cryptic 3' splice-site just upstream from the Neo gene which generates a splice RNA that serves as mRNA for the Neo gene. Human ADA fused to the mouse metalothionein I (MTI) and to the simian virus 40 (SV40) promoters have been cloned in N2 (MAX and SAX vectors). The titer of the virus generated is about $0.2-2x10^6$ CFU/ml. The presence of the MTI-ADA and of SV40-ADA into N2 has reduced from 5-10 fold the level of aberrantly spliced RNA compared to N2. For these reasons, we have constructed the vector ΔN2ADA deleting the Neo gene and replacing it by a human ADA cDNA under control of the LTR regulating

able to infect the target cell, integrates into the genomic DNA, expresses the harbor gene but is unable to package its genomic RNA into a virion. The development of such vector has been fostered by I. Verma and Richard Mulligan in 1983 and 1984 respectively (50,54). These two groups replaced the viral genes with an exogenous gene, transfected tissue culture cells and 24-48 hours later infected the same cells with a helper virus which provided the gene necessary for the formation of viral particles. The population of viruses that was obtained consisted of a mixture of defective virus and the helper virus. For the development of gene therapy the production of helper virus must be avoided and the knowledge of the localization of the Ψ sequence (55) has made possible the development of a system in which the recombinant defective virus can be produced but not the helper virus. The construction of an ecotropic recombinant virus missing the P sequence and integrated as provirus has permitted the creation of a cell line, called Ψ2, that produces all the viral proteins necessary for making an intact viral particle but lacks the property of viral packaging and viral production (56). When the cell line is transfected with the defective but Ψ+ retroviral recombinants, the cell produces defective viral particles containing the recombinant retrovirus within 48 hours. This particular cell line will produce replication defective viral particle that can infect mouse and rats cells through only one cycle. Utilizing the same basic principle, packaging cell line (57,58) that allows the production of helper free amphotropic retrovirus capable of infecting a broad range of cells, including human cells, have been developed.

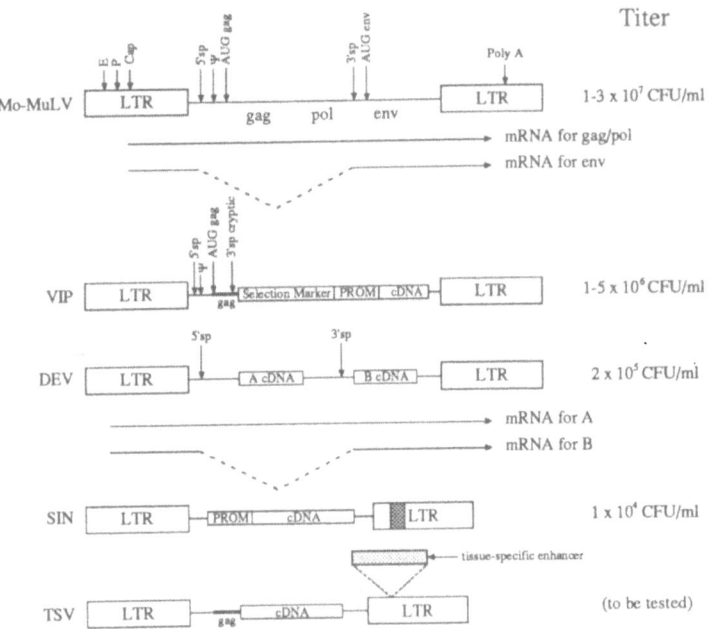

Titer

Mo-MuLV — 1-3 x 10⁷ CFU/ml

VIP — 1-5 x 10⁶ CFU/ml

DEV — 2 x 10⁵ CFU/ml

SIN — 1 x 10⁴ CFU/ml

TSV — (to be tested)

Figure 2: Different vectors used in gene transfer experiments.
Mo-MuLV: Moloney-Murine Leukemia Virus.
E: Enhancer, P: Promoter, Cap: Cap site.
5'sp: 5' splicing site (donor), Ψ: packaging signal
AUG gag: Translational starting codon.
3'sp: 3' splicing site (acceptor).
Poly A: Poly A sequence.
LTR: Long terminal repeat.
VIP: Vector with internal promoter.
DE: Double expression vector.
SIN: Self-inactivating vector.
TSV: Tissue-specific vector.
▮indicates the deletion of the enhancer
and promoter in the 3' LTR.
▨indicates the substitution of the
Moloney enhancer with a tissue-
specific enhancer in the 3' LTR.

sequences. The titer has been restored and after infection of murine bone marrow stem cells, mice expressing human ADA after 6 months have been obtained. N2 vector has been successfully used for gene transfer *in vitro* (62,63) and *in vivo* (64,65). One major problem of N2 was the production of helper virus after transfection of Ψ2 and PA12 packaging cell line, because of recombinational events due to the 418 bp sequence of the gag gene present in N2. This problem has been circumvented by the production of the PA317 and GP+E-86 packaging cell lines with a production of a high titer virus (2-8x10⁶ CFU/ml and 1-4x10⁶ CFU/ml respectively).

2) The double expression (DE) vectors.

The DE vectors (66), contain two foreign genes: the first gene replaces the gag/pol coding sequences and is expressed by an unspliced RNA species, the second gene replaces the coding sequences for the viral envelope and it is expressed from a spliced RNA species by the removal of the long intron. Furthermore, the expression of sequences inserted in this vector is achieved through the use of normal retrovirus transcriptional signals. Although good titer and expression were obtained using this type of vector (2x10⁵ and 10% of endogenous ADA respectively), one major problem reduced its usefulness for gene therapy experiment. In the construction of this vector, it was assumed that the viral splice junction sequences regulated the splicing process. In other words, the expression of the genes introduced into DE vectors was assumed to be dependent by the efficient formation of the two viral RNA species, which in turn is dependent on a properly regulating splicing process. But recently, DE vectors containing human β interferon and the mouse DHFR cDNAs in the env position, have shown that the efficient expression of the spliced gene product was dependent on the presence of additional intron sequences emphasizing the importance of internal sequences for the biogenesis of Mo-MLV RNA species (67).

3) The self-inactivating (SIN) and handicapped vectors.

These last vectors have been designed to minimize the risk of the promoter and enhancer sequences contained in the LTRs activating, flanking sequences (i.e. oncogenes) after their integration in the genomic DNA with potential harmful effects on the infected cells. The SIN vector (68) contains the deletion of 299 bp in the U3 region of the 3' LTR spanning the enhancer and the promoter sequences but leaving the TATA box intact. Since the U3 region of the 3' LTR, as a consequence of the mechanism of replication of retrovirus serves as template for the formation of both LTRs U3 regions in the progenery virus, any modification introduced in it is automatically transmitted to the 5' LTR. As a result, the proviral DNA in the infected cells becomes transcriptionally inactive, thus producing two consequences: the inhibited expression of the transduced gene, that is under the control of an internal promoter, and the reduction of insertional activation. The handicapped vector (69) is a derivative of the pAFVXM vector which differs from the conventional vectors by having two different LTR sequences. The 5' LTR has sequences from Abelson murine leukemia virus (A-MLV) and from Harvey murine sarcoma virus (HA-MSV), while the 3' LTR has Mo-MLV and A-MLV sequences. In addition a deletion of the 5' splicing donor (SD) site has been performed. Apart from the different composition of the LTRs, the handicapped vector differ from the SIN vectors since deletion of 327 bp in the U3 region of the 3' LTR contains not only the promoter and the enhancer but also the TATA box. The major disadvantage of these vectors is the low titer (~ 10⁴ CFU/ml) probably insufficient for use in gene therapy experiments. Nevertheless, the construction of these vectors has contributed to identify some of the fine mechanisms regulating the expression of transduced genes. Recently, for example, a disabled vector in which the expression of the HPRT gene was dependent from different promoters (human metalothionein IIA, human cytomegalovirus) has been constructed (70). This vector contains a deletion in the 3' LTR of the regulatory elements including the TATA and CAAT boxes, plus the SD site adjacent to the 5' LTR to prevent aberrant splicing events between the splice-donors site and any cryptic splicing acceptor present in the inserted DNA. The expression of the HPRT gene increased two fold using disabled vector indicating stabilization of viral transcript and interference between the two sets of transcriptional signals: the LTR regulation sequences and any internal promoter. Unfortunately the titer of this kind of virus dropped down to 10^2 CFU/ml.

4) The tissue-specific vectors

The last generation of vectors, that we could call tissue-specific vectors, was designed because although the titer of virus production with the other types of vectors is sufficiently high to efficiently infect rare subpopulation of cells, such as hematopoietic stem cells, the efficiency of expression is generally not satisfactory. Two potential explanations for this could be proposed: 1) the lack in the retroviral construct of elements specifically required in the hematopoietic stem cells to ensure gene expression at subsequent stage of hematopoietic cell development; 2) the production of unusual mRNA structures which are abnormally processed or degraded. In this regard, the construction of a vector harboring a cDNA sequence for the Bglobin gene and a 1.1 kb 3'flanking sequence which contains a putative enhancer element increased levels of human B globin expression in transgenic mice (71,72). Analysis of long term reconstituted animals has demonstrated that the human B globin mRNA expression was restricted to cells of the erythroid lineage. Only a very small amount of human B globin mRNA was found in B and T lymphocytes or in macrophages. Although a tissue specific expression was obtained, the absolute level of Bglobin transcription was low, 0.4-4% of endogenous mouse B globin mRNA levels per copy of provirus. Despite this low level of expression of inserted proviral sequence in reconstituted mice, the tissue specificity of this type of construct appears promising. We have designed a vector containing a 700 bp μ chain immunoglobulin enhancer to express the ADA cDNA specifically in B lymphocytes. This enhancer 700 bp fragment is attractive because it contains 4 protein binding sites: E1, E2, E3, E4 and an octamer sequence that is highly conserved in all human or murine immunoglobulin gene enhancers. The octamer sequence 5'ATGCAAAT is consistently found, at a position 60-80 bp upstream from the site of the transcriptional initiation of human and murine heavy chain variable region (V_H) genes. The reverse orientation 5'ATTTGCAT has been found in the light chain variable region (V) gene promoters. In addition, this octamer sequence constitutes a protein binding site and has been reported to specifically direct the expression of genes in B lymphocytes if cloned in position 0 to -70bp from the cap site (73). The octamer has also been cloned in the 3' LTR of a N2 derived vector in substitution of the Moloney enhancer and both vectors are currently being evaluated.

BONE MARROW AND GENE TRANSFER

At present bone marrow cells as delivery system for gene transfer experiments in human, appear to be the most readily tissue accessible tissue that one can manipulated, corrected and reimplanted. In the last several years, bone marrow transplantation (BMT) has been applied to several different classes of inherited disorders such as immunodeficiency diseases, hematological disorders, osteopetrosis and lysosomal storage diseases.

However, despite the extensive experience in BMT, the procedure is both costly and dangerous. Only a limited number of patients have HLA-compatible sibling donors available (allogenic BMT) and a high skilled and experienced team is required. Even in the most experienced centers, many short term and long term complications of the procedure arises. The short term complications include failure of engraftment, acute infections and toxicity associated with the use of preparative regimens and acute graft vs host disease (GVHD). The long term complications include chronic GVHD, incomplete engraftment leading to partial reconstitution of all hematological functions, and long term toxicity associated with chronic immunosuppression. Most of these problems might be circumvented with the application of autologous BMT which eliminates the lack of donors and avoids problems of immunological mediated rejection. Despite the characteristic of bone marrow tissue (accessibility, easy manipulation and standard protocols for the transplantation), one major problem that needs to be thoroughly investigated is the efficient infection of the primitive stem cells, the self renewable pluripotent cells that give rise to all hematopoietic lineages. To circumvent this problem, researchers are studying methods to increase the proportion of transduced stem cells before transplantation. One of the most interesting approaches is the application of immunological techniques to purify the stem cells population based on their cell surface differentiation antigenes. The application in mice of a negative selection for progenitors by removing all BM cells expressing selective markers characteristic of differentiated B cells, granulocytes, myelomonocytic cells and T cells has lead to the detection of a cell surface differentiation antigen, Thy-1, present on progenitors which, recognized by a monoclonal antibody, Scal, identifies a population of cells intermediate in size between bone marrow lymphocytes and large myeloid cells. Most of these are in Go-G1 phase of mitotic cycle and at least 97% bearing a 2N amount of DNA. Injections of only 30 of these cells were able to reconstitute 50% of lethally irradiated mice and to reconstitute all blood cell types in survivors (74). In the human system monoclonal antibodies against CD34 (eg My10, B1-3C5,12-8) have been found to purify a cell population enriched in stem cells and progenitors wells in human. CD34 is also expressed on a similar marrow population in baboons making it possible to test expression on pluripotent stem cells capable of hematopoietic reconstitution in vivo. After treatment of the bone marrow with antibody 12-8, an enriched population of cells CD34+ has been isolated and was capable of completely restoring hematopoiesis when transplanted into lethally irradiated baboons. After 6 months to one year the durable engraftment suggested that the stem cells are contained within the CD34+ population. Unfortunately, the possible contribution of endogenous radio resistant stem cells to hematopoietic reconstitution was not investigated (75-77).

These methods of purification of stem cells might have wide application to BMT. They could be used to infect stem cells in a very specific way with a retrovirus for gene therapy of an inherited disease or as an alternative approach to methods currently used to deplete tumor cells *ex vivo* from the marrow of the patient undergoing autologous BMT. In addition, application of gene transfer to purified stem cells could be very useful to treat certain form of cancer. The identification of bone marrow cells in cancer patient could permit their infection with retrovirus vector carrying drug resistance genes, therefore conferring a selective advantage in the treatment of various neoplasms. Experiments in mice have shown that hematopoietic cells infected with a retrovirus vector containing the gene for the methotrexate resistance have reconstituted bone marrow population cells treated with high dosage of MTX (78). A multidrug resistance gene (MDR) was also studied in murine experiments (79).

CONCLUSION

Modern recombinant DNA technology allows us to consider the feasibility of somatic gene therapy in humans. Problems which need to be addressed before this can be seriously envisioned are: (a) demonstration of proviral integration in a true population of self-renewable stem cells; (b) improvement in long-term expression in the progeny of the stem cells and (3) demonstration of long-term safety of those procedures.

We are confident that these progresses are likely to take place in a near future.

Supported in part by Cystic Fibrosis Foundation Fellowship # CCF F054 9-01 (M.S.), PHS grant # HD21452 and Howard Hughes Medical Institute.

REFERENCES

1. Anderson WF (1972) In: Hamilton N. (ed) The new genetics and the future of the man. Michigan: Eerdmans Grand Rapids
2. Presidnet's Commission for the Study of Ethical Problems in Medicine and Biomedical and Behavioral Research, 1982, *Splicing Life*, Government Printint Office, Washington, D.C.
3. Friedman T, (1983) Gene Therapy:Fact and Fiction, New York: Cold Spring Harbor Laboratory, Cold Spring Harbor
4. Motulsky AG, (1983) Impact of genetic manipulation on society and medicine. Science 219:135-140
5. Anderson WF, (1984) Prospects for human gene therapy. Science 226:401-408
6. Walters LR, (1986) The ethics of human gene therapy. Nature 320:225-227
7. Belmont JW, Caskey CT (1986) Developments leading to human gene therapy. In: Kucherlapati R (ed) Gene Transfer Plenum Publishing Corporation
8. Beaudet AL, Bowcock A, Buchwald M (1986) Linkage of cystic fibrosis to two tightly linked DNA markers: joint report from a collaborative study. Am J Hum Genet 39:681-693
9. Beaudet AL, Buffone GJ (1987) Prenatal diagnosis of cystic fibrosis. J of Ped 111(4):630-633
10. Caskey CT, Ward P, Hejtmancik F (1988) DMD carrier detection and prenatal diagnosis via recombinant DNA methods. Adv Neurol 48:83-91
11. Koenig M, Hoffman EP, Bertelson CJ, Monaco AP, Feener C, Kunkel LM (1987) Complete cloning of the Duchenne muscular dystrophy (DMD) cDNA and preliminary genomic organization of the DMD gene in normal and affected individuals. Cell 50:509-517
12. Long G, Chandra T, Woo S, Davies E, Kurachir K (1984) Complete sequence of the cDNA for human α-1-antitrypsin and the gene for the S variant. Biochem 23:4828-4837
13. Garver RI, Chytil A, Karlsson S, Fells GA, Brantly ML, Courtney M, Kantoff PW, Nienhuis AW, Anderson WF, Crystal RG (1987) Production of glycosylated physiologically "normal" human α_1-antitrypsin by mouse fibroblasts modified by insertion of a human α_1-antitrypsin cDNA using a retroviral vector. Proc Natl Acad Sci 84:1050-1054
14. Garver JR RI, Chytil A, Courtney M, Crystal RG (1987) Clonal gene therapy:transplanted mouse fibroblast clones express human α1-antitrypsin gene in vivo. Science 237:762-764
15. McKenzy WN (1985) In: Stainbury JB, Wyngarden JB, Frederickson DS, Goldstein JC, Bowman NS (ed) Metabolic basis of inherited diseases, 5th ed. 438 New York: McGraw-Hill
16. Horwich AL, Kraus JP, Williams K, Kalousek F, Konigsberg Wm Rosenberg LE (1983) Molecular cloning of the cDNA coding for rat ornithine transcarbamoylase. Proc Natl Acad Sci USA 80:4258-4262
17. Scherer SE, Veres G, Caskey CT (1988) The genetic structure of mouse ornithine transcarbamylase. Nucl Acid Res 15:1593-1601
18. Horwich AL, Fenton WA, Williams KR, Kalousek F, Kraus JP, Doolittle RF, Konigsberg W, Rosenberg LE (1984) Structure and expression of a complementary DNA for the nuclear coded precursor of human mitochondrial ornithine transcarbamylase. Science 224:1068-1076
19. Veres G, Craigen WJ, Caskey CT (1986) The 5' flanking region of the ornithine transcarbamylase gene contains DNA sequences regulating tissue-specific expression. J Biol Chem 261:7588-7591
20. DeMars R, LeVan SL, Trend BL, Russell LB (1976) Abnormal ornithine carbamoyltransferase in mice having the sparse-fur mutation. Proc Natl Acad Sci USA 73(5):1693-1697
21. Doolittle DP, Hulbert LL, Cordy C, (1974) A new allele of the sparse fur gene in the mouse. J Hercol 65:194-195
22. Veres G, Gibbs RA, Scherer SE, Caskey CT (1987) The molecular basis of the sparse fur mouse mutation. Science 237:415-417
23. Sorge J, West C, Westwood B, Beutler E (1985) Molecular cloning and nucleotide sequence of human glucocerebrosidase cDNA. Proc Natl Acad Sci USA 82:7289-7293
24. Tsuji S, Choudary PV, Martin BM, Winfield S, Barranger JA, Ginns EI (1986) Nucleotide sequence of cDNA containing the complete coding sequence for human lysosomal glucocerebrosidase, J Biol Chem 261:1,50-53
25. Sorge J, Kuhl W, West C, Beutler E (1987) Complete correction of the enzymatic defect of type I Gaucher disease fibroblasts by

retroviral-mediated gene transfer. Proc Natl Acad Sci USA 84:906-909

26. Valerio D, Duyvesteyn MGC, Dekker BMM, Weeda G, Berkvens ThM, van der Voorn L, van Ormondt H, Van der Eb AJ (1985) Adenosine deaminase: characterization and expression of a gene with a remarkable promoter. EMBO J 4:22, 437-443

27. Adrian GW, Wiginton DA, Hutton JJ (1984) Structure of adenosine deaminase mRNAs from normal and adenosine deaminase-deficient human cell lines. Mol Cell Bio 4:1712-1717

28. Palmer RD, Hock RA, Osborn WRA, Miller AD (1987) Efficient retrovirus-mediated transfer and expression of a human adenosime deaminase gene in diploid skin fibroblasts from an adenosine deaminase-deficient human. Proc Natl Acad Sci USA 84:1055-1059

29. Kantoff PW, Kohn DB, Mitsuya H, Armentano D, Sieberg M, Zwiebel JA Eglitis MA, McLachlin JR, Wiginton DA, Hutton JJ, Horowitz SD, Gilboa E, Blaese RM, Anderson WF (1986) Correction of adenosine deaminase deficiency in cultured human T and B cells by retrovirus-mediated gene transfer. Proc Natl Acad Sci USA 83:6563-6567

30. Valerio D, Duyvesteyn MGC, van der Eb AJ (1984) Intriduction of sequences encoding functional human adenosine deaminase into mouse cells using a retroviral shuttle system. Gene 34:163-168

31. Friedman R (1985) Expression of human adenosine deaminase using a transmissible murine retrovirus vector system. Proc Natl Acad Sci USA 82:703-607

32. Williams DA, Orkin SH, Mulligan RC (1986) Retrovirus-mediated transfer of human adenosine deaminase gene sequences into cells in culture and into murine hematopoietic cells in vivo. Proc Natl Acad Sci USA 83:2566-2570

33. Belmont, JW, Henkel-Tigges J, Chang SMW, Wager-Smith K, Kellems RE, Dick JE, Magli MC, Phillips RA, Bernstein A, Caskey CT (1986) Expression of human ademosine deaminase in murine haematopoietic progenitor cells following retroviral transfer. Nature 322(24):85-387

34. Nelson DL, Chang SMW, Henkel-Tigges J, Wager-Smith K, Belmont JW, Caskey CT (1986) Gene replacement therapy for inborn errors of purine metabolism. In Symposia on Quantitative Biology, Cold Spring Harbor Laboratory

35. Belmont JW, MacGregor GR, Wager-Smith K, Fletcher FA, Moore KA, Chang SMW, Hawkins D, Villalon D, Caskey CT (1988) Expression of human adenosine deaminase in murine hematopoietic cells. Molec Cell Biol in press

36. Goddard J, Caput D, Williams S, Martin D (1983) Cloning of human purine-nucleoside phosphorylase cDNA sequences by complementation in Escherichia coli. Proc Natl Acad Si USA 80:4281-4285

37. Wilson JM, Tarr GE, Mahoney WC, Kelley WN (1982) Human hypoxanthine-guanine phosphoribosyltransferase, complete amino acid sequence of the erythrocyte enzyme. J Biol Chem 257:18, 10978-10985

38. Melton D, Konecki D, Brennand J, Caskey CT (1984) Structure, expression and mutation of the hypoxanthine phosphoribosyltransferase gene. Proc Natl Acad Sci USA 81:2147-2151

39. Patel PI, Franson PE, Caskey CT, Chinault AC (1986) Fine structure of the human hypoxanthine phosphoribosyltransferase gene, Mol Cell Biol 6:393-403

40. Hughes SH, Wahl GM, Capecchi MR (1975) Purification and characterization of mouse hypoxanthine-guanine phosphoribosyltransferase. J Biol Chem 250:1, 120-126

41. Wilson JM, Young AB, Kelley WN (1983) Hypoxanthine-guanine phosphoribosyltransferase deficiency. The molecular basis of the clinical syndromes. N Engl J Med 309(15):900-910

42. Wilson JM, Tarr GE, Kelley WN (1983) Human hypoxanthine (guanine) phosphoribosyltransferase:An amino acid substitution in a mutant form of the enzyme isolated from a patient with gout. Proc Natl Acad Sci USA 80:870-873

43. Wilson J, Kelley WN (1983) Molecular basis of hypoxanthine-guanine phosphoribosyltransferase deficiency in a patient with the Lesch-Nyhan syndrome. J Clin Invest 71:1331-1335

44. Myers RM, Larin Z, Maniatis T (1985) Detection of single base substitutions by ribonuclease cleavage at mismatches in RNA:DNA duplexes. Science 230(4731):1242-1246

45. Myers RM, Maniatis T (1986) Recent advances in the development of methods for detecting single-base substitutions associated with human genetic diseases. In:Cold Spring Harbor symposia on quantitative bioloby 51:275-284

46. Gibbs RA, Caskey CT (1987) Identification and localization of mutations at the Lesch-Nyhan locus by robonuclease A cleavage. Science 236:303-305

47. Gossler A, Doetschman T, Korn Reinhard, Serling E, Kemler R (1986) Transgenesis by means of blastocyst-derived embryonic stem cell lines. Proc Natl Acad Sci USA 83:9065-9069

48. Kaufmann MH, Evans MJ, Robertson EJ, Bradley A (1984) Influence of injected pluripotential (EK) cells on haploid and diploid parthenogenetic development. J Embryol Exp Morphol 80:75-86

49. Munir I, Rossiter B, Caskey CT (1988) In:Antisense RNA: Antisense inhibition of HPRT in vitro and in vivo. New York, Cold Spring Harbor. In press

50. Miller AD, Jolly DH, Friedmann T, Verma IM (1983) A transmissible retrovirus expression human hypoxanthine phosphoribosyltransferase (HPRT): Gene transfer into cells obtained from humans deficient in HPRT. Proc Natl Acad Sci USA 80:4709-4713

51. Miller AD, Eckner RJ, Jolly DJ, Verma IM (1984) Expression of a retrovirus encoding human HPRT in mice. Science 225:630-632

52. Willis RC, Jolly DJ, Miller AD, Plent MM, Esty AC, Anderson PJ, Chang HC, Jones OW, Seegmiller JE, Friedmann T (1984) Partial phenotypic correction of human Lesch-Nyhan (hypoxanthine-guanine phosphoribosyltransferase-deficiency) lymphoblasts with a transmissible retroviral vector. J Biol Chem 259:12, 7842-7849

53. Chang SMW, Wager-Smith K, Tsao TY, Henkel-Tigges J, Vaishnav S, Caskey CT (1987) Construction of a defective retrovirus containing the human hypoxanthine phosphoribosyltransferase cDNA and its expression in cultured cells and mouse bone marrow. Mol

Cell Biol 7:2,854-863

54. Mulligan RC (1984) Construction of highly transmissible mammalian cloning vehicles derived from murine retroviruses. In: Inyouye M (ed) Experimental manipulation of gene expression New York: Academic Press

55. Watanabe S, Temin HM (1983) Construction of a helper cell line for avian retriculoendotheliosis virus cloning vectors. Mol Cell Biol 3:12,2241-2249

56. Mann R, Mulligan RC, Baltimore D (1983) Construction of a retrovirus packaging mutant and its use to produce helper-free defective retrovirus. Cell 33:153-159

57. Cone RD, Mulligan RC (1984) High-efficiency gene transfer into mammalian cells:Generation of helper-free recombinant retrovirus with broad mammalian host range. Proc Natl Acad Sci USA 81:6349-6353

58. Miller AD, Law MF, Verman IM (1985) Generation of helper-free amphotropic retrovirus that transduce a dominant-acting methotrexate-resistant dihydrofolate reductase gene. Mol Cell Biol 5:431-437

59. Miller AD, Buttimore C (1986) Redesign of retrovirus packaging cell lines to avoid recombination leading to helper virus production. Mol Cell Biol 6:2895-2902

60. Markowitz D, Goff S, Bank A (1988) A safe packaging line for gene transfer: separating viral genes on two different plasmids. J Virol 62:1120-1124

61. Keller G, Paige C, Gilboa E, Wagner EF (1985) Expression of a foreign gene in myeloid and lymphoid cells derived from multipotent haematopoietic precursors. Nature 318:149-154

62. Eglitis MA, Kantoff P, Gilboa E, Anderson WF (1985) Gene expression in mice after high efficiency retroviral-mediated gene transfer. Science 230:1395-1398

63. Hock RA, Miller AD (1986) Retrovirus-mediated transfer and expression of drug resistance genes in human haematopoietic progenitor cells. Nature 320:275-277

64. Kantoff PW, Kohn DB, Mitsuya H, Armentano D, Sieberg M, Zwiebel JA, Eglitis MA, McLachlin JR, Wiginton DA, Hutton JJ, Horowitz SD, Gilboa E, Blaese RM, Anderson WF (1986) Correction of adenosine deaminase deficiency in human T and B cells using retroviral-mediated gene transfer. Proc Natl Acad Sci USA 83:6563-6567

65. Kwok WW, Sceuning F, Stead RB, Miller AD (1986) Retroviral transfer of genes into canine hemopoietic cells in culture: A model for human gene therapy. Proc Natl Acad Sci USA, in press

66. Cepko CL, Roberts BE, Mulligan RC (1984) Construction and applications of a highly transmissible murine retrovirus shuttle vector. Cell 37:1053-1062

67. Armentano D, Yu SF, Kantoff PW, von Ruden T, Anderson WF, Gilboa E (1987) Effect of internal viral sequences on the utility of retroviral vectors. J Virol 61:1647-1650

68. Yu SF, von Ruden T, Kantoff PW, Garber C, Seiberg M, Ruther U, Anderson WF, Wagner EF, Gilboa E (1986) Self-inactivating retroviral vectors designed for transfer of whole genes into mammalian cells. Proc Natls Acad Sci USA

83:3194-3198

69. Hawley RG, Covarrubias L, Hawley T, Mintz B (1987) Handicapped retroviral vectors efficiently transduce foreign genes into hematopoietic stem cells. Proc Natl Acad Sci USA 84:2406-2410

70. Yee JK, Moores JC, Jolly DJ, Wolff JA, Respess JG, Friedmann T (1987) Gene expression from transcriptionally disabled retroviral vectors. Proc Natl Acad Sci USA 84:5197-5201

71. Cone RD, Weber-Benarous A, Baorto D, Mulligan RC (1987) Regulated expression of a complete human B-globin gene encoded by a transmissible retrovirus vector. Mol Cell Biol 7(2):887-897

72. Dzierzak EA, Papayannopoulou T, Mulligan RC (1988) Lineage-specific expression of a human B-globin gene in murine bone marrow transplant recipients reconstituted with retrovirus-transduced stem cells. Nature 331:35-41

73. Wirth T, Staudt L, Baltimore D (1987) An octamer oligonucleotide upstream of a TATA motif is sufficient for lymphoid-specific promoter activity. Nature 329:174-178

74. Spangrude GJ, Heimfeld S, Weissman IL (1988) Purification and characterization of mouse hematopoietic stem cells. Science 241:58-62

75. Civin CI, Strauss LC, Brovall C, Fackler MH, Schwartz JF, Shaper JH (1984) Antigenic analysis of hematopoiesis. A hematopoietic progenitor cell suface antigen defined by a monoclonal antibody raised against KG-1a cells. J Immunol 133:1,157-165

76. Lu L, Walker D, Broxmeyer HE, Hoffman R, Hy W, Walker E (1987) Characterization of adult human marrow hematopoietic progenitors highly enriched by two-color cell sorting with MY10 and major histocompatibility class II monoclonal antibodies. J Immunol 139:1823-1829

77. Berenson RJ, Andrews RG, Bensiger WI, Kalamasz D, Knitter G, Buckner CD, Bernstein IW (1988) Antigen CD34[+] marrow cells engraft lethally irradiated baboons. J Cli Invest 81:951-955

78. Williams DA, Hsieh K, DeSilva A, Mulligan RC (1987) Protection of bone marrow transplant recipients from lethal doses of methotrexate by the generation of methotrexate-resistant bone marrow. J Exp Med 166:210-218

79. Guild BC, Mulligan RC, Gros P, Housman DE (1988) Retroviral transfer of a murine cDNA for multidrug resistance confers pleitropic drug resistance to cells without prior drug selection. Proc Natl Acad Sci USA 85:1595-1599

80. Su TS, Bock HG, O'Brien WE, Beaudet AL (1981) Cloning of cDNA for argininosuccinate synthetase mRNA and study of enzyme overproduction in a human cell line. J Biol Chem 256(22):11826-11831

81. Lawn RM, Wood WI, Gitschier J, Wion KL, Eaton D, Vehar GA Tuddenham EGD (1986) Cloned Factor VIII and the molecular genetics of hemophilia. In Symposia on quantitative biology 51 New York: Cold Spring Harbor

82. Russell DW, Yamamoto T, Schneider WJ, Slaughter CJ, Brown MS, Golstein JL cDNA (1983) Cloning of the bovine low density lipoprotein receptor: Feedbach regulation of a receptor mRNA. Proc Natl Acad Sci USA

80:7501-7505

83. Woo SLC, DiLella AG, Marvit J, Ledley FD (1986) Molecular basis of phenylketonuria and potential somatic gene therapy. In Symposia on quantitative biology 51 New York: Cold Spring Harbor

84. Ledley FD, Lumetta M, Nguyen PN, Kolhouse JF, Allen RH (1988) Molecular cloning of L-methulmalonyl-CoA mutase: Gene transfer and analysis of *mut* cell lines. Proc Natl Acad Sci USA 85:3518-3521

14 Towards Gene Therapy for Adenosine Deaminase Deficiency

D. Valerio, V.W. van Beusechem, M.P.W. Einerhand, P.M. Hoogerbrugge, H. van der Putten, P.M. Wamsley, Th.M. Berkvens, I.M. Verma, R.E. Kellems, and D.W. van Bekkum

INTRODUCTION

Deficiency of adenosine deaminase (ADA) activity causes an autosomally inherited form of severe combined immunodeficiency (ADA⁻SCID) disease (1, 2). It has been suggested that this form of SCID is caused by a defect in T- and B-cell differentiation due to the accumulation of adenine nucleosides in the absence of functional ADA (2). The cloning of sequences encoding human ADA (3, 4, 5) opened new ways to investigate the molecular basis of ADA⁻SCID disease (6-10) and allowed studies aimed at the development of gene therapy protocols for ADA⁻SCID patients (11-15).

The objective of such a gene therapy protocol would be to repopulate the lymphoid compartment of the patients by introducing a functional ADA gene into their hemopoietic stem cells. Using retroviral vector technology, we and others were able to introduce the gene for ADA as well as a variety of other genes into hemopoietic stem cells. However, upon engraftment of these stem cells into irradiated recipients, expression of the newly introduced genes was usually very inefficient (12, 15, 16).

To overcome this problem we followed several strategies: (i) we studied the possibilities of employing the natural ADA-promoter to direct gene expression in the hemopoietic system. Transgenic mice were generated that harbor the human ADA-promoter linked to a marker gene. The expression patterns of the transgene suggested the presence of distinct regulatory elements in the ADA-gene that are required for its expression in hemopoietic cells. (ii) we designed retroviral vectors which mediate gene expression in primitive cells. This was achieved by altering the enhancer element present in the Long Terminal Repeat (LTR) that is known to be responsible for the tissue specificity of viral expression (17). With this newly constructed vector viral expression was obtained in the hemopoietic system of both the mouse and rhesus monkey. (iii) we developed a system that allows the selection of cells expressing additional ADA. Using this protocol we should be able to produce and titrate retroviruses that solely carry an ADA-gene without the presence of additional marker genes.

STUDY OF THE ADA PROMOTER IN TRANS-GENIC MICE

The promoter of the human gene for ADA is extremely G/C-rich, contains several G/C-box motifs (GGGCGGG), lacks any apparent TATA or CAAT boxes and has a significantly higher CpG/GpC ratio than is found in the remainder of the genome (see Fig. 1) (18). These features are commonly found in promoters of genes that lack a strong tissue specificity and are referred to as "housekeeping genes". Like other housekeeping genes, the ADA-gene is expressed in all tissues (19). In order to study the activity of the ADA promoter transgenic mice were generated that harbor a chimeric gene composed of a 132-bp ADA promoter fragment linked to a reporter gene encoding the bacterial enzyme Chloramphenicol Acetyl Transferase (CAT) (see also Fig. 1 for structure of the construct).

Expression of the ADA-CAT transgene was measured by performing CAT assays on lysates of a variety of different organs obtained from transgenic mice. Figure 2 shows representative results of such an experiment. As can be seen, CAT activity was detected in extracts from all organs tested. No great variations in expression levels were apparent when comparing the different tissues except for liver, testis and bone marrow in which only low levels of CAT activity were detected. We believe that this pattern of activity is a general characteristic of the 132-bp promoter fragment, since this expression pattern is shared by two different strains of ADA-CAT transgenic mice as well as by mice that harbor another fusion gene in which the ADA promoter was placed in front of the human ADA cDNA (to be published elsewhere).

Because of the potential application of the human ADA promoter to direct expression of the ADA-gene in hemopoietic cells, we were interested to know which cell types actually express CAT activity in the hemopoietic organs of the ADA-CAT mice. Therefore we isolated the hemo-

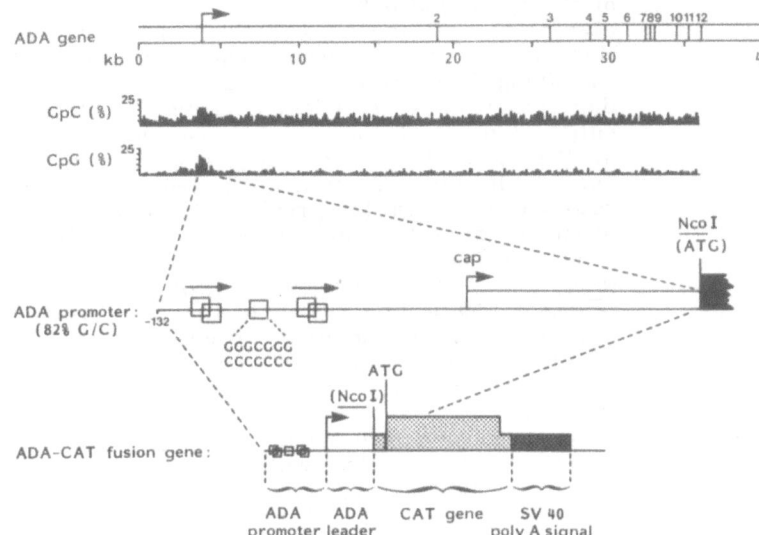

Figure 1
Structure of the human ADA-gene and construction of an ADA-CAT fusion gene. The bar at the top represents the ADA-gene, exons are indicated as vertical lines. The arrow denotes the transcription initiation site in front of exon 1. Exons 2 to 12 are indicated. The presence of a CpG-rich island in the 5' region is shown in the subsequent two lines. The number of GpC and CpG doublets present within 100-bp are depicted as vertical bars). In a fine map of the 5' region of the ADA-gene the G/C motifs and the 16-bp direct repeat in the promoter are depicted as boxes and arrows, respectively. The cap site as well as the NcoI site which coincides with the ATG initiation codon are indicated. The figure also shows the structure of the ADA-CAT fusion gene in which the 132-bp promoter of the ADA-gene as well as all of its 5' untranslated sequences are fused to the CAT-gene which is followed by the 3' untranslated region from the SV40-T-antigen.

poietic cells from thymus and spleen. In addition, these cell fractions were highly enriched for T- or B-cells by stimulation with Concanavalin A (Con A) and lipopolysaccharide (LPS), respectively. Surprisingly, no CAT expression was found in any of these lymphoid cells (not shown). Since CAT activity was readily detectable in extracts made from whole organs we suspected that ADA promoter activity was restricted to the stromal components. To substantiate this we studied CAT expression in hemopoietic tissues of mice after reconstitution with bone marrow from transgenic mice. A successful engraftment would result in chimeric mice in which the hemopoietic cells carry ADA-CAT sequences whereas the stromal components of thymus and spleen (epithelia, connective tissue) do not. Grafting was confirmed by Southern blot analysis of DNA from the hemopoietic organs of mice that had received a bone marrow transplantation. ADA-CAT sequences were detected in day-10 spleen colonies and in the spleen, the thymus and the bone marrow of reconstituted mice 30 days or more after transplantation (not shown). Despite successful engraftment of the transgenic bone marrow CAT activity was not detectable in inspected spleen colonies, nor in the bone marrow and the thymus of the long-term reconstituted mice (see Fig. 2). Quantification of this CAT assay revealed that the cells in the chimeric spleen expressed 80-fold less CAT than the spleen cells of the donor. Thus, CAT expression in the hemopoietic organs of the transgenic donor mice is restricted to non-hemopoietic cells. Apparently, the ADA promoter sequences tested are insufficient to promote expression in hemopoietic cells.

For many genes a causal relationship has been found between hypermethylation and gene inactivity, and vice versa (reviewed in 20). We therefore determined the methylation patterns of the transgenes in tissues in which the reporter gene was either expressed or not expressed. Our data clearly indicate that in hemopoietic cells the transgenes are largely hypermethylated whereas in tissues where CAT activity was

detected the transgenes were found to be hypomethylated (not shown). These data suggest that the mechanism responsible for the suppression of the ADA-CAT-gene in hemopoietic cells could be hypermethylation. Treatment of the hemopoietic cells with the demethylating agent 5-azacytidine will be performed to establish whether this indeed is the case.

CONSTRUCTION OF A RETROVIRAL VECTOR WITH HYBRID LTRs THAT MEDIATES GENE EXPRESSION IN HEMOPOIETIC CELLS OF BOTH MURINE AND PRIMATE ORIGIN

Retroviral vectors can be used to transfer new genes into a wide variety of cells. In addition, the fact that requirements for viral replication can be divided into cis- and trans-acting factors made it possible to generate systems that allow the production of helper-free replication defective viruses (21). These systems consist of a retroviral shuttle vector carrying the in-cis requirements for viral replication as well as the gene of interest under the transcriptional control of either the 5' LTR or an additional promoter. Transfection of such a vector into cells that produce the proteins required for virus production results in a stable helper-free virus-producing cell line. A drawback of this system is that some cells are refractory to viral expression. It has been reported for example that retroviral genes introduced in Embryonal Carcinoma (EC) cells, normal bone marrow cells or pre-implantation embryo's are not, or very poorly expressed (13, 16, 20, 22). In search for the basis of this lack of expression investigators have recently shown in undifferentiated EC cells that the enhancer of Moloney Murine Leukemia Virus (Mo-MLV) can act as an in cis repressor on heterologous promoters (23). Moreover, replacement of this Mo-MLV enhancer by the enhancer of a Polyoma virus mutant (PyF101) selected for growth on EC cells yielded an LTR (Δ Mo+PyF101) that could transiently promote the expression of a reporter gene in F9 EC cells

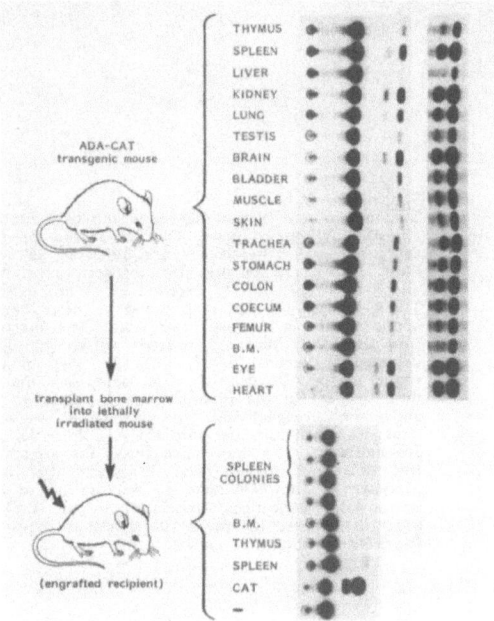

Figure 2
Transgene expression pattern in mice harboring the ADA-CAT fusion gene. CAT assays were performed with 15 µg of protein from different organs of a representative ADA-CAT transgenic mouse. After chromatography to separate the acetylated forms, autoradiography was performed for 48 hours and two weeks respectively. Both exposures are shown. Furthermore, bone marrow from transgenic mice was transplanted into lethally irradiated recipients. Hemopoietic tissues of successfully engrafted animals were monitored for CAT activity (20 µg/assay). For reference we used 0.5 units of pure enzyme (CAT) and spleen lysate from a non-transgenic mouse.

originally located only in U3 of the 3'-LTR is now present in both the 5'- and the 3'-LTR. To test this notion a Southern blot was prepared containing DNA from a virus producing cell line digested with a restriction enzyme that cuts either 5' or 3' of the inserted polyoma enhancer in the LTR (NheI and SstI, respectively). In Fig. 4 it can be seen that both digestions resulted in a 2.8 kb provirus fragment hybridizing to a neo^R probe. When probed with the polyoma enhancer the lanes containing either the NheI- or the SstI-digested DNA revealed a fragment identical in length to the provirus (2.8 kb) as well as an additional fragment containing flanking cellular sequences. These results show the polyoma enhancer is now present in both the 5'- and the 3'-LTR of the virus. Amphotropic virus producing cell lines were generated by infecting PA317 cells with this virus. Neo-resistant PA317 colonies were isolated and tested for the production of virus. A cell line designated P1-D was shown to produce amphotropic ΔMo+PyF101-neo virus in a titer of 3×10^5 G418^R CFU/ml. This virus-producing cell line was used in our subsequent studies.

In order to test whether ΔMo+PyF101-neo viruses could confer G418 resistance to EC cells F9 cells were infected with the ΔMo+PyF101-neo virus and selected with G418. This resulted in a titer of 5×10^3 CFU/ml in F9 cells that were refractory to expression of genes promoted from the Mo-MLV-LTR. We conclude therefore that, in the context of a replication defective retrovirus, the ΔMo+PyF101 LTR is active in EC cells.

(24). To test whether this LTR could be used to direct expression of retroviral vectors in the hemopoietic system we have produced a replication defective retrovirus in which a dominant selectable marker (neo^R) is under the control of the ΔMo+PyF101-LTR.

In Fig. 3 the retroviral shuttle vector p ΔMo+PyF101-neo is shown. A neo^R-gene was placed downstream from a 5'-LTR which contains the Mo-MLV enhancer (present in U3). As a 3'-LTR the ΔMo+PyF101-LTR was used. Viruses were generated with this construct by transfecting it into PA317 cells (25). These viruses were subsequently used to infect Ψ-2 cells (21). G418 resistant Ψ-2 colonies were isolated and further investigated. By virtue of this method the virus has undergone one round of replication. Consequently, the Polyoma F101 enhancer

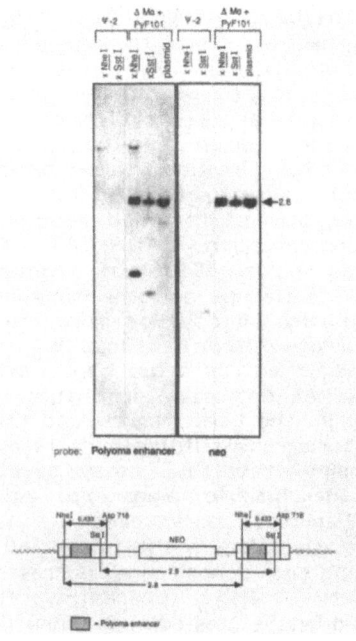

Figure 4
Structure of the ΔMo+PyF101-neo provirus. After one round of replication the polyoma enhancer element located in the 3' LTR of the shuttle vector, should be present in both the 3' and the 5' LTR of the provirus. Digestion of such a provirus with NheI and SstI should therefore both result in the generation of a 2.8 kb DNA fragment that will characteristically hybridize to a PyF101 enhancer probe as well as to a neo probe (see lower panel). In the upper panel the Southern blots are shown. Genomic DNA (10 µg) of Ψ-2 cells and a ΔMo+PyF101-neo virus-producing cell line were used. As control, plasmid DNA (10 pg) cut with SstI, was included. Restriction enzyme digestions and probes are indicated.

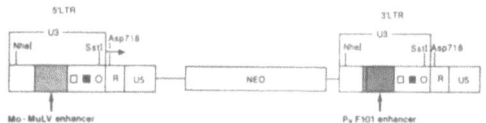

Figure 3
Structure of the retroviral shuttle vector p ΔMo+PyF101-neo. A neo^R-gene is situated in between an LTR in which U3 is derived from Mo-MLV (5') and the ΔMo+PyF101-LTR (3'). Enhancer elements are denoted. The TATA-box (open circle), CAAT-box (closed rectangular) and G/C motif (open rectangular) that are part of the Mo-MLV promoter as well as relevant restriction enzyme sites are indicated.

In primitive hemopoietic stem cells (CFU-S) we have reported high efficiency gene transfer using Mo-MLV based retroviral vectors. However, the results obtained in these experiments indicated that the expression of the introduced genes is often low or undetectable in the CFU-S derived spleen colonies (12). Here we examine the expression of the neo^R-gene in committed progenitors derived from CFU-S infected with ΔMo+PyF101-neo virus. Optimal infection efficiencies were obtained by cocultivating bone marrow cells with irradiated virus producing cells in the presence of recombinant human IL-1α and murine IL-3 followed by a preselection with G418. Since no CFU-S could be detected in bone marrow cells that were co-cultivated with control fibroblasts and preselected with G418, we conclude that all of the CFU-S that survived preselection expressed the neo^R-gene. In order to establish whether the neo^R-gene was still expressed in hemopoietic progenitors derived from the infected CFU-S we assayed spleen colonies for the presence of G418^R progenitors. Infected and preselected bone marrow cells were injected into lethally irradiated mice. After 10 or 12 days spleen colony containing spleens were dispersed into single cells and plated in methylcellulose with and without G418. The results shown in Table 1 indicate that G418 resistant progenitors were present in a frequency ranging between 70 and 100% in the cell lineages tested. This was found to be the case in two independent experiments. These findings sharply contrast with the results we obtained with a similar vector in which an unchanged Mo-MLV-LTR drives neo^R expression. When this virus was used in identical experiments, no significant increase in G418 resistant progenitors could be detected.

As a preclinical model for gene therapy, we plan to develop an autologous bone marrow transplantation protocol in rhesus monkeys using genetically modified hemopoietic stem cells. Initial studies were undertaken to infect rhesus monkey hemopoietic stem cells with ΔMo+PyF101-neo virus and to examine the expression of the neo^R-gene in myeloid progenitors derived from these stem cells in vitro.

Since cell division is thought to be essential for retroviral infection (17) we first established culture conditions that would stimulate division of rhesus monkey stem cells. The bone marrow suspension, obtained from an exsanguinized animal, was approximately 5- to 10-fold enriched for hemopoietic stem cells by albumin density gradient centrifugation (sp gr < 1.064 g/cm^3) (26). The stem cell stimulating conditions were essentially as described for murine stem cells (27, a modification of 28), with addition of recombinant human GM-CSF. As a possible source of stem cell stimulating factors normal rhesus monkey serum (NMS) was added in a 5% (v/v) concentration. The stem cell stimulation was compared to the same conditions with NMS substituted by FCS. Because of the lack of an in vitro assay for more primitive cells the effect of the stem cell stimulation was monitored for the presence of myeloid progenitor cells (CFU-C; 27), using recombinant human GM-CSF. As can be seen in Table 2, supplementing the culture with NMS significantly increased colony formation whereas supplementation with FCS did not increase the number of CFU-C. The NMS is therefore expected to provide essential growth factors. The data that we previously obtained in the murine system indicated that the most efficient procedure for infecting bone marrow stem cells was to co-cultivate them with irradiated virus producing fibroblasts. However, using this technique we also observed a variable cytotoxic effect on the hemopoietic stem cells. We therefore set out to establish whether irradiated NIH/3T3 cells exert a cytotoxic effect on rhesus monkey stem cells. It may be concluded from the results shown in Table 2 that co-cultivation with NIH/3T3 cells drastically decreases the number of CFU-C. We conclude from this experiment that infection of primate bone marrow cells should preferably not be performed by co-cultivation with virus producing fibroblasts. Therefore, we tried to establish culture conditions that would allow the infection of rhesus monkey stem cells using virus containing supernatants.

Bone marrow cells were infected after different periods of stem cell stimulating culture. At the day of infection the culture medium was replaced three times at two hour intervals by virus supernatant from the P1-D cell line with 0.4 µg polybrene/ml and supplemented for stem cell stimulation. The next day the cells were

Table 1

SUSTAINED RETROVIRAL EXPRESSION UPON ENGRAFTMENT OF MICE WITH ΔMo+PyF101-neo VIRUS-INFECTED AND PRESELECTED BONE MARROW CELLS

exp. no	progenitors	colonies/5x10^6 cells −G418	+G418
1	CFU-C$_1$*	195	195
	CFU-M**	99 ± 29	69 ± 19
	BFU-e***	60	70
2	CFU-M	11 ± 5	12 ± 2

Lethally irradiated mice were injected with infected and preselected bone marrow cells. At day 10 or day 12 the spleens were collected. The cells from 4 spleens (exp. 1) and 16 spleens (exp. 2) were pooled and plated in methylcellulose at densities of 1 × 10^6 (exp. 1) and 1.5 × 10^7 nucleated cells/ml (exp. 2) in the absence (−G418) and presence of 1 mg/ml G418 (+G418). Mock infected controls did not form colonies in the presence of G418. The number of colonies is given as mean ± S.E.M.
* CFU-C$_1$ - Granulocyte, monocyte and granulocyte/monocyte colony forming units
** CFU-M - colony forming unit-monocyte
***BFU-e - burst forming unit-erythrocyte.

Table 2

EFFECT OF STEM CELL STIMULATING CULTURE ON CFU-C FORMATION

culture period (days)	culture conditions	nr. of day 10 CFU-C per 10^5 enriched BM cells	percentage of untreated
0	−	292 ± 28	100
3	+ NMS	559 ± 46*	191
	+ FCS	282 ± 74**	97
	+ NMS/3T3	16 ± 5***	5
	+ FCS/3T3	20 ± 8***	7

Number of CFU-C given as mean ± S.E.M.
*p < 0.01; ** N.S.(p > 0.05); ***p < 0.001, as compared to untreated control value (Student's t-test).
−/3T3 = co-cultivated with irradiated (20 Gy) NIH/3T3 cells.

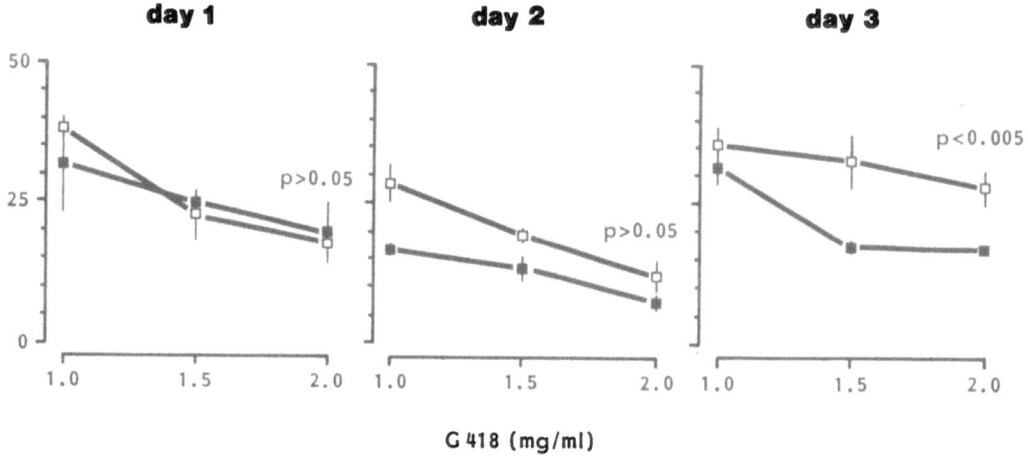

Figure 5
Infection of rhesus monkey BM cells after various periods of stem cell stimulating culture. Enriched rhesus monkey bone marrow cells were cultured under stem cell-stimulating conditions for 1, 2 or 3 days prior to infection in three two-hour hits with freshly harvested supernatant from □ a ΔMo+PyF101-neo virus producing cell line (P1-D) or ■ PA317 cells under the same conditions. The next day the cells were plated for CFU-C with or without G418 in different concentrations (1, 1.5 and 2 mg/ml). The percentage of colony formation is given in mean ± S.D. as compared to the unselected control values. Significant infection efficiency by P1-D virus was tested at 2 mg/ml G418 using Student's t-test.

plated for CFU-C. The expression of the neo^R-gene was examined by selection with G418 at concentrations up to 2 mg/ml. Figure 5 shows that in contrast to results obtained with murine bone marrow cells, mock infected rhesus monkey bone marrow cells exhibit a substantial background tolerance to G418 in all the concentrations tested. Infection with ΔMo+PyF101-neo virus could significantly increase colony formation in the presence of 2 mg/ml G418 only after three days of stem cell stimulating culture. This suggests that stimulation of the target cell for infection is not obtained before day three. Due to the limitations of the available in vitro assays infection of the stem cell will have to be assessed by in vivo studies.

SELECTION OF CELLS EXPRESSING ADDITIONAL ADA

We and others have previously employed retroviral vectors that harbor and express a selectable gene in addition to the gene of interest. This was not always an optimal situation because in some instances it was shown to result in unwanted and unexplained suppression mechanisms (29). Moreover, since our data indicate that tissue specificity of retroviral vector expression can be modulated by alterations in the LTR the employment of specialized internal promoters was less urgent. We have therefore set out to generate simple LTR-ADA-LTR viruses. In order to generate and titrate such viruses we adapted a selection procedure that allows the isolation of cells expressing human ADA in addition to the endogenous (murine) ADA (30). This is achieved by growing cells in the presence of the cytotoxic adenosine-analogue Xylofuranosyl-adenine (Xyl-A) as

well as the transition-state analogue deoxycoformicin (dCF) which inhibits ADA activity. Cells transfected with a functional ADA-gene are expected to survive the applied selection pressure by deaminating Xyl-A to its nontoxic inosine analogue.

To set up the selection procedure we first determined the seeding efficiency of NIH/3T3 cells without selection. This was shown to be 15-20% within a range of 10 to 1000 cells plated. Thus, colony formation correlated with cell survival in a range of 1 to 100%. At a cell concentration of 10^3/ml a Xyl-A concentration of 4 µM exerted a cytotoxic effect as indicated by a 20% decrease in colony formation. These conditions were used as a standard procedure. To determine the minimal concentration of dCF required to kill the Xyl-A treated NIH/3T3 cells a titration was performed with dCF-concentrations ranging from 1 to 20 nM. In Fig. 6 it can be seen that this resulted in a concentration-dependent survival pattern in which 3 nM dCF appeared to be the minimal concentration that completely inhibits colony formation. The cytotoxic effect was shown to be solely due to the activity of Xyl-A, since culturing the cells in medium containing 20 nM dCF did not lead to a significant decrease in cell survival (Fig. 6). Consequently, a combination of 4 µM Xyl-A and 3 nM dCF was initially chosen for the selection of cells expected to express additional ADA. The selective conditions were maintained for 7 to 10 days before colony formation was scored.

The ability to confer Xyl-A/dCF-resistance to NIH/3T3 cells by introducing a functional human ADA-gene was shown by transfection of a retroviral ADA shuttle vector (p ΔMo+PyF101-ADA). This construct harbors the human gene for ADA in the context of a ΔMo+PyF101 retroviral vector. In addition, outside of the viral

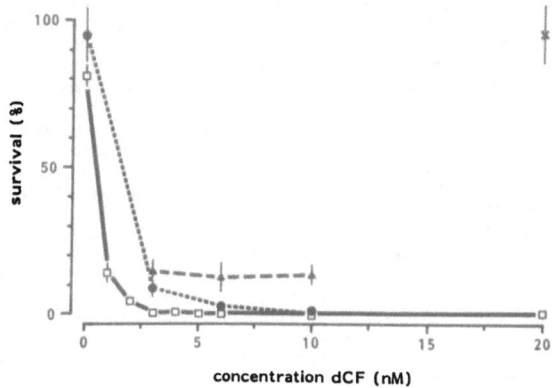

Figure 6
Xyl-A/dCF toxicity on NIH/3T3 cells. Percentage of cells surviving selection with 4 µM Xyl-A and different amounts of dCF as compared to unselected controls. Values given as mean ± S.D. from six determinations. □NIH/3T3 cells only expressing murine ADA, ▲NIH/3T3 cells transfected with p ΔMo+PyF101-ADA (results from independent experiment), ●NIH/3T3 cells cultured for two hours in medium harvested from NIH/3T3 cells expressing human ADA (mock infection), ×NIH/3T3 cells cultured in medium containing 20 nM dCF without addition of Xyl-A.

sequences, the plasmid carries a neo^R-gene which is driven by the SV40 early region promoter. Following transfection the cells were selected for neo^R expression. A mixed pool of G418^R cells exhibited weak human ADA expression as demonstrated by an ADA-specific staining of electrophoretically separated cell lysates (Fig. 7). Upon selection of these cells with Xyl-A/dCF about 13% were shown to be resistant to dCF-concentrations up to 10 nM (Fig. 6). Individual selected colonies expressed human ADA equal to or stronger than the endogenous (murine) ADA-activity (Fig. 7). This clearly demonstrates the usefulness of the Xyl-A/dCF culture as a selection method for NIH/3T3 cells expressing additional ADA.

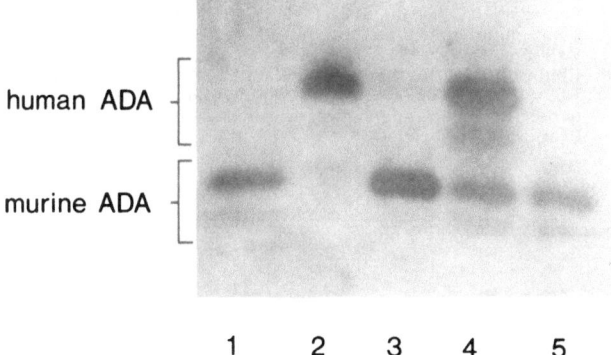

human ADA

murine ADA

1 2 3 4 5

Figure 7
ADA isozyme pattern of cell lysates after cellulose acetate electrophoresis. Lane 1: NIH/3T3 cells; lane 2: human red blood cells; lane 3: NIH/3T3 cells transfected with p Mo+PyF101-ADA; lane 4: NIH/3T3 cells transfected with p Mo+PyF101-ADA and resistant to 4 µM Xyl-A + 3 nM dCF; lane 5: mock infected NIH/3T3 cells resistant to 4 µM Xyl-A + 3 nM dCF.

Surprisingly, our data also showed that a mock infection of NIH/3T3 cells with culture supernatant can result in a significant increase in colony formation at lower dCF concentrations, although the resistant cells did not express human ADA (Fig. 7). This artefactual background could be eliminated by increasing the dCF-concentration to 10 nM (see Fig. 6). At this concentration all cells surviving the selection procedure observed sofar express additional ADA.

CONCLUSIONS

The ultimate goal of our study is to derive gene therapy protocols for ADA⁻SCID patients. Ideally, this should be conducted in such a way that expression of the introduced ADA-gene is directed by its natural transcriptional machinery. We therefore studied the activity of the human ADA promoter in transgenic mice. In summary, our data show that the 132-bp ADA promoter fragment can direct expression of a reporter gene in a great variety of different cell types as can be expected for a promoter of a typical housekeeping gene. Surprisingly, gene expression was not detected in blood cells. This is unexpected since apparently in ADA-deficient patients the only cell types that absolutely require ADA activity are of hemopoietic origin (8). We would therefore like to propose that hemopoietic ADA expression is controlled by a thusfar unidentified distinct regulatory element, which is not included in the transgenes described in this study. Our finding that the 132-bp ADA promoter fragment is not active in the T- and B-cell lineages is of considerable importance for experiments aimed at the reconstitution of ADA-gene activity in the hemopoietic system using retroviral vector technology. In particular, retroviral vectors to be used in gene therapy for ADA⁻SCID patients should preferably express ADA in lymphoid cells. Hence, the use of human ADA promoter sequences in retroviral vector systems (12) should await the identification of the relevant regulatory elements.

One of the major problems encountered when developing gene therapy protocols using retroviral gene transfer methods was the lack of sustained expression of the introduced genes. To alleviate this expression block we constructed a retroviral vector with mutant enhancer sequences capable of expression in EC cells and in primitive hemopoietic cells. In the murine hemopoietic system this expression was shown to be sustained when examined in CFU-S derived progenitors. Our _in vitro_ data with rhesus monkey hemopoietic progenitors suggest that indeed vectors based on the described retroviral construct are suitable for gene therapy purposes. In order to produce and titrate comparable ADA-containing viruses we developed a selection system for cells expressing additional ADA. Elaborate testing of these viruses in a preclinical rhesus monkey model is our next step towards gene therapy for ADA deficiency.

ACKNOWLEDGEMENTS

Part of this research was supported by the Netherlands Organization for Scientific Research (NWO).

REFERENCES

1. Giblett ER, Anderson JE, Cohen F, Pollara B, Meuwissen HJ (1972). Adenosine deaminase deficiency in two patients with severe impaired cellular immunity. Lancet II: 1067-1069.
2. Thompson CB, Seegmiller JE (1980). Adenosine deaminase deficiency and severe combined immunodeficiency disease. In: Meister A (ed) Advances in Enzymology, vol 51, New York: Academic Press.
3. Valerio D, Duyvesteyn MGC, Meera Khan P, Van Kessel GA, De Waard A, Van der Eb AJ (1983). Isolation of cDNA clones for human adenosine deaminase. Gene 25: 231-240.
4. Wiginton DA, Adrian GS, Friedman RL, Suttle DP, Hutton JJ (1983). Cloning of DNA sequences of human adenosine deaminase. Proc Natl Acad Sci 80: 7481-7485.
5. Orkin SH, Daddona PE, Shewach DS, Markham AF, Bruns GA, Goff SC, Kelly WN (1983). Molecular cloning of human adenosine deaminase gene sequences. J Biol Chem 258: 12753-12756.
6. Valerio D, Duyvesteyn MGC, Van Ormondt H, Meera Khan P, Van der Eb AJ (1984). Adenosine deaminase (ADA) deficiency in cells derived from humans with severe combined immunodeficiency is due to an aberration of the ADA protein. Nucl Acid Res 12: 1015-1024.
7. Valerio D, Dekker BMM, Duyvesteyn MGC, Van der Voorn L, Berkvens TM, Van Ormondt H, Van der Eb AJ (1986). One adenosine deaminase allele in a patient with severe combined immunodeficiency contains a point mutation abolishing enzyme activity. EMBO J 5: 113-119.
8. Berkvens TM, Gerritsen EJA, Oldenburg M, Breukel C, Wijnen JT, Van Ormondt H, Vossen JM, Van der Eb AJ, Meera Khan P (1987). Severe combined immune deficiency due to a homozygous 3.2-kb deletion spanning the promoter and first exon of the adenosine deaminase gene. Nucl Acid Res 15: 9365-9378.
9. Bonthron DT, Markham AF, Ginsburg D, Orkin SH (1985). Identification of a point mutation in the adenosine deaminase gene responsible for immunodeficiency. J Clin Invest 76: 894-897.
10. Adrian GS, Wiginton DA, Hutton JJ (1984). Structure of adenosine deaminase mRNAs from normal and adenosine deaminase-deficient human cell lines. Mol Cell Biol 4: 1712-1717.
11. Valerio D, Duyvesteyn MGC, Van der Eb AJ (1985). Introduction of sequences encoding functional human adenosine deaminase into mouse cells using a retroviral shuttle system. Gene 34: 163-168.
12. Valerio D, Visser TP, Wagemaker G, Van der Eb AJ, Van Bekkum DW (1986). The introduction of human ADA sequences into mouse haematopoietic stem cells. In: Vossen J, Griscelli C (eds). Progress in immunodeficiency research and therapy. Vol. II, Amsterdam: Elsevier.
13. Lim B, Williams DA, Orkin SH (1987). Retrovirus-mediated gene transfer of a human adenosine deaminase: expression of functional enzyme in murine hematopoietic stem cells in vivo. Mol Cell Biol 7: 3459-3465.
14. Belmont JW, Henkel-Tigges J, Chang SMW, Wager-Smith K, Kellems RE, Dick JE, Magli MC, Phillips RA, Bernstein A, Caskey CT (1986). Expression of human adenosine deaminase in murine haematopoietic progenitor cells following retroviral transfer. Nature 322: 385-387.
15. McIvor RS, Johnson MJ, Miller AD, Pitts S, Williams SR, Valerio D, Martin DW Jr, Verma IM (1987). Human purine nucleoside phosphorylase and adenosine deaminase: gene transfer into cultured cells and murine hematopoietic stem cells by using recombinant amphotropic retroviruses. Mol Cell Biol 7: 838-846.
16. Williams DA, Orkin SH, Mulligan RC (1986). Retrovirus-mediated transfer of human adenosine deaminase gene sequences into cells in culture and into murine hemato-poietic cells in vivo. Proc Natl Acad Sci USA 83: 2566-2570.
17. Weiss R, Teich N, Varmus H, Coffin J (1982). RNA tumor viruses. Molecular biology of tumor viruses, 2nd edn. Cold Spring Harbor: Cold Spring Harbor Laboratory.
18. Valerio D, Duyvesteyn MGC, Dekker BMM, Weeda G, Berkvens TM, Van der Voorn L, Van Ormondt H, Van der Eb AJ (1985). Adenosine deaminase: characterization and expression of a gene with a remarkable promoter. EMBO J 4: 437-443.
19. Dynan WS (1986). Promoters for housekeeping genes. Trends in Genet 8: 196-197.
20. Jaenisch R, Jaehner D (1984). Methylation, expression and chromosomal position of genes in mammals. Biochim Biophys Acta 782: 1-9.
21. Mann R, Mulligan RC, Baltimore D (1983). Construction of a retrovirus packaging mutant and its use to produce helper-free defective retrovirus. Cell 33: 153-159.
22. Van der Putten H, Botteri FM, Miller AD, Rosenfeld MG, Fan H, Evans RM, Verma IM (1985). Efficient insertion of genes into the mouse germ line via retroviral vectors. Proc Natl Acad Sci USA 82: 6148-6152.
23. Sassone-Corsi P, Duboule D, Chambon P (1986). Viral enhancer activity in teratocarcinoma cells. Cold Spring Harbor Symp on Quant Biol 50: 747-752.
24. Linney E, Davis B, Overhauser J, Chao E, Fan H (1984). Non-function of a Moloney Murine Leukaemia Virus regulatory sequence in F9 embryonal carcinoma cells. Nature 308: 470-472.
25. Miller AD, Buttimore C (1986). Redesign of retrovirus packaging cell lines to avoid recombination leading to helper virus production. Mol Cell Biol 6: 2895-2902.
26. Dicke KA, Tridente G, Van Bekkum DW (1969). The selective elimination of immunologically competent cells from bone marrow and lymphocyte cell mixtures. III. In vitro test for detection of immunocompetent cells in fractionated mouse spleen cell suspensions and primate bone marrow suspensions. Transplantation 8: 422-434.
27. Merchav S (1986). The effect of Rauscher leukaemia virus on haemopoietic cell differentiation in genetically defective W/W^v and normal mice. Thesis Erasmus University of Rotterdam, The Netherlands.
28. Löwenberg B, Dicke KA (1974). Studies on the in vitro proliferation of pluripotent haemopoietic stem cells. In: Leukemia and aplastic anemia. Proc Int Conf on Leukemia and Aplastic Anemia. Rome.
29. Emerman M, Temin HM (1986). Quantiative analysis of gene suppression in integrated retrovirus vectors. Mol Cell Biol 6: 792-800.
30. Kaufman RJ, Murtha P, Ingolia DE, Yeung C-Y, Kellems RE (1986). Selection and amplification of heterologous genes encoding adenosine deaminase in mammalian cells. Proc Natl Acad Sci USA 83: 3136-3140.

15 Abnormal Gene Expression as an Early Indicator of Relapse in Acute Myelogenous Leukemia (AML)

M.J. Evinger-Hodges, J.A. Spinolo, I. Cox, and K.A. Dicke

ABSTRACT

We have identified the abnormally high expression of two genes, MYC and SIS, which occur with high frequency in the leukemic cell population of untreated and relapsed acute myelogenous leukemia (AML) patients. With few exceptions, the presence of this abnormal population of cells was found to be in good agreement with the percentage of blast cells determined morphologically. Such an abnormal population of cells was also present in 70% of AML patients examined in clinical complete remission (CR). In these cases the frequency of this abnormal cell population is generally much lower than that present in untreated/relapse patients. Continued examination of this group of patients showed a good correlation between the presence of these abnormal cells and eventual relapse of the patient.

We have now examined 10 AML long-term survivors (median CR duration 38 months, range 14-78 months) after bone marrow transplantation. At this time we have found only 3 cases (30%) in which we are able to detect the presence of any abnormal bone marrow cells as described by gene expression studies. Unlike the short term remission patients studied, there is no overexpression of MYC detectable in this patient group. At this time even in the presence of as high as 80% of bone marrow cells overexpressing SIS, none of these patients have relapsed. Continued surveillance of these patients will show if this technology can be useful in the prediction of CR duration in AML.

INTRODUCTION

Numerous reports of oncogene overexpression in acute myelogenous leukemia (AML) have been published over the past decade (1-5). Recently several groups in addition to ourselves have identified two genes, MYC and SIS, which are present at unusually high levels in the bone marrow cells of untreated and/or relapsed AML patients (8-10). By Northern and dot blotting analyses we were unable to detect this abnormality in the bone marrow cells of any AML patient clinically classified as being in remission. Since the limitations of these techniques are in the range of 1-5% contamination of leukemic cells in the samples tested, any abnormality present at lower levels would remain undetectable.

Through the application of a modified RNA-in situ hybridization we have identified the presence of such abnormal cells as defined by gene expression in 7/10 AML patients studied recently after remission induction, but often at a much lower frequency than that found in untreated or relapse AML (9). Several, but not every patient, in which we found this abnormal group of cells have relapsed; in contrast, none of the patients whose bone marrow cells were found to be normal in their expression of MYC and/or SIS have relapsed since this study was completed. These results led us to question whether the presence of such an abnormal population of cells could be predictive of early relapse in acute leukemia.

The median disease-free survival in AML is 12-15 months and relapses occur after two years in only 15% of cases. If the abnormal cell population is composed of clonogenic leukemic cells, we would not expect to find them in our AML long term survivors after bone marrow transplantation. The results of this

TABLE 1
Frequency of Abnormal Gene Expression in Untreated/Relapsed AML

Patient	MYC	SIS	FMS	RAF
Untreated	10/10	8/10	2/10	0/10
Relapsed	15/15	12/15	4/12	2/12
Normal Untreated	0/15	0/15	0/15	0/15
Normal On chemo	0/3	0/3	0/3	0/3
Normal ABMT	0/3	0/3	0/3	0/3

TABLE 2
Comparison of % Blast with % Gene Overexpression in AML Bone Marrow

Patient	% Blast	% myc	% sis	% fms	% raf
1	7.2	70	70	0	0
2	24	25	25	0	0
3	40	90	90	90	0
4	49	30	50	25	0
5	51	50	50	0	0
6	60	75	75	-	-
7	63	85	90	87	0
8	65	65	90	70	70
9	76	70	75	0	0
10	78	75	80	70	75
11	81	85	0	-	-
12	96	>90	>90	-	-

study will help us determine what the significance might be of such a population of cells in the bone marrow of leukemia patients.

MATERIALS AND METHODS

Cells

Fresh human normal bone marrow cells were obtained from hematologically normal donors. Leukemia marrow samples were obtained from patients either at diagnosis or while in relapse, from remission patients whose marrow was being stored for future transplantation, or from follow-up examinations of the marrow of long term survivors after high dose chemotherapy and bone marrow transplantation. In all cases bone marrow buffy coat preparations were used for cytospin preparations.

RNA-in situ hybridization

A very sensitive and rapid RNA-in situ hybridization procedure was performed as described earlier (r). Briefly the buffy coat cells are suspended in medium containing 2% serum and then deposited on slides as a cytospin preparation. The cells are fixed with 75% ethanol/20% acetic acid for 15 minutes at room temperature. After a brief wash in phosphate-buffered saline (pH 7.2), the cells are permeabilized by treatment with 0.01% Triton X-100 (5 minutes, room temperature). Hybridizations were performed in 50% formamide at 52 C, followed by extensive washes in 1X SSC, 0.5X SSC and finally 0.1X SSC containing RNase. The slides were coverslipped in 50% glycerol/50% phosphate-buffered saline before being viewed under a fluorescence microscope.

Probes

The probes for both MYC and SIS were purchased from Amersham (Arlington Heights, IL) subcloned into an SP6 vector permitting the synthesis of full-length single-stranded RNA probes. These probes were sized between 200 and 400 base-pairs (r) and then labeled with Photobiotin™ (BRL, MD).

Detection of the biotin-labeled hybrids was performed by the addition of FITC-labelled streptavidin (BRL, MD). Unbound streptavidin was removed by large volume washes in 0.1X SSC containing 0.1% Triton X-100.

Quantification

We have used the level of fluorescence/cell as an initial estimate of expression levels in these studies. Earlier studies using cell lines have demonstrated that by this method we have the sensitivity to distinguish between the expression of a gene expressed at 5-10 copies per cell and one present at 20 copies per cell. In the experiments described in this paper overexpression of a gene was defined by a greater than 3-fold increase in the level of fluorescence present in a cell as compared to that found in the brightest cell present in normal bone marrow under the same hybridization conditions.

TABLE 3
Overexpression of MYC and SIS in AML Remission Marrow.

Patient	% MYC	% SIS	Present DX
1	75.0	80.0	Relapse
2	8.0	2.0	Relapse
3	5.0	2.0	Relapse
4	0.2	0.2	Relapse
5	0.1	1.0	Relapse
6	2.5	2.5	CR
7	0.05	0	CR
8	0	0	CR
9	0	0	CR
10	0	0	CR

TABLE 4
AML: Long Term Survivors

Pt. No.	%MYC	%SIS	Present Diagnosis
1	–	–	CR: 31 MOS.
2	–	–	CR: 34 MOS.
3	–	–	CR: 74 MOS.
4	–	–	CR: 44 MOS.
5	–	75	CR: 22 MOS.
6	–	–	CR: 20 MOS.
7	–	–	CR: 42 MOS.
8	–	–	CR: 63 MOS.
9	–	5	CR: 43 MOS.
10	–	80	CR: 18 MOS.

RESULTS

Expression levels of four genes, MYC, SIS, FMS and RAF, were studied at the single cell level in both untreated and relapsed AML patients (Table 1). Two of these genes, MYC and SIS, appear to be abnormally expressed in a variable percentage of bone marrow cells in nearly 100% of AML untreated and relapsed patients examined. In contrast, cells containing abnormally high levels of mRNA for FMS and/or RAF were present only sporadically in this patient group (Table 1). The expression of these genes at such high levels was not detectable in any normal bone marrow cell examined, nor could such cells be found in the highly proliferative cell population present after bone marrow transplantation or after chemotherapy treatment.

The presence of such abnormal cells, as defined by gene expression, in these AML patients was also correlated with the percentage of blast cells determined morphologically. As shown in Table 2, the percentage of cells overexpressing either MYC or SIS at least equals, and often exceeds, the number of blast cells present in the marrow.

AML remission

We reported earlier (9) the results of a study examining 10 AML patients morphologically determined to be in remission. In this study we demonstrated that in 7/10 patients a subpopulation of cells could be identified by RNA-in situ hybridization which contained a high expression level of MYC and/or SIS as compared to normal bone marrow.

The clinical status of these 10 patients has continued to be followed over the past year since this study was completed. As shown in Table 3, 5/7 patients in which an abnormal population of cells was identified have now relapsed. None of the three patients whose bone marrow cells appeared normal by this assay for gene overexpression have relapsed over this same time period.

All of these patients were examined within 2-8 months after remission induction. Those patients which relapsed did so within 5-6 months after testing positive for abnormal gene expression. At this time the two patients of this group in which the presence of an abnormal subpopulation of cells is detectable continue to be in remission 3 months and 1 year after testing.

AML long term survivors

To help us determine the significance of this abnormal cell population in the eventual clinical stability of the AML patient, we have also examined bone marrow cells of 10 AML patients who are long term survivors after bone marrow transplantation. The median CR duration at the time of examination for this group was 38 months with the individual remissions ranging from 14-78 months.

As seen in Table 4, the presence of a similarly abnormal cell population expressing both MYC and SIS at high levels in this patient group does not occur.

However, in 3/10 patients a high level of SIS expression alone was present in a variable percentage of cells occasionally as high as 80% At this time, none of the three patients identified with this abnormality at the RNA level have been classified as having a recurrence of leukemic cells in the bone marrow by conventional morphological criteria.

DISCUSSION

We had previously described the high frequency of bone marrow cells with MYC and/or SIS overexpression in untreated/relapsed AML patients as compared to normal bone marrow cells which express these genes at very low levels. The percentage of these abnormal cells parallels or exceeds the percentage of blast cells found in the marrow of these patients. The continued presence of such abnormal cells at very low frequency in greater than 70% of remission marrows examined from patients within a year after remission induction led us to question to significance of these cells in the eventual clinical stability of the patient. To help us determine this answer, we obtained bone marrow samples from 10 AML long term survivors (median CR duration 38 months; range 14-78 months) and examined these cells by RNA-in situ hybridization to see if any MYC or SIS overexpressing cells could be detected.

We find that the occurrence of such a population of abnormal cells is a rare event in this patient group. Whereas the presence of such abnormal cells was detectable in 70% of the AML patients examined shortly after remission induction, we found such an abnormality to be present in only 30% of the long term survivors. Interestingly, although we find MYC to be overexpressed in every short term remission patient in which we find SIS overexpression, this was not the case in any of the long term remission patients examined. Of the 3/10 patients in which SIS was expressed at high levels, none had MYC present at levels higher than normal; nor do we find any evidence of relapse in this patient group.

Since SIS encodes for the B-chain of the growth factor PDGF, it is conceivable that it is the unregulated response of these myeloid cells from AML patients to PDGF stimulation which is responsible for the high levels of MYC mRNA detected within these cells. In the long term survivor population, it is possible that this lack of high MYC expression is indicative of a regulated response of these cells to growth factor stimulation. In this case we would not expect these patients to relapse unless there is an increase in MYC expression similar to that found in AML relapse patients. These last patients will continue to be closely followed for any signs of relapse and possible correlation of relapse with increased MYC expression.

Initial evaluation of these long term survivors gives us an indication that detection of abnormal gene expression may well be a parameter which can be used to monitor remission duration in AML patients after BMT.

BIBLIOGRAPHY

1) Westin EH, Wong-Staal F, Gelmann EP, Dalla Favera R, Papas TS, Lautenberger J, Evaa A, Reddy EP, Tronick SR, Aaronson SA, Gallo RC (1982) Expression of cellular homologues of retroviral oncogenes in human hematopoietic cells. Proc Natl Acad Sci USA 79:2490-2424

2) Blick M, Westin E, Gutterman J, Wong-Staal F, Gallo R, McCredie K, Keating M, Murphy E (1984) Blood 64:1234-1239

3) Ferrari S, Torelli U, Selleri L, Donelli A, Venturelli D, Warni E, Moretti L, Torelli G (1985) Study of the level of expression of two oncogenes, c-myc and c-myb in acute and chronic leukemias of both lymphoid and myeloid lineage. Leuk Res 9:833-842

4) Slamon D, Boone TC, Murdock DC, Keith DE, Press MF, Larson RA, Souza LM (1986) Studies of the human c-myb gene and its product in human acute leukemias. Science 233:347-351

5) Evinger-Hodges MJ, Dicke KA, Gutterman JU, Blick M (1987) Proto-oncogene expression in human normal bone marrow. Leukemia 1:597-602

6) Thompson CB, Challoner P, Neiman PE, Groudine M (1985) Levels of c-myc oncogene mRNA are invariant throughout the cell cycle. Nature, Lond. 314, 363-366

7) Alitalo K, Koskinen P, Makela TP, Saksela K, Sistonen L, Winqvist R (1987) myc oncogenes: activation and amplification. Biochimica et Biophysica Acta 907:1-32

8) Preisler HD, Kinniburgh AJ, Wei-Dong G, Khan S (1987) Expression of the protooncogenes c-myc, c-fos, and c-fms in acute myelocytic leukemia at diagnosis and in remission. Cancer Research 47:874-880

9) Evinger-Hodges MJ, Bresser J, Brouwer R, Cox I, Spitzer G, Dicke KA (1988) Myc and sis expression in acute myelogenous leukemia. Leukemia 2(1):45-49

Z.S. Al-Lebban, J.B. Jones, M.A. Eglitis, W.F. Anderson, and C.D. Lothrop, Jr.

INTRODUCTION

The possibility that human genetic diseases may be treated by transfer of normal genes to correct abnormalities caused by defective genes is becoming more practical with recent advances in techniques for efficient gene transfer (1). The unique structure and mode of propagation of retroviruses make them well-suited for gene transfer. Bone marrow is a likely initial target for gene therapy, because extensive experience exists in manipulating and treating this tissue ex vivo and because some genetic diseases exert a primary tissue effect in the marrow. Several groups have reported successful transfer and expression of exogenous genes in hematopoietic progenitor cells (2-24). Most in vivo studies have focused on mice using vectors derived from the Moloney murine leukemia virus (11,14,15,17,18,20). The expression of retroviral vector transduced genes in hematopoietic stem cells in vivo has been variable in most species but is generally transient and only involves a small fraction of the hematopoietic cells (11,14,15,18,20-24). Stable long term expression of retroviral transduced genes in hematopoietic progenitor cells in species other than mice has not been reported. We found that canine and feline hematopoietic progenitor cells could also be infected with retroviral vectors carrying the neomycin phosphotransferase (neo^R) gene (2,12). Therefore, the dog and cat are excellent large animal models to test gene therapy because they have a number of genetic disorders similar to human diseases. Furthermore, the dog is an established preclinical model of bone marrow transplantation.

In this study, we have characterized gene transfer in hematopoietic progenitor cells from fetal, neonatal and adult dogs and cats using an in vitro CFU-GM assay. The methods developed in the in vitro studies were then applied to a canine and feline autologous bone marrow transplantation protocol to investigate the long term expression of retroviral transduced genes into hematopoietic stem cells in vivo.

MATERIALS AND METHODS

ANIMALS

Fetal, neonatal and adult dogs and cats of both sexes, were used in these studies. All adult animals were in apparent good health and had been previously vaccinated and dewormed before experimental study. Neonates were studied during the first 4 days of life. Dog and cat fetuses were obtained by closed ovariohysterectomy. Fetal development and gestation age were assessed by radiography and/or ultrasonography. All animals were housed in facilities fully accredited by the American Association for the Accreditation of Laboratory Animal Care.

CELL COLLECTION

Bone marrow was obtained from adult animals by percutaneous aspiration using pentabarbitol sedation of dogs and ketamine/xylazine sedation of cats. Bone marrow from neonatal animals was obtained after euthanasia by barbiturate overdosage.

Fetal livers were removed within half an hour of hysterectomy, freed of the extrahepatic bile duct and gall bladder and rapidly transferred into chilled Hank's balanced salt solution with Ca^{2+} and mg^{2+} as previously described (12). The mononuclear cells from the fetal, neonatal and adult hematopoietic cells were separated using ficoll-hypaque gradients, washed and counted as previously described (12).

CFU-GM assay: Canine or feline granulocyte- macrophage colony forming units (CFU-GM) were determined as previously described (2,12). Briefly, 7.5×10^5 nucleated marrow or fetal liver cells were plated in 1 ml Iscove's Modified Dulbecco's Medium (IMDM) (GIBCO, Grand Island, NY) containing 15% fetal calf serum (Whittaker Bioproducts; Walkersville, MD), 10% colony stimulating activity (neutropenic dog serum for dog cultures and conditioned medium from PHA stimulated feline mononuclear cells for cat cultures) in 0.8% methylcellulose. To determine the percentage of drug-resistant CFU-GM, the mononuclear cells were plated with and without the antibiotic G418; G418 when added, was present at a concentration of 1.7 or 2.0 mg/ml. The % infection is calculated as the the number of colonies that grow in the presence of G418 divided by the total number of colonies grown in the absence of G418 x 100. The cultures were maintained at $37°C$ in humidified air with 5% CO_2. The CFU-GM were scored after 6 to 8 days incubation. Statistical significance of all data from in vitro CFU-GM experiments was determined using the student's t test. The number of CFU-GM of each animal was calculated as the average of four individual cultures of that marrow sample. The data are summarized as CFU-GM (mean ± SD) for each marrow infection, where the "n" number refers to the number of different animals used in that experiment.

RETROVIRAL VECTOR AND INFECTION PROCEDURE

The retroviral vectors N2 or SAX were used in these experiments to characterize retroviral vector mediated gene transfer into hematopoietic progenitor cells of dogs and cats. The N2 vector is derived from the Moloney murine leukemia virus and contains the bacterial neo[R] gene. The SAX vector was derived from the N2 vector and contains the neo[R] gene as well as the human adenosine deaminase (AdA) gene driven by a simian virus 40 (SV40) promoter. The construction of these vectors has been described in detail previously (24,25).

For infection, bone marrow and fetal liver cells were incubated for 4 to 8 hours with retroviral vector-containing supernatants from fibroblast packaging cells at $37°C$ in the presence of 12 µg/ml polybrene (Aldrich Chemical Co., Milwaukee, WI). After infection, the bone marrow cells were recovered by low speed centrifugation, washed 3 times with complete media and plated in the CFU-GM assay.

AUTOLOGOUS BONE MARROW TRANSPLANTATION

Bone marrow was obtained by percutaneous aspiration and infected with the N2 or SAX retroviral vector as described above. After infection, cells were washed 3 times with complete media and infused into donor animals which had received 7.5 Gy (0.2 Gy/minute) total body irradiation from a cobalt 60 source. A portion of the infected bone marrow was plated in the CFU-GM assay at the time of infection in most animals. All animals were supported with fluid therapy and antibiotics, as necessary, until bone marrow reconstitution was complete. The complete blood count (CBC) was determined daily in transplanted animals until reconstitution was complete and then periodically thereafter. For each animal, the multiplicity of infection (MOI) was calculated as the ratio of viral particles (esimtated by the titer of drug-resistant CFUs) to bone marrow mononuclear cells.

Neomycin Phosphotransferase (NPT) Assay. Cells were recovered from methylcellulose cultures, washed 3 times and then lysed by five freeze/thaw cycles. The cleared cell lysates were separated using non-denaturing polyacrylamide gel electrophoresis (PAGE) and the NPT activity was measured using the method of Reiss et al (26).

RESULTS

Comparison of Gene Transfer into Fetal, Neonatal and Adult Hematopoietic Progenitor Cells: The efficiency of gene transfer into canine and feline hematopoietic progenitor cells of various developmental stages was determined with the in vitro CFU-GM assay. Cells were cultured in the presence or absence of 1.75 or 2.0 mg/ml G418 (The total amount present was actually 3.50 or 4.0 mg/ml since only about 50% of stock G418 is active). At a G418 dose of 1.75 mg/ml or greater there were no CFU-GM colonies in uninfected control cultures. The proportion of G418 resistant CFU-GM colonies from several canine and feline fetal, neonatal and adult marrow samples after infection with the N2 retroviral vector is summarized in Fig. 1. The mean ± SD G418-resistant CFU-GM after infection of feline fetal liver, neonatal and adult bone marrow with the N2 vector was 3.1±1.2, 11.1±13.9 and 3.4±3.2, respectively. The mean ± SD G418-resistant CFU-GM after infection of canine fetal liver, neonatal and adult bone marrow was 1.2±1.7, 32.0±18.0 and 3.2±0.8, repectively. The mean ± SD G418-resistant CFU-GM of neonatal dog and cat bone marrow after infection with N2 retroviral vector was significantly higher than that of adult dog (P<0.05) and adult cat (P<0.05) bone marrow. The number of G418-resistant CFU-GM obtained after infection of fetal liver hematopoietic progenitor cells was similar to that of adult bone marrow cells. The basis for increased infectivity of neonatal hematopoietic progenitor cells is not known but was not associated with an increased fraction of actively cycling cells as determined by tritiated thymidine suicide (data not shown).

The G418-resistant CFU-GM from neonatal animals produced neomycin phosphotransferase (Fig. 2) as previously reported for adult canine bone marrow infected with the N2 retroviral vector (2). This demonstrates that drug resistant CFU-GM were expressing the neo[R] gene.

Gene expression in dogs and cats reconstituted with autologous bone marrow.

To investigate the potential problems with gene therapy we reconstituted 2 dogs and 4 cats, which had been lethally irradiated, with autologous bone marrow that had been infected with the N2 or SAX retroviral vector. All dogs and cats used in this study were successfully engrafted with retroviral infected autologous bone marrow. All animals demonstrated prompt engraftment at approximately 15 to 30 days after transplantation based upon evaluation of the CBC and platelet counts (Table 1). In most animals, various degrees of gastrointestinal toxicity and hematological complications were observed. However, the severity of these pathological complications was not greater than that previously observed from lethal irradiation and autologous transplantation of uninfected bone marrow.

The pretransplant infection rate was between 3 and 12% in 4 of the 5 animals which were evaluated (Table 2). These results are similar to our previous findings on bone marrow infections and indicated that the "supernatant" infection protocol was adequate for large-scale bone marrow infections required in a large animal retroviral infection/autologous bone marrow transplantation protocol.

Bone marrow obtained from the transplanted animals from approximately 30 days to greater than 300 days post-transplant was tested for expression of the neo[R] gene with the in vitro CFU-GM assay. Expression of the neo[R] gene was evident in all animals post-transplantation based on the CFU-GM assay. Sequential evaluation of bone marrow for G418-resistant CFU over a period of several months showed that several animals still expressed the neo[R] gene after more than 200 days. However, the fraction of G418-resistant CFU did decline with time in all animals except for Yenkee. The fraction of bone marrow progenitor cells expressing the neo[R] gene in the autologous bone marrow transplant recipients was initially similar to that previously observed with in vitro infection experiments with a comparable MOI. The neo[R] gene was not detected in bone marrow from any of the transplanted animals by standard Southern hybridization, suggesting that the neo[R] gene was present in fewer than 1 gene copy per 100 bone marrow cells. Detection of neomycin phosphotransferase (NPT) activity in a transplant recipient confirmed that G418 resistance was associated with enzyme activity (data not shown). The presence of NPT activity in transplant recipients demonstrates, without question, the expression of the exogenous neo[R] gene in animals.

DISCUSSION

Retroviral vectors provide an efficient means for the transfer of exogenous genes into hematopoietic cells (2-24). The experiments described above demonstrated that retroviral vectors containing the dominant selectable neo[R] gene can infect the hematopoietic progenitor cells of fetal, neonatal and adult cats and dogs. The dramatic increase in infectability of some canine and feline neonatal marrow cells is unusual, but demonstrates that the rate of infectivity of hematopoietic progenitor cells depends not only on the retroviral vector but also on the developmental stage of the target hematopoietic cells. This increase, apparently, is not due to an increased fraction of actively cycling progenitor cells in neonatal bone marrow (2,9,12). It would seem that some factor(s) other than mitosis is creating this marked increase in retroviral gene transfer seen in some marrows of the neonatal animals. The basis for increased infectivity of neonatal bone marrow is unknown but could be due to 1) undefined qualities of hematopoietic progenitor cells/microenvironment which greatly affect retroviral infection. 2) the developmental timing or the types of hematopoietic progenitor cells in neonatal bone marrow. If so, this may vary between species since canine and feline fetal hematopoietic progenitor cells did not generate an increased number of G418-resistant CFU-GM as was seen with an in utero sheep transplantation model (23). 3) physiological changes which affect hematopoietic cells that take place during birth in both dogs and cats. Although the mechanism for the marked increase in gene transfer of neonatal hematopoietic progenitor cells is unknown this observation may be important for developing an optimal in vivo gene therapy protocol.

Retroviral vector-mediated gene transfer into canine and feline hematopoietic progenitor cells with autologous bone marrow transplantation is a useful model system to develop protocols for large-scale manipulation, retroviral infection and bone marrow reconstitution to investigate the prospect for treatment of human genetic diseases by gene therapy. This study also represents the first successful attempt of gene transfer into feline hematopoietic cells in vivo using a retroviral vector. We infected canine and feline bone marrow cells by incubating these cells with virus-containing medium from fibroblast producing cells. Our results and those of previously reported primate studies (24) contrast with a recent canine study (19) which suggested that infection of canine marrow by co-cultivation is more reliable in gene transfer and reconstitution than the exposure of marrow to virus-containing medium. Transient but significant expression of the neo[R] gene, after autologous bone marrow transplantation was evident in all animals based on the CFU-GM assay. Sequential evaluation of bone marrow over a period of several months showed that several animals continue to express the neo[R] gene after more than 200 days. The detection of neomycin phosphotransferase activity in a transplant recipient (Martina) confirmed that G418 resistance was due to the in vivo expression of the exogenous neo[R] gene in that animal. The N2 or SAX proviral sequences were not detected in marrows from any of the transplanted animals by standard Southern hybridization, suggesting the need for more sensitive methods to detect proviral sequences in bone marrow from transplanted animals. The low efficiency of gene transfer and expression in vivo may be due to the retroviral vectors used in these studies and/or the difficulty in isolating and infecting a large number of pluripotent stem cells. Improvement of current infection protocols, modification of vector design and the selection of infected pluripotent stem cells may be critical for efficient transfer of an exogenous genes to hematopoietic stem cells. Nevertheless, this study points out that expression of exogenous gene in large animals for several months is possible. However, it is obvious that gene transfer protocols must be improved before routine application of gene therapy to treatment of human genetic disease is possible.

SUMMARY

Retroviral vectors containing the selectable bacterial neo[R] gene (conferring resistance to the neomycin analogue G418) were used to demonstrate gene transfer into canine and feline hematopoietic progenitor cells in vitro and in vivo. The increased infectivity of neonatal hematopoietic cells relative to bone marrow from fetal and adult animals demonstrates that the efficiency of retroviral vector mediated gene transfer can be greatly influenced by the developmental stage of the target hematopoietic cells.

Transient but significant expression of the neo[R] gene after retroviral infection and autologous transplantation, was evident in 4 cats and 2 dogs based on the CFU-GM assay. Sequential evaluation of these animals over a period of several months showed that several animals continue to express the neo[R] gene after more than 200 days. These data show that retroviral vectors can be used to transfer exogenous genes into canine and feline bone marrow cells.

ACKNOWLEDGEMENTS

The Authors thank Janet Jolly for her technical assistance and Betsy Cagle for preparing this manuscript.

This work was supported in part by a generous gift form the G. Harold and Leila Mathers Charitable Foundation and by NIH Grant HL-15647.

REFERENCES

1. Anderson WF (1984): Prospects for human gene therapy. Science 226:401
2. Eglitis MA, Kantoff PW, Jolly JD, Jones JB, Anderson WF, Lothrop CD (1988): Gene transfer into hematopoietic progenitor cells from normal and cyclic hematopoietic dogs using retroviral vectors. Blood 71:717
3. Joyner A, Keller G, Phillips RA, Bernstein A (1983): Retrovirus transfer of a bacterial gene into mouse hematopoietic progenitor cells. Nature 305:556
4. Gruber HE, Finley KD, Hershberg RM, Katzman SS, Laikind PK, Seegmiller JE, Friedman T, Yee JK, Jolly DJ (1985): Retroviral vector-mediated gene transfer into hematopoietic progenitor cells. Science 230:1057
5. Rothstein L, Pierce JH, Klassen V, Greenberger JS (1985): Amphotrophic retrovirus vector transfer of the v-ras oncogene to human hematopoietic and stromal cells in continuous bone marrow cultures. Blood 65:744
6. Hock RA, Miller AD (1986): Retroviral-mediated transfer and expression of drug resistance genes in human hematopoietic progenitor cells. Nature 320:275
7. Kwok WW, Scheming F, Stead RB, Miller AD (1986): Retroviral transfer of genes into canine hemopoietic progenitor cells in culture: A model for human gene therapy. Proc. Natl. Acad. Sci. USA. 83:4552
8. Belmont JW, Henkel-Tigges J, Chang SM, Wager-Smith K, Kellens RE, Dick JE, Magli MC, Phillips RA, Bernstein A, Caskey CT (1986): Expression of human adenosine deaminase in murine hematopoietic progenitor cells following retroviral transfer. Nature 322:385
9. Hogge DE, Humphries RK (1987): Gene transfer to primary normal and malignant human hemopoietic progenitors using recombinant retroviruses. Blood 69:611
10. Magli MC, Dick JE, Hugzar D, Bernstein A, Phillips RA (1987): Modulation of gene expression in multiple hematopoietic cell lineages following retroviral vector gene transfer. Proc. Natl. Acad. Sci. USA. 84:7989.
11. Keller G, Paige C, Gilboa E, Wagner EF (1985): Expression of a foreign gene in myeloid and lymphoid cells derived from multipotent hematopoietic precursors. Nature 318:149
12. Al-Lebban ZS, Henry MJ, Jones JB, Eglitis MA, Anderson WF, Lothrop CD Jr. (1988): Increased efficiency of gene transfer with retroviral vectors in neonatal hematopoietic progenitor cells. Blood (submitted)
13. Anklesaria P, Sakakeeny MA, Klassen V, Rothstein L, Fitzgerald TJ, Appel M, Greenberger JS, Holland CA (1987): Expression of selectable gene transferred by a retroviral vector to hematopoietic stem cells and stromal cells in murine continuous bone marrow cultures. Exp. Hematol. 15:195
14. Dick JE, Magli MC, Huszar D, Phillips RA, Bernstein A (1985): Introduction of a selectable gene into primitive stem cells capable of long-term reconstitution of the hematopoietic of w/w mice. Cell 42:71
15. Willams DA, Lemischka IR, Nathan DG, Mulligan RC (1984): Introduction of new genetic material into pluripotent hematopoietic stem cells of the mouse. Nature 310:476
16. Lothrop CD Jr., Al-Lebban ZS, Jones JB, Smith JR, Baker HJ, Eglitis MA, Anderson WF (1988): Gene expression in dogs and cats reconstituted with autologous bone marrow infected with retroviral vectors. Science (submitted)
17. Miller AD, Eckner RJ, Jolly DJ, Friedmann T, Verma IM (1984): Expression of a retrovirus encoding human HPRT in mice. Science 225:630
18. Dzierzak WA, Papayannopoulou T, Mulligan RC (1988): Lineage-specific expression of a human B-globin gene in murine bone marrow transplant recipients reconstituted with retrovirustransduced stem cells. Nature 331:35
19. Stead RB, Kwok WW, Storb R, Miller AD (1988): Canine model for gene therapy: Inefficient gene expression in dogs reconstituted with autologous marrow infected with retroviral vectors. Blood 71:742
20. Eglitis MA, Kantoff P, Gilboa E, Anderson WF (1985): Gene expression in mice after high efficiency retroviral mediated gene transfer. Science 230:1395
21. Williams DA, Orkin SH, Mulligan RC (1986): Retrovirus mediated transfer of human adenosine deaminase gene sequences into cells in culture and into murine hematopoietic cells in vivo. Proc. Natl. Acad. USA. 83:2566
22. McIvor RS, Johnson JJ, Miller AD, Pitts S, Williams SR, Valerio D, Martin DW, Verma IM (1987): Human purine nucleoside phosphorylase and adenosine deaminase: Gene transfer into cultured cells and murine hematopoietic stem cells by using recombinant amphotropic retroviruses. Mol. Cell. Biol. 7:838
23. Kantoff PW, Flake AW, Eglitis MA, Harrison MR, Gilboa E, Zanjani ED, Anderson WF (1988): In utero gene transfer and expression: A sheep transplanation model (submitted)
24. Kantoff PW, Gillio AP, McLachlin JR, Bordignon C, Eglitis MA, Kernan NA, Moen RC, Kohn DB, Yu S-F, Karson E, Karlsson S, Zwiebel J, Gilboa E, Blaese RM, Nienhuis A, O'Reilly RJ, Anderson WF (1987): Expression of human adenosine deaminase in nonhuman primates after retrovirus-mediated gene transfer. J. Exp. Med. 166:219
25. Armentano D, Yu SF, Kantoff PW, Von Ruden T, Anderson WF, Gilboa E (1987): Effects of internal viral sequences on the utility of retroviral vectors. J. Virol. 61:1647
26. Reiss B, Sprengel R, Will H, Schaller H (1984): A new sensitive method for qualitative and quantitative assay of neomycin phosphotransferase in crude cell extracts. Gene 30:211

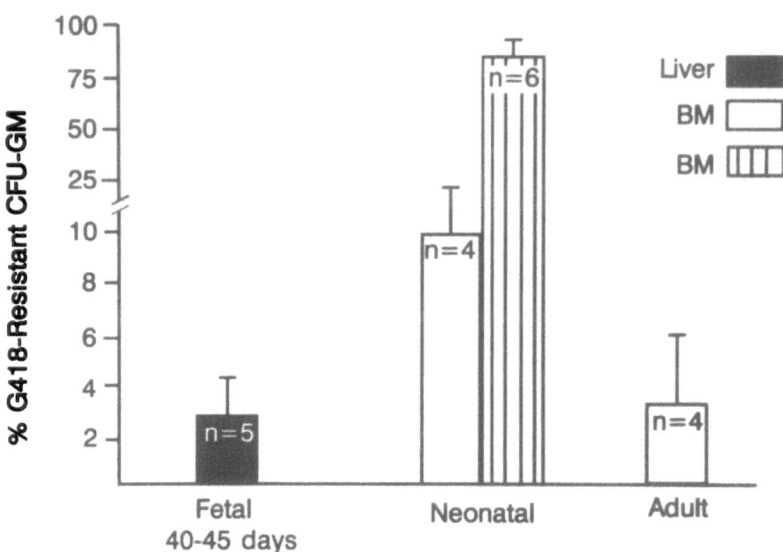

Fig. 1 a. Hematopoietic progenitor cells from cat liver and bone marrow were obtained, infected with the N2 retrovirus and G418-resistant CFU-GM were determined after 6-8 days culture. Bone marrow cells from neonatal cats were incubated with the N2 retrovirus for either 4 (A) or 8 (B) hours. The data are summarized as mean ± SD.

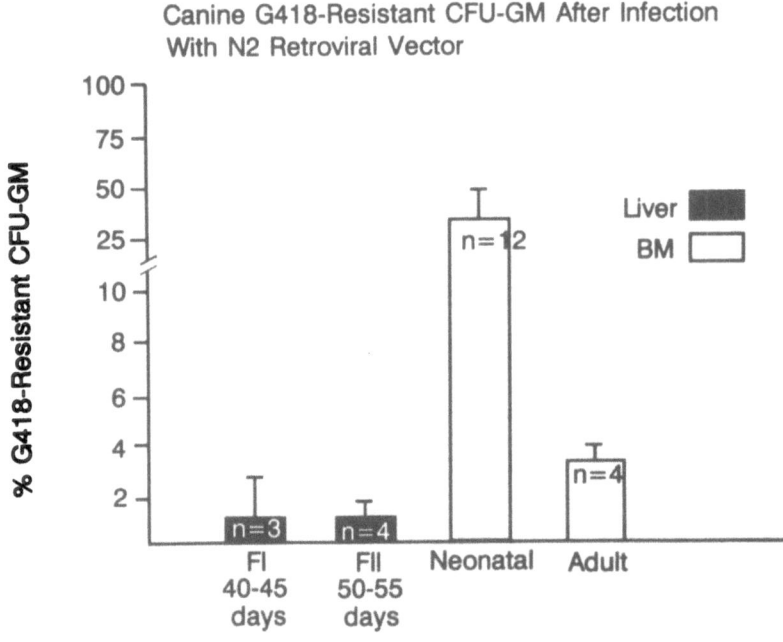

Fig. 1 b. Hematopoietic progenitor cells from dog liver and bone marrow were obtained, infected with the N2 retrovirus and G418-resistant CFU-GM were determined after 6-8 days culture. The data are summarized as mean ± SD.

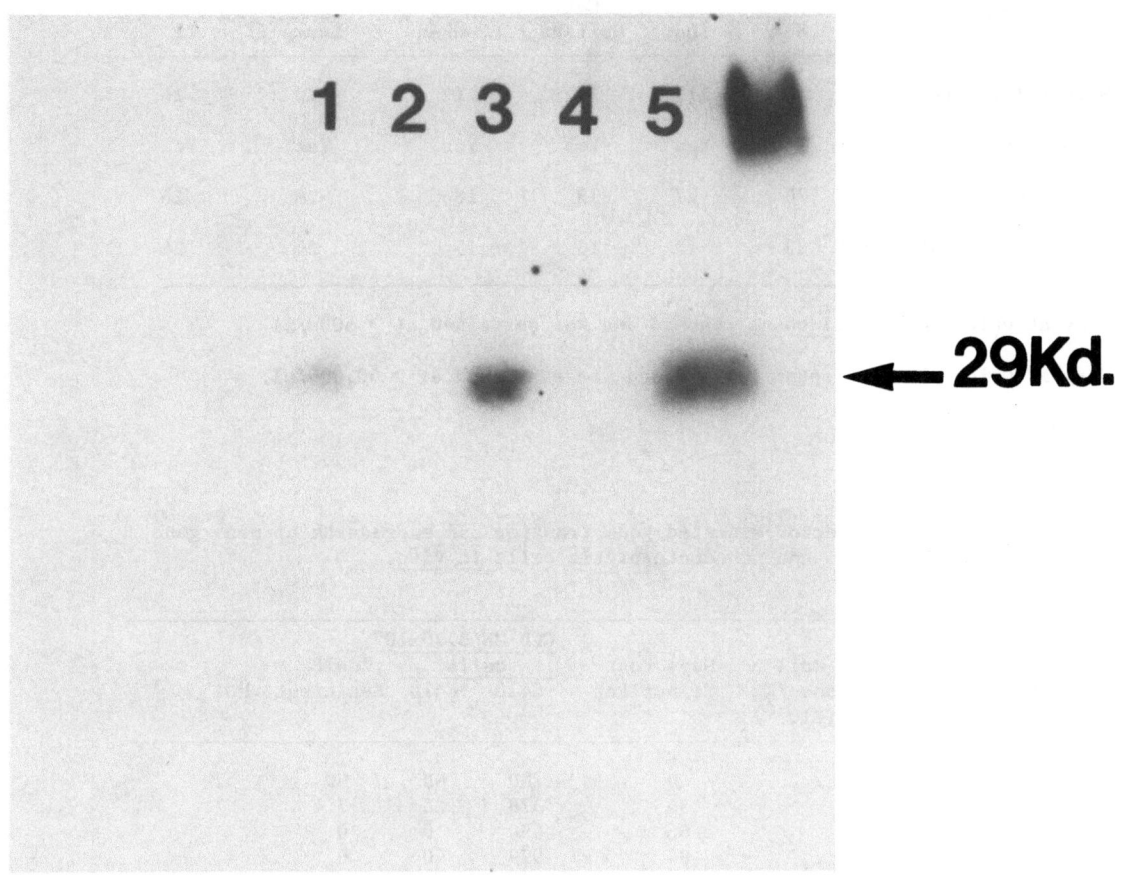

Fig. 2.   Neomycin phosphotransferase activity in G418-resistant neonatal feline CFU-GM.   Colonies re-
          sistant to 2000 G418/ml were recovered from methyl cellulose cultures, washed extensively
          and assayed.   The arrow indicates the position of the expected 29Kd NPT protein: Lane 1 10^5
          V6 cells; Lane 2 - 10^5 uninfected feline CFU-GM cells; Lane 3 - 10^6 neonatal canine
          G418-resistant CFU-GM cells; Lane 4 - 10^6 uninfected feline CFU-GM cells; Lane 5 10^6 V6
          cells.   V-6 cells are the fibroblas producing cells from which the N2 rich media
          supernatants were prepared.

108

Table 1. Bone marrow reconstitution of dogs and cats with infected autologous bone marrow.

Parameter	Cat				Dog	
	Sam	Andy	Martina	Yenkee	Scamp II	ZB
Survival (months)	>3*	>12	>12	>12	>10	>10
Reconstitution	Yes	Yes	Yes	Yes	Yes	Yes
PMN >5000/µl (day)[a]	26	17	13	18	28	26
Plts > 50,000/µl (day)[b]	23	13	13	15	24	28

[a]Day at which neutrophil count reached and was sustained at $\geq$ 5000/µl.

[b]Day at which platelet count reached and was sustained at $\geq$ 50,000/µl.

*Died on day 100.

Table 2. Retroviral-vector-mediated gene transfer and expression of neo[R] gene in feline and canine hematopoietic cells in Vivo.

Animal	Vector	Cell dose/kg ($\times 10^{-8}$)	Days Post-Transplant	CFU-GM/3.75×10^6 Cells -G418	+G418	%G418 Resistant	MOI
Sam (f)	N2	2.2	0	ND	ND	ND	>2
			30	378	28	7.4	
			66	490	0	0	
			94	979	0	0	
Andy (f)	N2	2.2	0	932	55	5.9	>2
			30	627	35	5.6	
			66	404	0	0	
			94	1336	5	0.4	
			154	551	19	3.5	
			194	563	15	2.7	
			223	480	16	3.3	
			348	210	1	0.5	
Martina (f)	N2	5.5	0	1211	40	3.3	>2
			26	540	30	5.5	
			67	349	4	1.1	
			96	240	0	0	
			214	390	4	1.0	
Yenkee (f)	SAX	3.7	0	468	0	0	>2
			26	420	13	3.1	
			67	386	18	4.7	
			96	414	9	2.2	
			214	975	35	3.5	
Scamp II (d)	N2	0.93	0	167	8	4.7	>2
			48	1263	56	4.4	
			85	739	4	0.5	
ZB (d)	N2	1.3	0	160	20	12.5	>2
			48	530	3	0.7	
			85	625	11	1.8	

Cells were cultured in the presence or absence of 2 mg/ml G418. At 2 mg/ml G418 there were no CFU-GM in uninfected control cultures.

ND - not done; f - feline; d - canine

K. Akai, M. Ueda, G. Kawanishi, Y. Miura, and T. Suda

SUMMARY

Erythropoietin (Ep) is a glycoprotein which is required for the proliferation and differentiation of late erythroid progenitor cells. The Ep gene was cloned and large quantities of recombinant Ep is available for the investigation of the action mechanism of Ep. Recently, we found that a small fraction (5%) of an interleukin-3 dependent murine myeloid cell line, FDC-P2 was able to respond to Ep. Moreover, we could establish a subclone EP-FDC-P2, which is dependent on Ep for proliferation and survival, from parental cell FDC-P2. To investigate the role of Ep in cell growth, the Ep gene was introduced into FDC-P2 or EP-FDC-P2 with a retrovirus vector, pZIPNeoSV(X)1. Infected FDC-P2 were selected for G-418 resistance, plated in methylcellulose medium and individual single colonies were cloned. Cell lines (FDC-P2/Ep or EP-FDC-P2/Ep) growing in the absence of exogenous growth factors were established. The Southern blot analysis using the probe of Ep cDNA showed that FDC-P2/Ep contained a single copy of the integrated pZIPNeoSV(X)1-Ep provirus. The Northern blot analysis revealed the Ep specific RNA transcript in the cell line. The growth of the FDC-P2/Ep cell line was inhibited by a monoclonal antibody against recombinant Ep. However, the amount of the antibody to inhibit the cell growth was much larger than that needed to neutralize Ep activity in the culture medium, suggesting that stimulation of growth occurred not only by external interaction of the secreted Ep with receptors, but also by immediate interaction at cell surface or internal interaction. FDC-P2/Ep was not able to induce tumor in nude mice, while transfection of IL-3 gene caused FDC-P2 to become tumorigenic. These results indicate that Ep acts as a growth factor in this system as well as IL-3 and that autocrine stimulation with Ep did not result in tumorigenicity.

INTRODUCTION

Normal hemopoietic precursors are able to proliferate and differentiate in the presence of hemopoietic growth factors. Erythropoietin (Ep) is a polypeptide growth factor required for the growth and terminal erythroid differentiation. Since human Ep gene has been molecularly cloned [1,2], a large amount of recombinant, purified Ep became available [3]. In order to clarify the action of Ep, it is essential to prepare the pure population of Ep-responsive cells.

Dexter et al. established immortalized cell lines derived from murine hemopoietic cells which are characterized by absolute dependency on exogenous CSF [4]. FDC-P2 is able to proliferate depending on IL-3 but not on GM-CSF or G-CSF. Recently, we found that a small population (5%) of FDC-P2 can respond to Ep for proliferation without erythroid differentiation. Moreover, to obtain high responder to Ep, we subcloned Ep-dependent cell line (EP-FDC-P2) from parental FDC-P2. To determine if the constitutive production of Ep by factor dependent cell lines is sufficient to induce cell transformation, we introduced a human Ep gene into FDC-P2 and EP-FDC-P2. Although the infected cells synthesized Ep and grew autonomously in vitro, they remained non-tumorigenic. However, they did induce tumor in irradiated nude mice.

MATERIALS AND METHODS

Mice: Nude mice (BALB/cA, nu/nu) were purchased from Japan Clea, Tokyo.
Cell Culture: FDC-P2 were routinely cultured in RPMI 1640 medium (GIBCO Laboratories, Grand Island, New York)

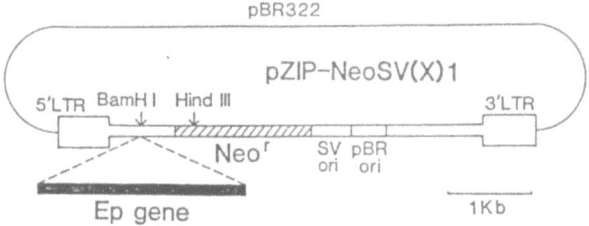

Fig.1 Diagram of Ep Expression Vector, pZIPNeoSV(X)1-Ep.

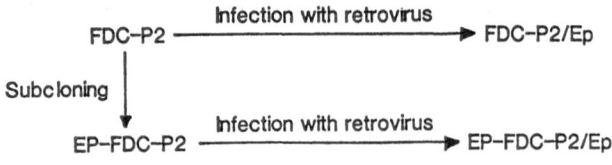

Fig.2 Establishment of Cell Lines.

supplemented 10% fetal calf serum (FCS, Flow Laboratories, North Ryde, Australia) in the presence of 100 U/ml murine recombinant (r) IL-3 at 37°C in 5% CO_2 95% in air. Recombinant murine IL-3 was obtained from the serum-free culture supernatants of COS-1 cells transfected with a cloned IL-3 gene using pCD-X vector as described before [5], specific activity being 10^5 units/mg protein. Serum-free pokeweed mitogen spleen cell conditioned medium (PWM-SCM) was occasionally used as a source of IL-3 [6]. Recombinant human Ep was obtained from the culture supernatants of BHK cells transfected with the Ep gene, and purified using anti-Ep antibody-coupled affinity column, specific activity being 96,000 units/A_{280} protein [3]. Psi 2 cells were maintained in Dulbecco's modified Eagle's medium (DMEM) supplemented with 5% FCS.

Establishment of Ep-dependent Cell Line: Parental FDC-P2 was able to proliferate in response to Ep but had trouble surviving in the presence of Ep alone. Parental FDC-P2 was cultured in the presence of 10% PWM-SCM and 2U/ml rEp. The concentration of PWM-SCM was reduced by half each week. After 8 weeks of acclimatization, the cells were cultured in the presence of 2 U/ml rEp, plated in the methylcellulose medium and single colonies were cloned. These sublines were dependent on Ep for the proliferation and survival, and one of them was designated as EP-FDC-P2.

Colony Formation in Methylcellulose Medium: Culture of 500 transfected or non-transfected FDC-P2 cells were prepared in 35-mm non-tissue culture dishes (Falcon, Oxnard, California) containing methylcellulose medium. One milliliter of 1.2% methylcellulose (Fisher Scientific, Norcross, Georgia) in alpha medium contained 30% FCS, 10 mg bovine serum albumin (BSA; Sigma Chemical Company, St. Louis, Missouri) with appropriate rEp or rIL-3 [6].

Infection of Psi2 with a Retrovirus Encoding Ep: A 2.4 Kb human Ep gene fragment [3] was inserted at the cloning site of a murine retrovirus vector, pZIPneoSV(X)1, and the plasmid DNA was transfected into the psi2 mouse packaging cell line [7,8]. The psi 2 cells contain a mutant Moloney leukemia provirus that is defective in its packaging sequences and cannot encapsidate its genomic RNA into virions. The psi2 provirus provides retroviral replicative functions in trans, so that vector genomic RNA is efficiently packaged into viral particles produced by the transfected cells. These helper free defective virus can infect other murine cells, leading to efficient integration of the vector provirus. Since the vector lacks genes necessary for virion production, the target cells do not release infectious particles. Figure 1 shows the vector containing the Ep and neomycin-resistance genes. Transfected psi2 cells were selected in 2mg/ml G-418 (Geneticin;GIBCO Laboratories) and G-418-resistant cells expressing Ep were selected. All cell lines were cultured in RPMI 1640 medium containing 10% FCS.

Infection to FDC-P2 or EP-FDC-P2 with Recombinant Virus: Virus producing psi2 cells (3×10^5 cells) were seeded in an 80 cm^2 culture flask and cultured for 2 days. FDC-P2 or EP-FDC-P2 (1.5×10^6 each) were cocultured with the selected psi2 cells in RPMI 1640 supplemented with rIL-3 or rEp in the presence of 2 µg/ml polybrene (Sigma Chemical) for 48 hr. Non adherent cells were harvested and selected in the media containing 2 mg/ml G-418. After three weeks of selection, the G418-resistant cells were maintained in complete medium containing rIL-3 or rEp. Cells were selected in the presence of G-418 and plated in the methylcellulose medium in the absence of growth factors. Factor independent colonies derived from single cells were grown up and subjected to further analysis. Independent clones were tested for the properties and tumorigenicity in nude mice. We designated FDC-P2 and EP-FDC-P2 infected with recombinant retrovirus containing Ep gene as FDC-P2/Ep and EP-FDC-P2/Ep, (Fig.2).

Monoclonal Antibodies to Ep: Monoclonal antibodies to rEp, R2 and R6 , used in this experiment, were purified from ascitic fluid of the mice which received Ep-directed hybridoma [9].

Measurement of Ep Activity in Culture Supernatant: Ep titers in the culture supernatants of each cell line were determined by radioimmunoassay or enzyme immunoassay. Competitive radioimmunoassay was carried out according to the method we described in 1987 [10], using purified human rEp and rabbit anti-Ep serum. Enzyme immunoassay was done by the sandwich method using two kinds of monoclonal antibodies to rEp [9]. The primary antibody was immobilized on the surface of a 96-well microplate and the sample Ep solution was reacted with immobilized antibody. After removing unreacted Ep, the secondary antibody which was labeled with alkaline phosphatase reacted to Ep. The amount of Ep was determined by measuring the activity of alkaline phosphatase observed in the micro wells.

Binding Assay of Ep to Receptors: Binding of ^{125}I-labeled Ep was measured by a modification of the method previously described [11].

Southern and Northern Blot Analyses: High molecular weight DNA was isolated [12] and 5 μg of the DNA was digested with HindIII. The fragments were separated on 0.8% agarose gel, transferred to a nylon membrane (Hybond-N, Amersham Japan, Tokyo) by the method of Southern [13] and hybridized with a ^{32}P-labeled Ep cDNA probe. The cDNA probe was the 0.8 Kb of Sau3A1 fragment of rescued proviral DNA from virus producing psi2 cells by the method of Cepko et al. [7]. For Northern analysis cellular RNA was isolated by guanidium thiocyanate method and poly (A[+]) RNA was selected on an oligo(dT) cellulose column [12]. Two ug of the poly (A[+]) RNA was electrophoresed in a formaldehyde gel transferred to a nylon membrane and hybridized with the ^{32}P-labeled Ep cDNA probe.

Transplantation Experiments: Transfected or non-transfected cells were washed with alpha-medium and embedded in a plasma clot. Approximately 1×10^6 cells were transplanted into untreated 3-month-old nude mice. In some experiments, the nude mice were irradiated with 400 rad whole body irradiation at 107 rad/min prior to injection of cells using Gammacell 40 with a 133TBq (3600Ci) ^{137}Cs-source (Atomic Energy of Canada Ltd., Ottawa, Ont). Mice were monitored twice weekly and when moribund, they were killed and tested.

RESULTS

Response of FDC-P2 to Ep, IL-3, G-CSF, and GM-CSF: IL-3 and Ep supported the colony formation of FDC-P2 in methylcellulose cultures, but G-CSF and GM-CSF did not. Figure 3 shows the relationship of the number of colonies to the concentration of each hemopoietic growth factor. The maximum number of colonies supported by rEp or rIL-3 was around 25 or 300 per 500 plated cells. Approximately 5% of FDC-P2 cells could respond to Ep for forming colonies. To analyze the morphology of constituent cells in colonies, individual colonies were lifted from the methylcellulose medium on day 7 of culture. They remained at the stage of blasts or promyelocytes. They showed no differentiation signs and were negative for benzidine staining.

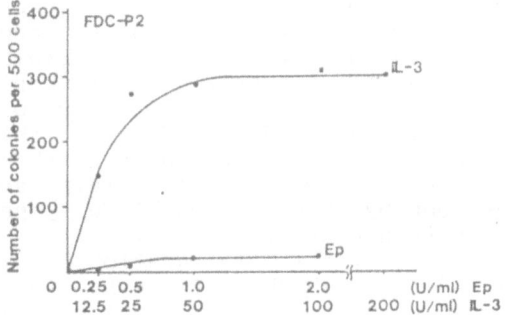

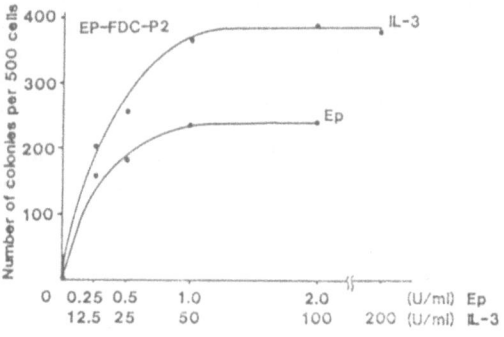

Fig.3 Response of FDC-P2 and EP-FDC-P2 to rEp or rIL-3.

Response of EP-FDC-P2 to IL-3 and Ep: Ep-dependent cell line was established from parental FDC-P2 after continuing culture of FDC-P2 in the presence of rEp as decreasing the concentration of PWM-SCM. This subclone was designated as EP-FDC-P2. This cell line was maintained by the support of rEp instead of rIL-3. EP-FDC-P2 formed around 250 or 400 colonies by the support of rEp or rIL-3 (Fig.3). The incidence of colony formation by EP-FDC-P2 in the presence of Ep was 10 times higher than that of FDC-P2, plating efficiency being 50%. There was no morphological difference between parental FDC-P2 and EP-FDC-P2.

Establishment of FDC-P2/Ep and EP-FDC-P2/Ep: After infection of FDC-P2 and EP-FDC-P2 with the retrovirus containing the Ep gene, we could establish FDC-P2/Ep and EP-FDC-P2/Ep. FDC-P2/Ep was selected and cultured in the presence of 2 mg/ml G-418 and 10% PWM-SCM, and plated in the methylcellulose medium in the presence of G-418 and in the absence of exogenous growth factors. Factor independent colonies derived from single cells were grown up and subjected to further analyses. We checked the Ep gene integration sites in 4 clones derived from individual colonies. They showed the same integration site of the recombinant virus, indicating that they were derived from the same clone. EP-FDC-P2/Ep was selected in the methylcellulose medium deprived of growth factors. Factor independent colonies were lifted up and grown in liquid culture. We could obtain 12 subclones which showed different integration sites. Since the incidence of colony formation by EP-FDC-P2 in the presence of Ep was 10 times higher than that of FDC-P2, EP-FDC-P2 could obtain an autocrine loop by introduction of Ep gene more easily than FDC-P2.

<u>Analyses of Viral Integration and Transcripts in FDC-P2/EP</u> : To make sure that Ep producing provirus was integrated into FDC-P2/Ep , high molecular weight DNA of FDC-P2/Ep was isolated, digested with restriction endonuclease HindIII and analyzed by Southern blotting using the Ep cDNA probe (Fig.4). FDC-P2/Ep showed a 4.0 Kb fragment hybridized with the Ep cDNA probe, while parental FDC-P2 cells showed no detectable band. Since HindIII cleaved the recombinant provirus genome only at one site, FDC-P2/Ep has a single proviral integration. FDC-P2 cell line should contain the endogenous murine Ep gene, but the murine Ep gene was not detected with human Ep cDNA probe in this condition.

We next performed Northern blot analysis of FDC-P2/Ep cells. While FDC-P2 did not express Ep mRNA, FDC-P2/Ep produced a 5.1 Kb transcript which corresponds to the expected genomic RNA of introduced provirus (Fig.4). Southern blot analysis of EP-FDC-P2/Ep confirmed viral integration. In 8 of 12 clones, a single band was detected; 3 clones showed a higher density band than the others. In 4 out of 12 clones, 2 or 3 bands were detected.

<u>Biological Properties of FDC-P2/Ep and EP-FDC-P2/Ep:</u> In the methylcellulose culture system, FDC-P2/Ep and EP-FDC-P2/Ep formed colonies without exogenous hemopoietic factors. Addition of 2 U/ml Ep did not increase the number of colonies. By the addition of 100 U/ml IL-3, the number of colonies increased significantly (Table 1). Plating efficiencies of EP-FDC-P2/Ep were different from each clone; clone 3 was 23% and clone 4 was 6%. The difference of colony formation may be caused by the different number of copies and integration site of the retrovirus.

Ep activities in supernatants of each clone were also shown in Table 1. Parental FDC-P2 showed no detectable Ep activity in the supernatant. After infection of the retrovirus containing the Ep gene, FDC-P2 produced a slight of Ep activity in the supernatant, while EP-FDC-P2/Ep produced a higher Ep activity.

Scatchard analysis of the ^{125}I-labeled Ep binding to FDC-P2 revealed the existence of a single class of binding site in extremely low abundance. It is very difficult to compare the number of binding sites between FDC-P2 and FDC-P2/Ep, because Ep responsive population of FDC-P2 was very low. However, because Ep responsive population was nearly 50% in EP-FDC-P2, decrease of binding sites-per-cell of EP-FDC-P2/Ep was significant. After introduction of Ep gene, down

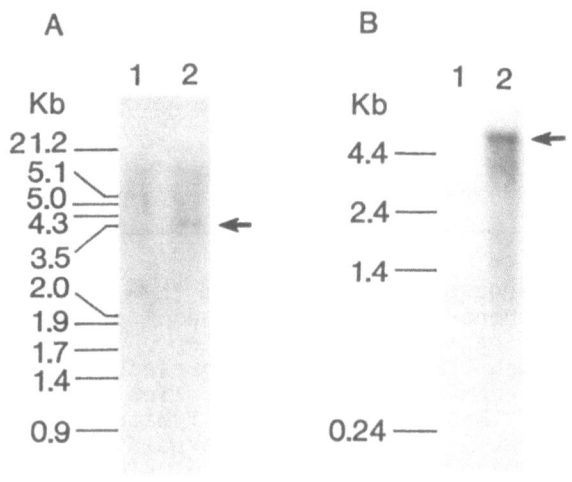

Fig.4 Southern and Northern Blot Analyses. DNA samples(A) and poly(A$^+$) RNA samples(B) were probed with Ep cDNA fragment. Lane 1;FDC-P2, Lane 2;FDC-P2/Ep.

Table 1   Properties of Each Clone

Cell line	Colony formation in methylcellulose[1]			Ep activity[2] (U/ml)	Ep binding site per cell
	medium	Ep (2U/ml)	IL-3 (100U/ml)		
FDC-P2	0	25±2	287±12	0	67
FDC-P2/Ep	210±20	220±20	395±5	0.015	56
EP-FDC-P2	0	239±15	382±14	-	306
EP-FDC-P2/Ep Cl.3	117±3	117±17	213±20	2.1	20
EP-FDC-P2/Ep Cl.4	30±3	33±3	69±4	2.2	35

1) Number of colonies per 500 cells/dish. Colonies were counted on day 7.
2) Culture supernatants were prepared from 3 day cultures. The initial cell concentration was 1x10^5 cells/ml.

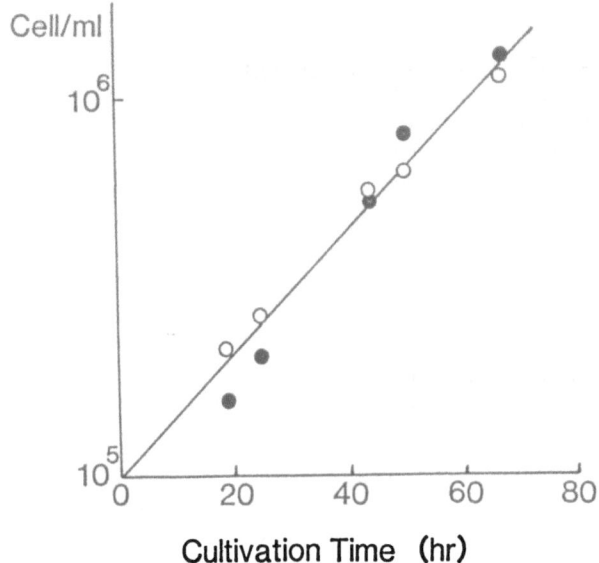

Fig.5 Effect of Anti-Ep Antibody on the Growth of FDC-P2. FDC-P2 (10^5/ml) were cultured with(○) or without(●) 200μg/ml of anti-Ep monoclonal antibody R6.

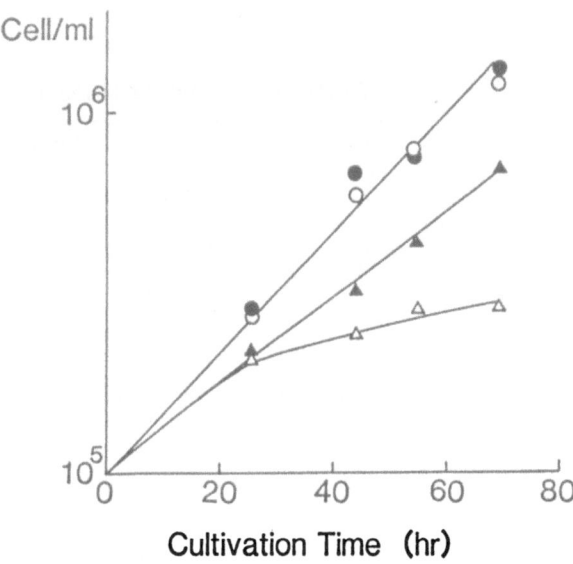

Fig.6 Effect of Anti-Ep Antibody on the Growth of FDC-P2/Ep. FDC-P2/Ep (10^5/ml) were cultured with 100μg/ml(▲) or 200μg/ml of R6(△), 200μg/ml of mouse IgG (○) as a control or without IgG(●).

modulation of Ep receptor might be mediated by its ligand.

The Effect of Anti-Ep-Monoclonal Antibody to the Growth FDC-P2/Ep : We first examined the effect of the anti-Ep monoclonal antibody on the growth of FDC-P2 supported by rIL-3 to check the specificity of the antibody. As shown in Fig. 5, the growth of parental FDC-P2 was not influenced by the anti-Ep monoclonal antibody at all. Moreover, the 200 μg/ml anti Ep antibody did not inhibit granulocyte-macrophage colony formation by normal bone marrow cells (data not shown). Next, we checked the neutralizing activity of this monoclonal antibody using the methylcellulose culture system of human bone marrow cells. One μg/ml of monoclonal antibody neutralized the activity of 2 U/ml rEp (Table 2). Using this monoclonal antibody, we checked if the growth of FDC-P2 depended on growth factor mediated by external stimulation. As Figure 6 shows, the growth of FDC-P2/Ep was inhibited by the addition of 200 μg/ml anti-Ep monoclonal antibody. This 200 μg of the monoclonal antibody can neutralize 400 U/ml of Ep activity. To inhibit the growth of FDC-P2/Ep about 25,000 fold more of the Ep monoclonal antibody is required than to neutralize the Ep activity in the culture supernatant from FDC-P2/Ep. The same amount of mouse IgG, as a control of the anti Ep antibody, did not inhibit the growth of FDC-P2/Ep.

The results prove autocrine stimulation of growth occurred by external interaction

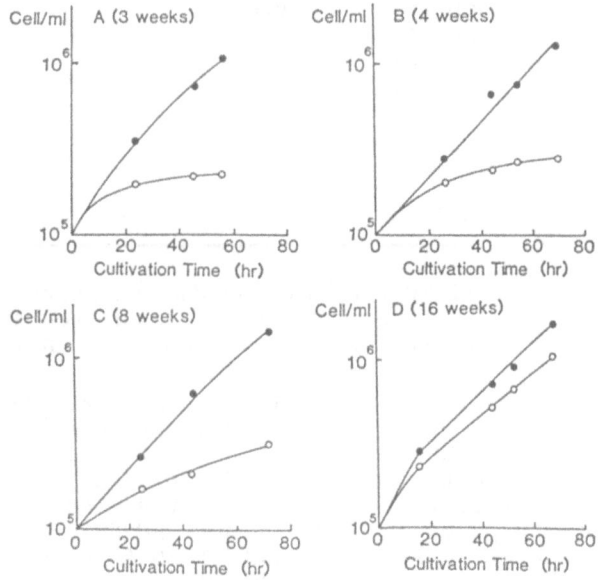

Fig.7 Sequential Studies on the Effect of Anti-Ep Antibody on the Growth of FDC-P2/Ep. Cells were cultured with(○) or without(●) 200μg/ml of R6.

of the secreted Ep with receptors. The growth inhibition of FDC-P2/Ep by monoclonal antibody became negligible in subsequent experiments after continuing the culture of FDC-P2/Ep in the absence of Ep. The growth was clearly inhibited by the monoclonal antibody at 3 weeks and 4 weeks after cell cloning. However, at 8 weeks of culture, the inhibition was

Table 2   Neutralized Activity of Anti Ep-Monoclonal Antibody

Ep	Anti-Ep	Number of erythroid colonies (% control)[1]
2 U/ml	0 µg/ml	155 (100)
2	0.1	96 (62)
2	1.0	4 (4)
2	10.0	0 (0)
	mouse IgG	
2	10.0	157 (101)

1) Number of colonies per 5×10^4 human bone marrow mononuclear non-phagocytic cells. Colonies were counted on day 7.

Table 3   Tumorigenicity of Hemopoietic Cells

Cells	Gene		Autocrine growth in vitro	Tumorigenicity[1]	Reference
FDC-P2			-	- (0/3)	
FDC-P2	Ep		+	- (0/5)[2]	
EP-FDC-P2	Ep	Cl.3	+	- (0/5)	
EP-FDC-P2	Ep	Cl.4	+	- (0/5)	
FDC-P2	IL-3		+	+ (25/25)	14)
Normal	IL-3		+	- (0/6)	14,22)
FDC-P2	GM-CSF		-	-	20)
FDC-P1	-		-	-	18,22)
FDC-P1	IL-3		+	+	22)
FDC-P1	GM-CSF		+	+	18,20)
Macrophage Cell Line	CSF-1		+	-	23)
32D	IL-3		+	+	21,22)

1) Numbers in parentheses represent the numbers of animals developing tumors/number of inoculated animals.
2) The cells induced tumors in irradiated mice (3/3).

partial and at 16 weeks of culture, the addition of monoclonal antibody did not affect the growth of FDC-P2/Ep (Fig.7). The growth of EP-FDC-P2/Ep was not completely inhibited even by the addition of 200 µg/ml anti Ep. Since EP-FDC-P2/Ep produced a larger amount of Ep than FDC-P2/Ep, 200 µg/ml anti-Ep was not enough.

Tumorigenicity: FDC-P2, FDC-P2/Ep and EP-FDC-P2/Ep were examined for the ability to induce tumors in nude mice. Cells (1×10^6) were injected subcutaneously. After 3 months, FDC-P2, FDC-P2/Ep, and EP-FDC-P2/Ep clones 3 and 4 had not induced tumors, while introduction of IL-3 gene caused FDC-P2 cells to become tumorigenic [14]. When FDC-P2/Ep was transplanted into 400 rad irradiated mice, tumors were developed in all mice within 2 to 3 weeks (Table 3).

DISCUSSION

Ep is a glycoprotein required for the proliferation and differentiation of late erythroid progenitors. In this study, we showed that a small population of IL-3 dependent myeloblastic cell line (FDC-P2) was able to respond to Ep and form colonies. We and other investigators reported that IL-3 dependent myeloid cell lines such as DA-1 and NFS-60 also respond to Ep and that NFS-60 showed a hemoglobin synthesis in the presence of Ep [15,16,17]. These cell lines provide us with an opportunity to study the growth effect mechanism of Ep. Moreover, we succeeded in subcloning the Ep-dependent cell line from parental FDC-P2, which can survive in the presence of Ep.

To investigate the role of Ep in cell-growth, human Ep gene was introduced into FDC-P2 and EP-FDC-P2 with a retrovirus vector. The plasmid in the transfected cells contains the Ep gene under the transcriptional regulation of the promoter in LTR. When infected with this retrovirus vector, FDC-P2/Ep and EP-FDC-P2/Ep showed the autonomous growth. While FDC-P2/Ep was selected for G-418 resistance, EP-FDC-P2/Ep was selected for factor-independent growth. FDC-P2/Ep secreted a very low titer of Ep, and EP-FDC-P2/Ep produced a high titer of Ep. The growth of FDC-P2/Ep was inhibited by a monoclonal antibody against rEp, suggesting an autocrine growth by Ep. However, the amount of the antibody to inhibit the cell growth was much larger than that needed to neutralize Ep activity in the culture medium. To our knowledge, this result represents one of very few instances in which autocrine growth was inhibited by a monoclonal antibody against the growth factor concerned. In contrast, the growth of EP-FDC-P2/Ep was not inhibited clearly even by the addition of large amounts of monoclonal antibody. It may be suggested that stimulation of growth occurred not only by external interaction of the secreted Ep with receptors, but also by immediate interaction at the cell surface or internal interaction. Lang et al. also reported that the growth of FDC-P1 transfected with the GM-CSF gene was not inhibited by anti GM-CSF antibody [18]. Similarly the growth of the T cell line transfected by the IL-2 gene was not completely inhibited by the anti IL-2 antibody [19]. In this study, we showed that growth inhibition by the anti Ep antibody was altered for a long period of time. Since FDC-P2/Ep was cultured in the absence of an external source of growth factor, it is likely that a subpopulation of cells has developed, which were no longer requiring growth factors. Recently Laker et al. proposed the following hypothesis: Two separate events are included in the acquisition of autonomous growth of factor dependent cells. The first event after the growth factor gene transfer to the cells was the acquisition of growth factor independence by external stimulation by endogenously synthesized growth factor. This initial step was followed by a second step that abrogated the requirement for external stimulation [20].

Table 3 summarizes our and previously reported data on tumorigenicity of transformed hemopoietic cells. Autocrine synthesis of colony stimulating factor can induce the malignant transformation of established factor dependent cell lines. For example, the introduction of genes encoding the granulocyte-macrophage colony stimulating factor (GM-CSF) or IL-3 into the myeloid cell line FDC-P1,FDC-P2 or 32D, or insertion of the interleukin-2 (IL-2) gene into a T cell line, abrogated their factor dependence and rendered the cells tumorigenic [14,18-22]

Tumorigenicity seems to depend upon three factors; (1) the gene transferred (2) target cells and (3) host to which cells were transferred. These results are not always comparable in separate experiments because the kind of cells and retrovirus vector was different from each other. However, we can deduce the following principles. (1) The capacity of the factor dependent cell lines (FDC-P1 or FDC-P2) producing IL-3 or GM-CSF confer a tumorigenic phenotype on these cells. Gene transfer of Ep and CSF-1 [23] did convert the factor dependent cells to independent cells but did not confer the tumorigenicity on these cells. Ep or CSF-1 support the late stage of differentiation and proliferation of erythroid or macroghagic lineage. They are lineage specific hemopoietic factors. In contrast, IL-3 and GM-CSF support the earlier stage of hemopoietic differentiation. Myelogenous leukemic cells are occasionally able to respond to IL-3 or GM-CSF but not to Ep or CSF-1 [23]. (2) Normal hemopoietic cells infected with a recombinant retrovirus containing IL-3 gene, mast cell-like permanent cell line was established which could proliferate without exogenous growth factors[15,21]. However, these cells remained non-tumorigenic. Although autocrine mechanisms may contribute to leukemogenesis, possibly as a part of multistep process involving other genetic events, our results suggested that additional regulatory mechanisms can limit the tumorigenic potential of immortalized, autocrine-stimulated hematopoietic cells. (3) Tumorigenicity also depends upon host factors. These FDC-P2/Ep are generally not tumorigenic in a non-irradiated group but they can induce tumor in the irradiated group, indicating that an irradiation-altered host environment played an important role in inducing tumorigenic transformation. Tumor immunity including cellular immunity and antibody formation may be suppressed by the whole body irradiation. Duhrsen and Metcalf have described the consistent development of leukemic transformation of FDC-P1 cells when injected intravenously into irradiated animals [24]. Naparstek et al. generated leukemogenic factor independent clones by cocultivation of CSF-dependent hemopoietic cell lines with irradiated stromal cells [25].

In conclusion Ep acts as a growth factor on the factor dependent myeloid cell line as well as IL-3, but autocrine stimulation with Ep did not result in tumorigenicity. Abrogation of growth-factor dependency by autocrine mechanism in an established hemopoietic cell line did not result in a tumorigenic phenotype.

REFERENCE

1. Jacobs K, Shoemaker C, Rudersdorf R, Neill SD, Kaufman RJ, Mufson A, Seehra J, Jones SS, Hewick R, Fritsch EF, Kawakita M, Shimizu T, Miyake T (1985) Isolation and characterization of genomic and cDNA clones of human erythropoietin. Nature 313:806-809
2. Lin F-K, Suggs, Lin C-H, Browne JK, Smalling R, Egrie JC, Chen KK, Fox GM, Martin F, Stabinsky Z, Badrawi SM, Lai P-H, Goldwasser E (1985) Cloning and expression of the human erythropoietin gene. Proc Natl Acad Sci USA 82:7580-7584
3. Goto M, Akai K, Murakami A, Hashimoto

C, Tsuda E, Ueda M, Kawanishi G, Takahashi N, Ishimoto A, Chiba H, Sasaki R (1988) Production of recombinant human erythropoietin in mammalian cells: host-cell dependency of the biological activity of the cloned glycoprotein. Biotechnology 6:67-71

4. Dexter TM, Garland J, Scott E, Scolnick E, Metcalf D (1980) Growth of factor-dependent hemopoietic precursor cell lines. J Exp Med 152:1036-1047

5. Yokota T, Lee F, Rennick D, Hall C, Arai N, Mosmann T, Nabel G, Cantor H, Arai K (1984) Isolation and characterization of a mouse cDNA clone that express mast-cell growth-factor activity in monkey cells. Proc Natl Acad Sci USA 81:1070-1074

6. Suda T, Suda J, Ogawa M, Ihle JN (1985) Permissive role of interleukin 3 (IL-3) in proliferation and differentiation of multipotential hemopoietic progenitors in culture. J Cell Physiol 124:182-190

7. Cepko CL, Roberts BE, Mulligan RC (1984) Construction and applications of a highly transmissible murine retrovirus shuttle vector. Cell 37:1053-1062

8. Mann R, Mulligan RC, Baltimore D (1983) Construction of a retrovirus packaging mutant and its use to produce helper-free defective retrovirus. Cell 33:153-159

9. Goto M, Murakami A, Akai K, Ueda M, Kawanishi G, Sasaki R, Chiba H (1988) New hybridomas for production of monoclonal antibodies to human erythropoietin. Nippon Nogeikagaku Kaishi 62:331

10. Mizoguchi H, Ohota K, Suzuki T, Murakami A, Ueda M, Sasaki R, Chiba H (1987) Basic conditions for radioimmunoassay of erythropoietin, and plasma levels of erythropoietin in normal subjects and anemic patients. Acta Haematol Jpn 50:15-24

11. Sasaki R, Yanagawa S, Hitomi K, Chiba H (1987) Characterization of erythropoietin receptor of murine erythroid cells. Eur J Biochem 168:43-48

12. Davis LG, Dibner MD, Battey JF (1986) Basic methods in molecular biology, Elsevier Science Publishing Co., Inc. NY

13. Southern EM (1975) Detection of specific sequences among DNA fragments separated by gel electrophoresis. J Mol Biol 98:503-517

14. Suda T, Ohno M, Suda J, Saito M, Miura Y, Kitamura Y (1988) Transfection of IL-3 gene to factor-dependent murine myeloid cell line(FDC-P2). Acta Haematol Jpn in press

15. Tsao C-J, Tojo A, Fukamachi H, Kitamura T, Saito T, Urabe A, Takaku F (1988) Expression of the functional erythropoietin receptors on interleukin-3-dependent murine cell lines. J Immunol 140:89-93

16. Sakaguchi M, Koishihara Y, Tsuda H, Fujimoto K, Shibuya K, Kawakita M, Takatuki K (1987) The expression of functional erythropoietin receptors on an interleukin-3 dependent cell line. Biochem Biophys Res Commun 146:7-12

17. Hara K, Suda T, Suda J, Eguchi M, Ihle JN, Nagata S, Miura Y, Saito M (1988) Bipotential murine hemopoietic cell line(NFS-60) that is responsive to IL-3, GM-CSF, G-CSF and erythropoietin. Exp Hematol 16:256-261

18. Lang RA, Metcalf D, Gough NM, Dunn AR, Gonda TJ (1985) Expression of a hemopoietic growth factor cDNA in a factor-dependent cell line results in autonomous growth and tumorigenicity. Cell 43:531-542

19. Yamada G, Kitamura Y, Sonoda H, Harada H, Taki S, Mulligan RC, Osawa H, Diamantstein T, Yokohama S, Taniguchi T (1987) Retroviral expression of the human IL-2 gene in a murine T-cell line results in cell growth autonomy and tumorigenicity. EMBO J 6:2705-2709

20. Laker C, Stocking C, Bergholz U, Hess N, De Lamarter JF, Ostertag W (1987) Autocrine stimulation after transfer of the granulocyte/macrophage colony-stimulating factor gene and autonomous growth are distinct but independent steps in the oncogenic pathway. Proc Natl Acad Sci USA 84: 8458-8462.

21. Jirik FR, Burstein SA, Treger L, Sorge A (1987) Transfection of a factor-dependent cell line with the murine interleukin-3(IL-3) c-DNA results in autonomous growth and tumorigenesis. Leukemia Research 11:1127-1134

22. Wong PMC, Chung S-W, Nienhuis AW (1987) Retroviral transfer and expression of the interleukin-3 gene in hemopoietic cells. Genes Devel 1:358-365

23. Roussel MF, Rettenmier CW, Sherr CJ (1988) Introduction of a human colony stimulating factor-1 gene into a mouse macrophage cell line induces CSF-1 independence but not tumorigenicity. Blood 71:1218-1225

24. Duhrsen U, Metcalf D (1988) A model system for leukemic transformation of immortalized hemopoietic cells in irradiated recipient. Leukemia 2:329-333

25. Naparstek E, Pierce J, Metcalf D, Shadduk R, Ihle J, Leder A, Sakakeeny MA, Wagner K, Falco J, FitzGerald TJ, Greenberger JS (1986) Induction of growth alteration in factor-dependent hematopoietic progenitor cell lines by cocultivation with irradiated bone marrow stromal cell lines. Blood 67:1395-1403

Part V. Regulation of Leukemogenesis

Chairperson: F.W. Ruscetti

18 Transforming Growth Factor β: A Negative Regulator of Normal and Leukemic Hematopoietic Progenitor Cell Growth

F.W. Ruscetti, G.K. Sing, L.R. Ellingsworth, and J.R. Keller

ABSTRACT

We have recently demonstrated that TGF-β1 and TGF-β2 are equipotent selective inhibitors of the growth and differenti- ation of murine and human hematopoietic progenitor cells. The proliferation of fresh unfractionated murine bone marrow by IL-3 and human bone marrow by IL-3 or GM-CSF was inhibited by TGF-β1 and TGF-β2 while the proliferation of murine bone marrow by GM-CSF and both murine and human marrow by G-CSF was not inhibited. Both mouse and human hematopoietic colony formation was differentially affected by TGF-β1. In particular, CFU-GM, CFU-GEMM, BFU-E and HPP-CFC, the most immature colonies, were inhibited by TGF-β1, whereas the more differentiated unipotent colonies CFU-G, CFU-M and CFU-E, were not affected. TGF-β1 inhibits IL-3-induced growth of murine leukemic cell lines within 24 hours. The cells were still viable and subsequent removal of the TGF- β1 results in the resumption of normal growth. Also, TGF-β1 directly inhibited the growth of isolated Thy-1-positive progenitor cells. TGF-β1 inhibited, in a dose dependent manner, the growth of factor-dependent NFS-60 cells in response to IL-3, GM-CSF, G-CSF, CSF-1, IL-4, or IL-6. TGF-β1 inhibited the growth of a variety of murine and human myeloid leukemias while erythroid and macrophage leukemias were insensitive. Some of these leukemic cells like HL-60 have little or no detectable TGF-β receptors on their cell surface, while other TGF-β resistant cells possess receptors. Thus, TGF-β may be an important negative regulator of normal and leukemic hematopoietic cell growth.

INTRODUCTION

TGF-β1 was originally identified in the supernatants of transformed fibro- blasts, by its ability to augment the anchorage independent growth of NRK 49F fibroblasts in soft agar. Since then, it has been recognized that TGF-β belongs to a family of polypeptide growth factors that regulates cell growth and function. These include: 1) two distinct forms of transforming growth factor beta (TGF-β), TGF-β1 and TGF-β2 (1,2,3), 2) inhibin, which suppresses follicle-stimulating hormone secretion, 3) Mullerian inhibitory substance, which causes regression of the Mullerian duct in the male embryo and 3) the predicted product of the Drosophilian decapentaplegic gene complex (2).

TGF-β1 has been purified to homogeneity from platelets, placenta, kidney and bone and is a 25-kd disulfide-linked homodimeric protein whose sequence is remarkably conserved with a single amino acid substitution between mouse and man (1). The cDNA sequences for this family of growth factors indicate that the mature proteins are synthesized from the carboxyl-terminal half of larger precursors, and share homologies to each other within this region and, in particular, show conservation in 7 out of 9 cysteines (1).

More recently, it has been shown to be multifunctional in that it enhances the proliferation of two mesenchymal cell types, osteoblasts, responsible for new bone formation, and Schwann cells, involved in the repair of peripheral nerves and synthesis of myelin (3). On the other hand, TGF-β1 has also been described as a potent inhibitor of proliferation of a variety of cell types in vitro including epithelial cells, embryonic fibroblasts, endothelial cells and T and B-lymphocytes (3). Thus, it has been postulated that TGF-β1 may act as an autocrine or paracrine negative regulator of cell growth.

TGF-β is also known to exert biological effects unrelated to cell proliferation, including the induction of extracellular matrix proteins such as collagen and fibronectin in human, rat, mouse and chicken fibroblasts (4), and the inhibition of the degradation of extracellular matrix proteins. Local injection of TGF-β1 in vivo results in a strong fibrotic response including activation of fibroblasts to produce collagen. Also, TGF-β1 accelerates the healing of incisional wounds in rats. Therefore, TGF-β1 has been proposed to play a role in inflammation and wound healing (4).

Immunohistochemical studies using antibodies to the N-terminus of TGF-β1 to localize the production of TGF-β1 showed that TGF-β1 is produced by cells in centers of active hematopoiesis including fetal liver and bone marrow, and active lymphopoiesis in the Hassall's corpuscles (5), suggesting a role for TGF-β in leucocyte differentiation. Therefore, experiments were designed to determine whether TGF-β1 might be involved in regulating normal hematopoietic cell growth and differentiation. In addition, the role of TGF-β in the autonomous growth of leukemic cell lines was evaluated.

For these studies we employed the bone marrow soft agar colony assay, which measures the clonal growth of hematopoietic progenitors and has historically been used to determine the factors and conditions required for normal hematopoietic cell growth and differentiation. This assay has facilitated the identification and cloning of the colony stimulating factors (CSFs) (6), which are polypeptide growth factors that enhance the proliferation and differentiation of hematopoietic cells in vitro.

RESULTS

EFFECT OF TGF-β ON NORMAL MURINE BONE MARROW PROGENITOR CELL GROWTH

The initial experiments to examine the effect of TGF-β1 on the growth of hematopoietic cells used 3H-thymidine proliferation assays with freshly aspirated murine bone marrow cells cultured in microtiter wells for 72 h in the presence of the various CSFs. For these studies, homogeneous TGF-β1 purified from bovine bone was used. In contrast to two recent reports (7,8)

stating TGF-β2 had no effect on hematopoiesis, our comparative study of the effects of TGF-β1 and TGF from two different sources showed little or no difference between the two. Thus, all the results presented here will have used TGF-β1. The CSFs employed were 3 (IL-3), which promotes the colony formation (CFU) of mixed lineage colonies that include granulocytes (G), macrophages (m), megakaryocytes (M), and erythroid cells (E), (CFU-GEMM), granulocyte/macrophage-CSF (GM-CSF), which promotes the growth of granulocytes and macrophages (CFU-GM), granulocyte-CSF (G-CSF), which promotes granulocyte growth, and erythropoietin (EPO), which promotes the terminal differentiation of erythroid cells.The results demonstrate that TGF-β1 alone has no effect on the growth of hematopoietic cells, but can inhibit between 50 to 70 percent of the IL-3-driven bone marrow proliferation in a dose dependent manner with an ED-50 of 5-10 pM. In contrast, G-CSF, GM-CSF and EPO driven proliferation is unaffected by TGF-β1. To determine what progenitors and which lineages may have been effected by TGF-β1, bone marrow cells were plated in soft agar in the presence of the CSFs. IL-3-induced colony formation was inhibited in a dose dependent manner with an ED50 of 5-10 pM, while G-CSF and GM-CSF driven colony formation were unaffected. Finally, even though the colony formation induced by IL-3 was significantly inhibited by TGF-β1, clusters of differentiated myeloid cells were consistently observed.

Since the inhibition of IL-3-driven proliferation was incomplete and ranged from 50 to 70, percent and since clusters of differentiated cells were observed in soft agar assays containing IL-3 and TGF-β1, we hypothesized that TGF-β1 might selectively regulate early hematopoietic progenitor cell growth. To test this hypothesis, we studied the effect of TGF-β1 on the growth of the most primitive CFU found in normal marrow populations, the multipotential CFU, CFU-GEMM. The CFU-GEMM develops in response to two signals, namely IL-3, which provides the early signal, and EPO which promotes the terminal differentiation of erythroid cells. In addition, we examined the primitive erythroid lineage colony, the burst forming unit (BFU-E), which also uses the same two growth signals, IL-3 and EPO. TGF-β1 is a potent inhibitor of CFU-GEMM and BFU-E formation; however, TGF-β1 had no effect on erythropoietin-driven proliferation of splenocytes from phenylhydrazine-treated mice (spleen cells are highly enriched for erythroblasts (9)) or erythroid colony formation, CFU-E, from normal bone marrow. Taken together, the results suggest that TGF-β1 selectively inhibits the growth of early progenitor cells, while the proliferation and differentiation of the more committed or differentiated progenitors are unaffected.

To determine if TGF-β1 could also regulate human hematopoietic cells, comparative studies were performed on freshly aspirated human bone marrow. In proliferation assays, TGF-β1 inhibited 50 to 70 percent of the IL-3- and GM-CSF-driven proliferation in a dose dependent manner with an ED50 of 5 to 15 pM. In contrast, G-CSF driven proliferation was unaffected. Similar results were observed in the colony assays, in that TGF-β1 inhibited GM-CSF and IL-3 induced CFU-GM, CFU-GEMM, and BFU-E progenitor cell growth, while G-CSF-induced colony formation was unaffected. In addition, similar to results seen with murine hematopoietic cells, clusters of terminally differentiated progeny of a single lineage were consistently seen in cultures containing TGF-β1 and GM-CSF or IL-3. However, in contrast to results with murine hematopoietic cells, human GM-CSF-induced colony formation was inhibited by TGF-β1. This may reflect differences in the two species, since human GM-CSF unlike murine GM-CSF is a potent inducer of the more primitive progenitors including CFU-GEMM (10). Therefore, these results are consistent with the concept that TGF-β1 selectively regulates early hematopoietic progenitor cells while the more differentiated and committed progenitors are unaffected.

EFFECT OF TGF-β1 ON PRIMITIVE MURINE PROGENITOR POPULATIONS

To determine whether the effect of TGF-β1 on hematopoietic cells was direct or indirect, pure populations of hemato-poietic progenitors were isolated from heterogeneous populations of murine bone marrow cells. This was accomplished using a previously described system which demonstrated that IL-3 was required for the induction of Thy-1 antigen expression on Thy-1-negative (Thy-1-) bone marrow cells. Single cell analysis of FACS sorted Thy-1+ cells has shown that these cells have the potential to differentiate to the various hematopoietic cell types, and contain progenitors with multipotent, bipotent and unipotent differentiation potential (11). Therefore, two aspects of this system were studied; first, what are the effects of TGF-β1 on IL-3-induced Thy-1 antigen expression, and second, does TGF-β1 inhibit the growth of isolated Thy-1+ progenitor cells in agar colony assays? TGF-β1 inhibited IL-3-induced Thy-1 antigen expression in a dose dependent manner with an ED_{50} of 5 to 10 pM. When these cultures were examined morphologically using a Jenner's

stain, there was an absence of proliferating myeloid blast cells normally present in cultures supplemented with IL-3. Next, to demonstrate the direct action of TGF-β1 on hematopoietic progenitor cell growth, Thy-1+ progenitors were isolated by fluorescent activated cell sorting (FACS) from bone marrow cultures grown for seven days in medium supplemented with IL-3, and then plated into soft agar colony assays. As previously demonstrated (12), IL-3 induces the formation of mixed lineage colonies that contain granulocytes, macrophages, and mast cells as well as pure monolineage colonies of each cell type. However, in addition to the previous results, Thy-1+ progenitors also give rise to CFU-GEMM in cultures supplemented with IL-3 and EPO. The results show that TGF-β1 inhibits greater than 50 per cent of the total colony formation which includes all of the multipotential progenitors, CFU-GEMM and CFU-GM, while colonies and clusters containing single lineages of granulocytes and macrophages are unaffected. TGF-β1 inhibits the induction of Thy-1 antigen on hematopoietic progenitors and can directly act on isolated Thy-1 progenitors to selectively inhibit multipotential cell growth and differentiation.

To determine at what stage in hematopoietic development progenitor cells acquire the ability to respond to TGF-β, bone marrow cells were obtained from the femurs of mice injected with 5-fluorouracil (5-FU), a compound that enriches for non-cycling primitive hematopoietic progenitors (13). One class of hematopoietic progenitor cell in post 5-FU bone marrow is the high proliferative potential colony forming cell (HPP-CFC) that proliferates in vitro to generate large macroscopic colonies (greater than 0.5mm). Therefore, we examined the effect of TGF-β1 on the growth of the three classes of HPP-CFC which have been defined. The first class of HPP, HPP-CFC-1, which can be induced to proliferate in response to interleukin [IL-1 (previously known as hematopoietin-1)] and CSF-1, represents the most primitive HPP progenitor capable of reconstituting the hematopoietic system (14). It can be inhibited by TGF-β1 in a dose dependent manner, with an ED-50 of 10-30 pm. The second class of HPP, HPP-CFC-2, which has been shown to be derived from HPP-CFC-1 (15) and is induced to proliferate by IL-3 and CSF-1, was also inhibited by TGF-β1 with similar dose dependent kinetics. Finally, a third class of HPP, HPP-CFC-HLGF-1, induced to proliferate from normal bone marrow cells in response to hemato-lymphopoietic growth factor (HLGF-1) and CSF-1 (16), was also inhibited. Thus, TGF-β1 can inhibit the growth of the most primitive hematopoietic progenitors detectable in vitro. Since the HPP progenitors are quiescent cells, the mechanism of negative regulation by TGF-β probably involves maintaining the progenitor cells in a non-cycling state.

EFFECT OF TGF-β1 ON LEUKEMIC MYELOID CELL GROWTH

Since the results obtained with normal hematopoietic progenitors suggest that TGF-β1 plays a role in the regulation of hematopoietic cell growth, experiments were designed to determine whether TGF-β1 also inhibits the growth of leukemic cell lines. Thus, we examined a variety of murine cell lines, including growth factor-dependent "preleukemic" progenitor cell lines blocked in early stages of differentiation as shown by phenotype, function and morphology and factor independent myeloid leukemic cells (17,18,19). TGF-β1 inhibits the proliferation of all the IL-3-dependent murine progenitor cell line tested to date, regardless of its derivation. In addition, the growth of NFS-60 cells, which proliferate in response to IL-3, GM-CSF, G-CSF, CSF-1, IL-6 and IL-4 was inhibited by TGF-β1 in a dose dependent manner with an ED50 of 5-10 pM regardless of the growth factor employed. TGF-β1 did not induce the differentiation of those lines that were inhibited. The inhibitory effects of TGF-β1 on the proliferation of factor dependent cell lines could be specifically neutralized with a polyclonal antiserum to TGF-β1 (5).

Examination of factor independent murine leukemic cell lines revealed that cell lines with an early blastic phenotype were inhibited by TGF-β but those with a more mature phenotype were not. WEHI-3B (myelomonocytic) and GG2EE (promonocytic) were inhibited, while the macrophage cell lines P388D1 and J774, the erythroid cell lines TP-3 and DS-19 and the mastocytoma P815 were insensitive to the effects of TGF-β1.

We next explored the question of whether factor dependence or independence plays a role in TGF-β sensitivity. To address this question, two IL-3 dependent cell lines, NFS-60 and 32D clone 23 were infected with oncogene containing retroviruses to abrogate their requirement for factor dependence (previously demonstrated 19). In separate experiments, retroviruses containing v-abl, v-src and v-fms were used to make factor independent variants of NFS-60 and 32D-cl23. In all cases, the factor independent cell lines obtained retained the ability to be inhibited by TGF-β suggesting that the response to TGF-β was not related to factor dependence.

TGF-β RECEPTOR DIFFERENCES ON HUMAN LEUKEMIC CELLS

A study of the effect of TGF-β on the growth of human myeloid lines revealed that all myelomonocytic cell lines except the promyelocytic HL-60 cell line were inhibited by TGF-β. In contrast to our data with myeloid leukemic cells, 4 out of 4 mature B-lymphoid leukemic cell lines and 4 out of 4 mature T-lymphoid leukemic cell lines were not inhibited by TGF-β while their normal counterparts were sensitive to TGF-β (20,21). To determine whether the differential response of these cell lines to TGF-β could be at the level of receptor expression, ligand binding and affinity labeling studies were performed. Three structurally distinct cell surface glycoproteins - a 280 Kd, an 85 Kd and a 65 Kd species - have been previously described as being high affinity TGF-β receptors (22). These receptor molecules were identified on all cell lines sensitive to TGF-β growth inhibition. When the TGF-β insensitive HL-60 cells were examined, they had no detectable TGF-β receptors. These results suggest that the loss of TGF-β receptor expression might play a role in the growth of some leukemias by allowing cells to escape negative growth regulation by TGF-β.

DISCUSSION

We have found that TGF-β is a potent selective inhibitor of early murine (23,24) and human (25) hematopoietic progenitor cell growth. It inhibits CFU-GEMM, BFU-E and CFU-GM colony formation but has no effect on CFU-G, CFU-E and CFU-M colony formation. Using purified IL-3 induced Thy-1 positive progenitors, we have shown that multipotent but not unipotent cell growth is directly inhibited by TGF-β (26). In addition, all three classes of the HPP progenitor cells, which are the earliest measurable cells in vitro, were inhibited by TGF-β. Since these cells are quiescent and have to be induced to enter the cell cycle in vitro by IL-1, it seems likely that TGF-β acts by interfering with the ability of cells to enter the cell cycle. Two forms of TGF-β, TGF-β1 and TGF-β2 were found to have identical effects in all the assays employed.

In summarizing the effects of TGF-β on leukemic cell growth, it appears that sensitivity to TGF-β is determined by the state of cellular differentiation and not the growth factor responsiveness of a particular cell type. For example, while

G-CSF and GM-CSF-induced normal bone marrow proliferation, and colony formation were not inhibited by TGF-β1, G-CSF and GM-CSF driven NFS-60 cell proliferation was inhibited. In addition, it was found that the ability to respond to TGF-β was not due to growth factor dependence or independence. The TGF-β1 resistant leukemic erythroid and macrophage cells may have been transformed at a stage in their maturation where they are insensitive to TGF-β1. On the other hand, the resistance to TGF-β inhibition of some leukemic T and B lymphocytes as well as HL-60 is due to the absence of detectable receptors for TGF-β. Thus, it appears that some leukemic cells can gain a growth advantage by escaping from the negative growth regulation of TGF-β.

The capacity to selectively inhibit early marrow progenitor cell as well as leukemic cell proliferation could have clinical applications. A variety of hematopoietic tumors may be responsive in vivo to TGF-β inhibition, which could be a useful adjunct to cytoreductive chemotherapy. In addition, if growth-arrested, quiescent marrow stem cells prove to be less sensitive to the effects of cycle active drugs, TGF-β could be used as a protective agent to ameliorate the dose-limiting toxicity of many chemotherapeutic agents.

REFERENCES

1. Sporn MB, Roberts AB, Wakefield LM, Assoian RK (1987) Transforming growth factor-β: Biological function and chemical structure. Science 232:532-534

2. Massague' J (1987) The TGF-β family of growth and differentiation factors. Cell 49:437-438

3. Sporn MB, Roberts AB (1988) Peptide growth factors are multifunctional. Nature 332:217-220

4. Roberts AB, Anzano MA, Wakefield LM, Roche NS, Stern DF and Sporn MB (1985) Type β transforming growth factor: A bifunctional regulator of cellular growth. Proc. Natl. Acad. Sci. USA 82:119-123

5. Ellingsworth, LR, Brennan JE, Fok, K, Rosen DM, Bentz H, Piez KA, Seyedin, SM (1986) Antibodies to the N-terminal portion of cartilage-inducing factor-A and transforming growth factor β. J. Biol. Chem. 261:12362-12371

6. Metcalf D (1985) The granulocyte-macrophage colony stimulating factors. Science 229:16-22

7. Ohta M, Greenberger JS, Anklesaria P, Bassols A, Massague' J (1987) Two forms of transforming growth-β distinguished by multipotential hematopoietic progenitor cells. Nature 329:539-541

8. Ottman OG, Pelus LM (1988) Differential proliferative effects of transforming growth factor-β on human hematopoietic progenitor cells. J. Immunol. 140:2661-2665

9. Ruscetti SK (1986) A 3H-thymidine incorporation assay to distinguish the effects of different friend erythroleukemia-inducing retroviruses on erythroid cell growth. J. Natl. Can. Inst. 77:241-245

10. Burgess AW, Metcalf D (1980) The nature and action of granulocyte macrophage colony stimulating factors. Blood 56:947-955

11. Keller JR, Weinstein Y, Hursey M, Ihle JN (1985) Interleukins 2 and 3 regulate the in vitro proliferation of two distinguishable populations of 20-α-hydroxysteroid dehydrogenase positive cells. J. Immunol. 135:1864-1871

12. Ihle JN, Keller J, Oroszlan S, Henderson LE, Copeland TD, Fitch F, Prystowsky MB, Goldwasser E, Schrader JW, Palaszynski E, Dy M, Lebel B (1983) The induction of hematopoietic colony formation from Thy-1 positive progenitors by IL-3. J. Immunol. 131:282-287

13. Hodgson GS, Bradley TR, Radley JM (1982) The organization of hemopoietic tissue as inferred from the effects of 5-fluorouracil. Exp. Hematol. 10:26-35

14. McNiece IK, Williams NT, Johnson GR, Kriegler AB, Bradley TR, Hodgson GS (1987) The generation of murine hematopoietic precursor cells from macrophage high proliferative potential colony forming cells. Exp. Hematol. 15:972-984

15. McNiece IK, Bradley TR, Kriegler AB, Hodgson GS (1986) Subpopulations of mouse high proliferative potential colony forming cells. Exp. Hematol. 14:856-865

16. Quesenberry P, Zenqxuan S, McGrath E, McNiece I, Shadduck R, Waheed A, Baber G, Kleeman E and Kaiser D (1987) Multilineage synergistic activity produced by a murine adherent marrow cell line. Blood 69:827-835

17. Greenberger JS, Eckner RJ, Sakakeeny M, Marks P, Reid D, Nabel G, Hapel A, Ihle JN, Humphries KC (1983) Interleukin-3 dependent hematopoietic progenitor cell lines. Fed. Proc. 42:2672-2681

18. Dexter TM, Garland J, Scott D, Scolnick E, Metcalf D (1980) Growth of factor-dependent hemopoietic precursor cell lines. J. Exp. Med. 152:1036-1049

19. Ihle JN, Morse HC, Keller J, Holmes K (1984) In Current Topics in Microbiology and Immunology. IL-3 dependent retrovirus induced lymphomas: Loss of the ability to terminally differentiate. New York Plenum Publishing Corp. 113:85-94

20. Kehrl JH, Wakefield LM, Roberts AB, Jakowlew S, Alvarez-Mon M, Derynck R, Sporn MB, Fauci AS (1986) Production of transforming growth factor-β by humans T-lymphocytes and its potential role in the regulation of T-cell growth. J. Exp. Med. 163:1037-1051

21. Kehrl JH, Roberts AB, Wakefield LM, Jakowlew S, Sporn MB, Fauci AS (1986) Transforming growth factor-β is an important immunomodulatory protein for human B lymphocytes. J. Immunol. 137:3855-3863

22. Cheifetz S, Weatherbee JA, Tsang ML-S, Andersen JK, Mole JE, Lucas R, Massague' J (1987) Complex patterns of TGF-β receptors. Cell 48:409-419

23. Ishibashi T, Miller SL, Burnstein SA (1987) Type β transforming growth factor is a potent inhibitor of murine megakaryocytopoiesis in vitro. Blood 69:1737-1741

24. Keller JR, Mantel C, Sing GK, Ellingsworth LR, Ruscetti SK, Ruscetti FW (1988) Transforming growth factor β1 selectively inhibits early murine hemato-poietic progenitors and inhibits the growth of IL-3 dependent myeloid leukemia cell lines. J. Exp. Med., in press

25. Sing GK, Keller JR, Ellingsworth LR, Ruscetti FW (1988) Transforming growth factor-β selectively inhibits human hematopoietic progenitor cells. Blood, in press.

26. Keller JR, Ellingsworth LR, McNiece Quesenberry PJ, Sing GK, Ruscetti, FW (1988) Transforming growth factor-β directly regulates primitive murine hematopoietic cell proliferation, submitted

19 The Regulation of Blast Stem Cell Self-Renewal in Myeloblastic Leukemia

E.A. McCulloch

ABSTRACT

The blast cells of Acute Myeloblastic Leukemia are considered as a novel lineage, composed of segments of normal differentiation programs assembled abnormally, reflecting abnormal gene expression. AML blast cells can be grown in culture; the earliest blast progenitors have been shown to have the stem cell capacity of self-renewal. This paper considers certain aspects of self-renewal regulation. First, based on drug survival curves measured in two culture techniques, the suggestion is made that the balance between self-renewal and differentiation is regulated by specific genes. Second evidence is presented that blast proliferative patterns are affected by the growth factors IL-3, GM-CSF, G-CSF and CSF-1. The most consistent pattern is seen with CSF-1, where cells show decreased self-renewal when exposed to CSF-1 added to cultures or to the products of the CSF-1 gene as detected by Northern analysis. Finally, the importance of self-renewal is emphasized. It is proposed that drugs might be classified by their effect on this crucial cell function; further, that the chemosensitivity of self-renewal is an attribute contributing to clinical outcome.

Introduction

The blast cells of acute myeloblastic leukemia (AML) are arranged as a cellular hierarchy; stem cells at the apex of the hierarchy maintain abnormal clones by their capacity for self-renewal and, through a differentiation-like step, produce terminally-dividing progeny[1,2]. It would be desirable to know whether the balance between renewal and determination is regulated and particularly if the balance can be affected by changing conditions either in culture or in vivo. The issue has theoretical importance because it is central to the behaviour of cell-renewal systems generally; it is of practical significance since therapy favóring determination might be beneficial by limiting division potential; in contrast, stimulation of renewal might be harmful.

Two complementary culture methods are available for the study of the balance between blast self renewal and determination. The first is a clonogenic assay[3]; blast stem cells form colonies consisting largely of proliferatively inert cells, although some self-renewal events can be detected by replating experiments[4]. Secondly blast stem cells increase in suspension cultures[5] as a consequence of self-renewal. In addition, non-dividing adherent cells are generated by some blast populations cultured in suspension[6]; thus, this assay provides evidence of both self-renewal and terminal divisions.

These techniques have yielded data that can be combined into a consistent operational model of the regulation of the balance between renewal and determination. It is the purpose of this paper to state the model, the data that supports it and some of the implications flowing from it.

The Model

The model is an amplification of the scheme proposed for AML blast cells generally[7]. Genetic evidence is available that leukemic transformation occurs in pluripotent stem cells[8]. In chronic myeloblastic leukemia (CML) differentiation programs lead to morphologically-normal and function end cells, although with a disturbed balance between lineages. In acute myeloblastic leukemia (AML), before determination, blast cells appear as a novel lineage. The earliest cells of this lineage have the stem property of self-renewal; they are the origins of programs assembled abnormally from normal components and ending in functionally ineffective morphologically abnormal blast cells (Lineage Infidelity[9]). We now propose that the balance between renewal and differentiation is regulated by mechanisms that include genetic factors determining the probability that cycles end in new stem cells or programmes leading to inert end cells. Other genes code for growth factors and their receptors and these may alter the probabilities of renewal or determination. Further we propose that the balance between self-renewal and determination is an important attribute contributing to outcome variation in AML and that the self-renewal process itself is the appropriate target for therapy. The experimental evidence follows.

Survival Curves

First, construction of drug survival curves using the two methods has shown three patterns [10,11]. For adriamycin both procedures yield curves with the same slope, expressed as the drug concentration required to reduce survival to 10% of control (D_{10})[12,13]. In contrast, cytosine arabinoside (ara-C) was more toxic when tested in suspension than in methylcellulose. The third pattern was observed with 5-azacytidine (5-aza) or 5-deoxyazacytidine; D_{10} values for these agents were greater when determined using the clonogenic assay in methylcellulose than when measured in suspension.

In each of experimental procedures, drugs act directly on stem cells, decreasing their growth capacity. The model contains the proposal that the differences observed between the two assays may be explained as a differential toxicity in suspension or in methylcellulose depending on the probability of either stem cell renewal or determination; thus, where conditions favor self-renewal (that is, in suspension) ara-C will be more toxic and 5-aza less toxic. In contrast, where terminal divisions are favored (that is, in methylcellulose), the reverse will be seen with 5-aza more toxic than ara-C. All three drugs act on DNA; the anthracyclines intercalate in DNA and may be cytotoxic by inhibiting preribosomal RNA synthesis; there is no known specificity for particular sequences in DNA[14]. In contrast ara-C is a cytidine analogue that is converted enzymatically to ara-CTP and then incorporated into DNA[15]. 5-aza is also a cytidine analogue, processed by different enzymes prior to incorporation into both RNA and DNA. It is considered to have functional specificity based on nitrogen substituted for carbon in the 5 position of the cytidine ring. This carbon is the site of methylation of DNA and and the 5-substituted base cannot accept methyl groups. Hypomethylation following 5-aza treatment has been shown to alter gene-expression[16] and to affect differentiation[17]. We have suggested that the observed effects of 5-aza on self renewal might be explained if it were integrated more efficiently into DNA bearing information related to self-renewal than into DNA generally[10]. By analogy a similar suggestion may be made for ara-C. These data and their proposed interpretations are the basis for the inclusion within the model of a genetic mechanism for the regulation of self renewal. The suggestion implies that specific genes exist whose activity determines the balance between self-renewal and determination.

Clinical Correlations

The model contains the proposal that self-renewal is both an important determinant of outcome and the proper target for treatment. Evidence for the first comes from studies of blast self renewal as estimated by determining the plating efficiencies of cell suspensions obtained from colonies derived from the blast progenitors of individual patients (PE2)[4]. PE2 varies greatly from patient-to-patient although it usually is stable with time in individuals[18]. We have observed regularly a statistically significant association between high PE2 values and poor outcome of treatment in AML[4,19,20,21]. Remission-induction clearly depends on aggressive chemotherapy; nonetheless, using the methylcellulose assay, D_{10} values for ara-C or adriamycin were not associated with response to chemotherapy regimens that included both agents[19]. In contrast, for both ara-C and 5-aza, a significant association was found between sensitivity in suspension and successful response to chemotherapy[12,22]. If, as postulated, the suspension assay detects self-renewal events, the association between drug sensitivity measured in suspension and successful outcome is consistent with the postulate in the model that self-renewal is the appropriate target for chemotherapy.

The patients used in these studies were treated with a single agent, high dose ara-C[21]; it was reasonable, therefore, to find that the ara-C sensitivity of the self-renewal function was an attribute contributing to successful therapy. No such argument can be made for the association between 5-aza sensitivity and response to treatment with ara-C. Rather, it might be postulated the the self-renewal function varies in its sensitivity to agents that are incorporated into DNA, a view consistent with the hypothesis that there are self-renewal specific genes. Further, that clones with "sensitive" renewal genes respond to chemotherapy, while those with "resistant" genes fail. The data support the suggestion that each AML patient has an intrinsically-determined response to commonly-used treatment regimens, provided these regimens contain active drugs. From this point of view, sensitivity in suspension is associated with response because it helps to identify those patients with chemotherapy sensitive self-renewal genes. Regardless, the observations are consistent with the model.

Effects of Growth Factors

Normal myelopoietic precursors respond to a variety of growth factors; these are usually glyoproteins secreted by endothelial cells, fibroblasts of continuous cancer cell lines[23]. Molecular clones have been obtained for many of these and recombinant factors are available[24]. Some recombinant factors, such as erythropoietin and CSF-1 are lineage specific, acting on late erythropoietic precursors or macrophage progenitors respectively. Others, such as GM-CSF and G-CSF were considered specific for granulocytes and/or macrophages; however, the availability of the recombinant products has shown that GM-CSF also acts at early stages of myelopoiesis, including erythropoiesis. The targets of IL-3 are considered to be very primitive cells, although activity on more mature elements has also been demonstrated.

IL-1, IL-4, IL-5 and IL-6 were first identified by their activities on lymphoid cells. Further, work with recombinant proteins has shown that they also have myelopoietic activity; IL-4 and IL-5 stimulate basophil and eosinophil precursors respectively; IL-1 and IL-6 are themselves inactive but potentiate the effects of others factors. IL-6 is synergistic with early acting factors that stimulate stem cells; IL-1, first described as hemopoietin-1, potentiates the response of macrophages to CSF-1 and increases secretion of G-CSF and GM-CSF. Features of these myelopoiesis-active factors, together with references, are summarized in Table 1.

Most AML blast populations depend for survival and growth in culture on the same factors that act on normal myelopoiesis. The methylcellulose and suspension assays have been used together to assess the effects of certain of the factors on blast cell renewal and determination. Variation from patient-to-patient was seen, and the factors had differing effects on the balance between self-renewal and determination. Usually, IL-3, GM-CSF and G-CSF[25,26,27,28] stimulated growth by increasing both self-renewal and post-deterministic divisions, although to varying degrees. There was a trend for IL-3 to stimulate renewal and G-CSF to favor determination. The most consistent observations were related to the effects of CSF-1; for sensitive populations, CSF-1 increased differentiation and often this was evident by the production of adherent cells incapable of further divisions[29].

Table 1

GROWTH FACTORS ACTIVE ON MYELOPOIESIS

NAME	TARGET	COMMENT	REF.
IL-3	EARLY CELLS	LIMITED ACTION LATE LYMPHOID ACTION	39,27, 40,41 42,43
GM-CSF	EARLY CELLS	OBVIOUS EFFECTS ON G/M ACTIVE THROUGHOUT LINEAGE	44,45 46,47, 48,49 50
G-CSF	GRANULOPOIETIC PRECURSORS	LINEAGE RESTRICTED	51,52, 53,54, 55,56, 57,28
CSF-1	MACROPHAGES MONOCYTES	LINEAGE RESTRICTED	58,59, 35,60, 61
ERYTHRO-POIETIN	ERYTHROPOIETIC PRECURSORS	LINEAGE RESTRICTED	62,63, 64,65
IL-1	SYNERGISTIC INDUCES FACTOR-PRODUCTION	MANY LINEAGES	66,67, 68,69, 70
IL-4	LYMPHOCYTES BASOPHILS	MYELOPOIETIC-LYMPHOID	71,72, 73
IL-5	LYMPHOCYTES EOSINOPHILS	MYELOPOIETIC-LYMPHOID	74
IL-6	MANY CELL CLASSES	VERY WIDE ACTIVITY SPECTRUM SYNERGISTIC	75,76

These data raise the issue of the role of growth factors in the regulation of blast cell growth; particularly are autocrine or paracrine mechanisms operative[30]? The genes for growth factors have been shown by Northern analysis to be expressed in blasts from some AML patients; however, secretion of active factors is much rarer than expression and often factor-expressing populations remain dependent on the factors that they express[31,32]. While not excluding autocrine mechanisms, these finding are not evidence that they are frequent or important.

CSF-1 is the most frequently expressed growth factor gene[33,34], and, as noted earlier, usually increased the probability of differentiation. In studies using growth of blast cells in suspension, CSF-1 expression positive clones were found to renew themselves significantly less well than expression negative populations[33]. Thus expression of CSF-1, like exposure to the hormone exogenously, had an inhibitory effect on blast growth. Although CSF-1 protein has yet to be found in most expression positive populations, studies of of the CSF-1 receptor, now known to be encoded by the fms proto-oncogene[35] are revealing. The distribution of fms expression among AML clones supports the view that the mechanism of inhibition by CSF-1 is similar for endogenous expression and exogenous exposure; as expected only fms expression positive clones responded to CSF-1. Most CSF-1 expression positive clones were also fms-expression positive. Regrettably, too few CSF-1 positive, fms negative examples were identified to test whether cells with this phenotype had different growth characteristics than the commoner fms positive type, although a trend towards better growth was seen in CSF-1 expression positive clones when fms was not expressed. Taken together, these data support the view that CSF-1 acts directly to inhibit blast growth by binding with its receptor, and that such interaction may take place intracellularly. Thus, the CSF-1 gene has an effect on the balance between renewal and differentiation. The mechanism linking CSF-1 gene expression and a shift towards differentiation remains conjectural. It is unlikely, however, that the CSF-1 gene has a direct regulatory effect on self-renewal; rather the effect may be secondary to intracellular signals for the activation of programmatic components that are normally associated with macrophage/monocyte differentiation.

Conclusion

Three major points related to a model of blast cell renewal are addressed in this paper. First, the suggestion is made that specific genes act to determine the probabilities of self-renewal and determination. Second that drugs can be identified that have specificity for these two central stem cell functions. Third, the probabilities of renewal and differentiation can be altered by growth factors. CSF-1, in particular, specifies a product that increase differentiation regardless of whether the factor is supplied exogenously or endogenously.

The issue of the regulation of self-renewal has always been central to stem cell physiology since it is this function that underlies the maintenance of independent cellular clones. The value of the postulates contained in the model is that suggestions are made for genetic control of processes that are often considered to be stochastic[36,37]. It follows that deliberate manipulations are feasible. These may be clinically important since a change in the balance towards increased differentiation may alter the aggressive behaviour of leukemic cells in vivo.

Growth factors are potentially useful for such manipulations. Factors have been shown to be active in vivo and have already been employed in some therapeutic settings (see references in Table 1). It is unlikely, however, that complete terminal differentiation can be induced often in blast populations, inspite of the observations of such effects in continuous cell lines[38]. Rather, within the abnormal blast lineage, determination might be increased relative to self-renewal; such a change might not be powerful enough in itself to serve as treatment. However, in combination, growth factors and chemothera-peutic drugs, the activities of each self-renewal and differentiation might be synergistic; the optimal combination of agents might be different for remission induction and remission maintenance. Regardless, the patient-to-patient variation observed in blast properties makes it likely that the properties of each clone should be determined in order to exploit their characteristic features in regimens designed for individual patients.

References

1. McCulloch EA, Till JE (1981) Blast cells in acute myeloblastic leukemia: A model. Blood Cells 7:63-77.

2. Griffin JD, Lowenberg B (1986) Clonogenic cells in acute myeloblastic leukemia. Blood 68:1185-1195.

3. Buick RN, Till JE, McCulloch EA (1977) Colony assay for proliferative blast cells circulating in myeloblastic leukaemia. Lancet 1:862-863.

4. Buick RN, Minden MD, McCulloch EA (1979) Self-renewal in culture of proliferative blast progenitor cells in acute myeloblastic leukemia. Blood 54:95-104.

5. Nara N, McCulloch EA (1985) The proliferation in suspension of the progenitors of the blast cells in acute myeloblastic leukemia. Blood 65: 1484-1493.

6. Langley GR, Smith LJ, McCulloch EA (1986) Adherent cells in cultures of blast progenitors in acute myeloblastic leukemia. Leukemia Res 10:953-959.

7. McCulloch EA (1984) The blast cells of acute myeloblastic leukemia. In: Clinics in Hematology. McCulloch EA: ed. Saunders, London, pp 503-515.

8. Fialkow PJ (1982) Cell lineages in hematopoietic neoplasia studied with glucose-6-phosphate dehydrogenase cell markers. In: J Cell Physiol (supplement 1). J. Cell Physiol Alan Liss. New York.

9. McCulloch EA (1987) lineage infidelity or lineage promiscuity? Leukemia 1: 235.

10. Motoji T, Hoang T, Tritchler D, McCulloch EA (1985) The effect of 5- azacytidine and its analogues on blast cell renewal in acute myeloblastic leukemia. Blood 65:894-901.

11. Wang C, McCulloch EA (1987) The sensitivity to 5-azacytidine of blast progenitors in acute myeloblastic leukemia. Blood 69:553-559.

12. Nara N, Curtis JE, Senn JS, Tritchler DL, McCulloch EA (1986) The sensitivity to cytosine arabinoside of the blast progenitors of acute myeloblastic leukemia. Blood 67:762-769.

13. Nara N, Yamashita Y, Murohashi I, Tanikawa S, Imai Y, Aoki N (1987) The effects on leukemic clonogenic cells in murine myeloid leukemia of 1-beta-D-arabinofuranosylcytosine and the anthracyclines adriamycin, daunomycin, aclacinomycin and 4'-epidoxorubicin. Cancer Research 47:2376-2379.

14. Myers CE (1982) Anthracyclines. In: Pharmacological Principles of Cancer Treatment. Chabner B: ed. W.B. Saunders Co., Philadelphia, pp 416-434.

15. Chabner BA (1982) Cytosine Arabinoside. In: Pharmocological Principle of Cancer Treatment. Chabmner BA: ed. W.B. Saunders, Philadephia, pp 387-401.

16. Riggs AD, PA Jones (1983) 5-methylcytosine, gene regulation, and cancer. Adv Cancer Res 40:1-25.

17. Taylor SM, Jones PA (1979) Multiple new phenotypes induced in 10 1/2 and 3T3 cells treated with 5-azacytidine. Cell 7:771-779.

18. McCulloch EA, Buick RN, Curtis JE, Messner HA, Senn JS (1981) The heritable nature of clonal characteristics in acute myeloblastic leukemia. Blood 58:105-109.

19. McCulloch EA, Curtis JE, Messner HA, Senn JS, Germanson TP (1982) The contribution of blast cell properties to outcome variation in acute myeloblastic leukemia (AML). Blood 59:601-608.

20. Curtis JE, Messner HA, Hasselback R, Elhakim TM, McCulloch EA (1984) Contributions of host- and disease-related attributes to the outcome of patients with acute myelogeneous leukemia (AML). J Clin Oncol 2:253-259.

21. Curtis JE, Messner HA, Minden MD, Minkin S, McCulloch EA (1987) High dose cytosine arabinoside in the treatment of acute myeloblastic leukemia: Contributions to outcome of clinical and laboratory attributes. J. Clin. Oncol. 5:532-543.

22. Wang C, Curtis JE, Senn JS, Tritchler DL, McCulloch EA (1987) Response to 5-azacytidine of leukemic blast cells in suspension: a biological parameter associated with response to chemotherapy. Leukemia 1:753-756.

23. Metcalf D (1984) The hemopoietic colony-stimulating factors. Elsvier. Amsterdam.

24. Clark SC, Kamen R (1987) The human hematopoietic colony-stimulating factors. Science 236:1229-1237.

25. Kelleher C, Miyauchi J, Wong G, Clark S, Minden MD, McCulloch EA (1987) Synergism between recombinant growth factors, GM-CSF and G-CSF, acting on the blast cells of acute myeloblastic leukemia. Blood 69:1498-1503.

26. Griffin JD, Young D, Herrman F, Wiper D, Wagner K, Sabbath DK (1986) Effects of recombinant GM-CSF on the proliferation of clonogenic cells in acute myeloblastic leukemia. Blood 67:1448-1453.

27. Delwel R, Dorssers L, Touw I, Wagemaker ER, Lowenberg B (1987) Human recombinant multilineage colony stimulating factor

(Interleukin-3): stimulator of acute myelocytic leukemia progenitor cells in vitro. Blood 70: 333-336.

28. Nara N, Murohashi I, Suzuki T, Yamashita Y, Maruyama Y, Aoki N, Tanikawa S (1987) Effects of recombinant human granulocyte colony-stimulating factor (G-CSF) on blast progenitors from acute myeloblastic leukaemia patients. Br J Cancer 56:49-51.

29. Miyauchi J, Wang C, Kelleher CA, Wong GG, Clark SC, Minden MD, McCulloch EA (1988) The effects of recombinant CSF-1 on the blast cells of acute myeloblastic leukemia in suspension culture. J Cell Physiol 135:55-62.

30. Sporn MB, Roberts AB (1985) Autocrine growth factors and cancer. Nature 313:745-747.

31. Cheng GYM, Kelleher CA, Miyauchi J, Wang C, Wong G, Clark S, McCulloch EA, Minden MD (1987) Structure and expression of genes of GM-CSF and G-CSF in blast cells from patients with Acute Myeloblastic Leukemia. Blood 71:204-208.

32. Young DC, Wagner K, Griffin JD (1987) Constitutive expression of the granulocyte-macrophage colony-stimulating factor gene in acute meyloblastic leukemia. J Clin Invest 79:100-106.

33. Wang C, Kelleher CA, Cheng GYM, Miyauchi J, Wong GG, Clark SC, Minden MD, McCulloch EA (1988) Expression of the CSF-1 Gene in the Blast Cells of Acute Myeloblastic Leukemia: Association with Reduced Growth Capacity. J Cell Physiol 135:133-138.

34. Young DC, Demetri GD, Ernst TJ, Cannistra SA, Griffin JD (1988) In vitro expression of colony-stimulating factor genes in human acute myeloblastic leukemia cells. Exp Hematol 16:378-382.

35. Sherr CJ, Rettenmier CW, Sacca R, Roussel MF, Look AT, Stanley ER (1985) The c-fms proto-oncogene product is related to the receptor for the mononuclear phagocyte growth factor, CSF-1. Cell 41:665-676.

36. Till JE, McCulloch EA, Siminovitch L (1964) A stochastic model of stem cell proliferation, based on the growth of spleen colony forming cells. Proc Natl Acad Sci (USA) 51:29-36.

37. Nakahata T, Gross AJ, Ogawa M (1982) A stochastic model of self-renewal and committment to differentiation of the primitive hemopoietic stem cells in culture. J Cell Physiol 113:455-458.

38. Sachs L (1978) Control of normal cell differentiation in the phenotypic reversion of malignancy in myeloid leukemia. Nature 274:535-539.

39. Fung MC, Hapel AJ, Ymer S, Cohen DR, Johnson RM, Campbell HD, Young IG (1984) Molecular cloning of cDNA for murine interleukin-3. Nature 307:233-237.

40. Donahue RE, Seehra J, Norton C, Turner K, Metzger M, Rock, B, Carbone S, Seghal, R, Yang YC, Clark SC (1987) Stimulation of hematopoiesis in primates with human interleukin-3 and granulocyte-macrophage colony stimulating factor. Blood 70:133a (abstr).

41. Messner HA, Yamasaki K, Jamal N, Minden MD, Wong GG, Clark SC (1987) Growth of human hemopoietic colonies in response to recombinant gibbon interleukin 3: Commparison with human recombinant granulocyte-macrophage colony-stimulating factor(GM-CSF). Proc Natl Acad Sci USA 84:1-5.

42. Yang Y-C, Ciarletta AB, Temple PA, Chung MP, Kovacic S, Witek-Gianotti JS, Leary AC, Kritz R, Donahue RE, Wong G-G, Clark SC (1986) Human IL- 3(Multi-CSF): Identification by expression cloning of a novel hemopoietic growth factor related to murine IL-3. Cell 47:3-10.

43. Mufson RA, Gesner TG, Turner K, Norton C, Yang Y-C, Clark S (1987) Characterization of IL-3 receptors on human acute myelogenous leukemia cell line KG-1. Blood 70:suppl 1, 118a.

44. Gough NM, Gough J, Metcalf D, Kelso A, Grail D, Nicola N, Burgess AW, Dunn AR (1984) Molecular cloning of cDNA encoding a murine hematopoietic growth regulator, granulocyte-macrophage colony stimulating factor. Nature 309:763-767.

45. Metcalf D (1986) The molecular biology and functions of the granulocyte-macrophage colony-stimulating factors. Blood 67:257-267.

46. Miyatake S, Otsuka T, Yokota T, Lee F, Arai K (1985) Structure of the chromosomal gene for granulocyte-macrophage colony-stimulating factor: comparison of the mouse and human genes. Embo J 4:2561-2568.

47. Sieff CA, Emerson SG, Donahue RE, Nathan DG, Wang EA, Wong GG, Clark SC (1985) Human recombinant granulocyte-macrophage colony-stimulating Factor: A Multilineage Hematoprotein. Science 230:1171.

48. Burgess AW, Begley, GC, Johnson GR, Lopez AF, Williamson DJ, Mermod JJ, Simpson RJ, Schmitz A, DeLmarter JF (1987) Purification and properties of bacterially synthesized human granulocyte-macrophage colony stimulating factor. Blood 69:43-51.

49. Groopman JE, Mitysyasu RT, Deleo MJ, Oette DH, Golde DW (1987) Effect of recombinant human granulocyte-macrophage colony-stimulating factor on myelopoiesis in the acquired immunodeficiency syndrome. N Eng J Med 317:593-598.

50. Vadhan-Raj S, Keating M, LeMaistre, A, Hittleman WN, McCredie K, Trujillo JM,

Broxmeyer HE, Henney C, Gutterman JU (1987)
Effects of recombinant human
granulocyte-macrophage colony-stimulating
factor in patients with myelodysplastic
syndromes. New Engl. J. Med. 25:1545-1552.

51. Souza LM, Boone, TC, Gabrilove J, Lai PH,
Zsebo KM, Murdock DC, Chazin VR, Bruszewski
J, Lu H, Chen KK, Barendt J, Platzer E, Moore
MAS, Mertelsmann R, Welte, K (1986)
Recombinant human granulocyte colony-
stimulating factor: effects on normal and
leukemic myeloid cells. Science 232:61-65.

52. Nagata S, Tsuchiya M, Asano S, Kaziro Y,
Yamazaki T, Yamamoto O, Hirata Y, Kubota N,
Oheda M, Nomura H, Ono M (1986) Molecular
cloning and expression of cDNA for human
granulocyte colony colony-stimulating factor. Nature
319:415-417.

53. Nagata S, Tsuchiya T, Asano S, Yamamoto O,
Hirata Y, Kubota N, Yasmazaki T (1986) The
chromosome gene structure and two mRNAs for
human granulocyte colony-stimulating factor.
Embo J 5:575-581.

54. Nomura H, Imazeki I, Oheda M, Kubota N,
Tamura M, Ono M, Ueyama Y, Asano S (1986)
Purification and characterization of human
granulocyte colony stimulating factor
(G-CSF). EMBO J 5:871.

55. Shimamura M, Kobayashi Y, Yuo A, Urabe A,
Okabe T, Komatsu Y, Itoh S, Takaku F (1987)
Effect of human recombinant granulocyte
colony-stimulating factor on hemopoietic
injury in mice induced by 5-fluorouracil.
Blood 69:353-355.

56. Welte K, Bonilla MA, Gillo AP, Boone TC,
Potter GK, Gabrilove JL, Moore MAS, O'Reilly
RJ, Souza LM (1987) Recombinant human
granulocyte colony-stimulating factor:
effects on hematopoiesis in normal and
cyclophosphamide-treated primates. J Exp Med
165:941-948.

57. Yuo A, Kitagawa S, Okabe T, Urabe A,
Komatsu Y, Itoh S, Takaku F (1987)
Recombinant human granulocyte
colony-stimulating factor repairs the
abnormalities of neutrophils in patients with
myelodysplastic syndromes and chronic
myelogenous leukemia. Blood 70:404-411.

58. Stanley ER, Guilbert LJ, Tushinski RJ,
Bartelmez SH (1983) CSF-1: a mononuclear
phagocyte lineage-specific hemopoietic growth
factor. J Cell Biochem 21:151-159.

59. Das SK, Stanley ER (1982) Structure-funtion
studies of a colony stimulating factor
(CSF-1). J Biol Chem 257:13679-13684.

60. Wong GG, Temple PA, Leary A,
Witek-Giannotti JS, Yang Y, Ciarletta AB,
Chung M, Murtha P, Kritz R, Kaufman RJ,
Ferenz CR, Sibley BS, Turner KJ, Hewick RM,
Clark SC, Yanai N, Yokota H, Yamada M (1987)
Human CSF-1: Molecular cloning and expression
of 4-kb cDNA encoding the human urinary

protein. Science 235:1504-1508.

61. Rambaldi A, Young DC, Griffin JD (1987)
Expression of the M-CSF (CSF-1) gene by human
monocytes. Blood 69:1409-1413.

62. Goldwasser E (1976) Erythropoietin and the
differentiation of red blood cells. Fed Proc
34:2285-2292.

63. Jacobs K, Shoemaker C, Rudersdorph R,
Neil SD, Kaufman RJ, Mufson A, Seehra J,
Jones SS, Hewick R, Fritsh,EF, Kawakita M,
Shimizu,T, Miyake T (1985) Isolation and
characterization of genomic and cDNA clones
of human erythropoietin. Nature 313:806-810.

64. Eschbach JW, Egrie JC, Downing MR, Browne JK,
Adamson JW (1987) Correcton of the anemia of
end-stage renal disease with recombinant
human erythropoietin. N Engl J Med
316:73-78.

65. Lin FK, Suggu S, Lin CH, Browne JK,
Smalling R, Egrie JC, Chen KK, Fox GM,
Martin F, Stabinsky Z, Badrawi SM, Lai PH,
Goldwasser E (1985) Cloning and expression
of the human erythropoietin gene. Proc Natl
Acad Sci (USA) 82:7580.

66. Mochizuki DY, Eisenman JR, Conlon PJ,
Larsen AD, Tushinski RJ (1987) Interleukin-1
regulates hematopoietic activity, a role
previously ascribed to hemopoietin-1. Proc
Natl Acad Sci.84:5267-5271.

67. Bird TA, Skalatvala J (1986) Identification
of a common class of high affinity receptors
for both types of interleukin-1 on connective
tissue cells. Nature 324:263-266.

68. Dower SK, Kronheim SR, Hopp TP, Cantrell M,
Deely M, Gillis S, Henney CS, Urdal Dl
(1986) The cell surface receptors for
Interleukin-1 alpha and interleukin-1 beta
are identical. Nature 324:266-268.

69. Zucali JR, Dinarello CA, Oblon DJ, Gross MA,
Anderson L, Weiner RS (1986) Interleukin 1
stimulates fibroblasts to produce
granulocyte-macrophage colony stimulating
activity (GM-CSA) and prostoglandin E2
(PGE2). J Clin Invest 77:1837.

70. Broudy VC, Kaushansky K, Harlan JM,
Adamson JW (1987) Interleukin-1 stimulates
human endothelial cells to produce
granulocyte-macrophage colony-stimulating
factor and granulocyte colony-stimulating
factor. J Immunol 139:464-468.

71. Lee F, Yokota, T, Otsuka T, Meyerson P,
Villaret D, Coffman R, Mosman T, Rennick D,
Roehm N, Smith C, Zlotnick A, Arai K-I
(1986) Isolation and characterization of a
mouse interleukin cDNA clone that expresses
BSF-1 activities and T cell and mast cell
sttimulating activities. Proc Natl Acad
Sci (USA) 83:2061-2065.

72. O'Garra A, Warren DJ, Holman M, Popham AM,
Sanderson CJ, Klaus GBB (1986) Interleukin 4

(B-cell growth factorII/eosinophil differentiation factor) is a mitogen and differentiation factor for preactivated murine B lymphocytes. Proc Natl Acad Sci (USA) 83:5228-5232.

73. Rennick D, Yang G, Muller-Sieburg C, Smith C, Arai N, Takabe Y, Gemmell L (1987) Interleukin 4 (B-cell stimulatory factor 1) can enhance or anatagonize the factor dependent growth of hemopoietic progenitor cells. Proc Natl Acad Sci USA 84:6889-6893.

74. Campbell HD, Tucker WQJ, Hort Y, Martinson ME, Mayo G, Clutterbuck EJ, Sanderson CJ, Young IG (1987) Molecular cloning, nucleotide sequence, and expression of the gene encoding human eosinophil differentiation factor (interleukin 5). Proc Natl Acad Sci USA 84:6629-6633.

75. Ikebuchi K, Wong GG, Clark SC, Ihle JN, Hirai Y, Ogawa M (1987) Interleukin 6 enhancement of interleukin 3-dependent proliferation of multipotential hemopoietic progenitors. Proc Natl Acad Sci USA 84:9035-9039.

76. Garman RD, Jacobs KA, Clark SC, Raulet DH (1987) B-cell-stimulatory factor 2 (beta 2 interferon) functions as a second signal for interleukin 2 production by mature murine T cells. Proc Natl Acad Sci USA 84:7629-7633.

20 Role of Hemopoietic Growth Factors inn the Proliferation of Acute Myeloid Leukemia

P. Mannoni, M. Lopez, N. Maroc, C. Fay, H. Torres, D. Razanajaona, A. Tabilio, and F. Birg

Abstract : Selective growth advantage of the leukemic clone over normal populations can either result from growth factor independency or, on the opposite, from an increased cell response to haemopoietic growth factors. In order to define the relationship between growth factors and leukemic cell proliferation, we studied the growth patterns of 44 Acute Myeloid Leukemia (AML) cell samples in vitro, with or without exogenous growth factors. As spontaneous proliferation was observed in 64% of the cases, growth factor gene expression and secretion of the factor in the culture medium were studied in some of these specimens. Primary AML cells were found to express several growth factor genes : IL-1α and ß, GM-CSF, CSF -1, and sometimes G-CSF and IL-6. Protein excretion in the culture supernatant was documented for GM-CSF, CSF-1 and IL-1α using specific functional assays. AML cells were shown to respond by an induction or an amplification of their proliferation to the addition of human recombinant IL-3, GM-CSF and G-CSF. The existence of an autocrine loop involving GM-CSF gene expression, excretion of the protein and inhibition by specific antibodies was demonstrated in some AML with high indexes of in vitro proliferation. These results suggest that haemopoietic growth factors play a major role in the proliferation of AML cells, since these cells are able to synthesize and to respond to the major CSF or Interleukins involved in normal haemopoiesis.

Key Words : AML, CSF and Interleukins, Autocriny.

The occurence of a leukemia is the result of several biological events resulting in the development of a clonal population with a selective growth advantage over normal tissues. However, even though leukemic cells have acquired the capacity to overgrow and sometimes to inhibit normal haemopoiesis, growth factors are still necessary, at least in vitro, to support their survival and their growth (7, 11, 16, 25, 31, 32, 37, 39, 42). A complex network of growth factors, coupled to cellular interactions, is involved in the control of haemopoiesis (1, 6, 10, 24, 36). It has been postulated that interruption of this hormonal regulation leads to the autonomous growth of some haematopoietic progenitors, responsible for the occurence and the progression of the tumor clone.

Oncogene activation, abnormal response to growth factors and, more recently, autocrine mechanisms have been sometimes recognized in such leukemic cells. However none of these factors, by itself, is sufficient to explain cell transformation resulting in the occurence of the tumor. A multistep carcinogenesis theory has been developed, which received experimental support from transfection experiments with cloned oncogenes, on one side, and the development of transgenic mouse strains, on the other side. For example, in gene transfer experiments utilizing retroviral vectors, expression of some oncogenes and growth autonomy in haematopoietic cells have been documented. On the opposite, loss of factor dependency and acquisition of tumorigenicity have been observed without detectable factor production in the same kind of models (18, 45). Moreover, it has been found that only a certain percentage of transgenic mice expressing human proto-oncogenes will develop tumors and that human GM-CSF transgenic mice will develop a systemic disease without malignancy (19). Looking at human tumors or leukemic cells, both genomic amplification of proto-oncogenes and increased expression of growth factor genes have been reported. In AML cells, activation of N-ras by specific point mutations has been reported in about 25% of the cases (3). In the same pathology, autonomous in vitro growth, release of growth factors by leukemic cells, and expression at the RNA level of genes coding for GM-CSF, CSF-1 and IL-1 have been observed (5, 12, 46 ,47). As regards growth factor dependency, contradictory observations have been reported, e-g spontaneous in vitro growth of leukemic cells and on the contrary, requirement on CSF for cell survival and growth. Facing these apparently discordant results, we undertook to define the frequency of autonomous growth and growth factor requirement for survival, growth and differentiation of primary leukemia cells, isolated from patients with AML. Involvement of known growth factor genes was analyzed at the transcriptional and protein synthesis levels. Examination of the results shows that haemopoietic growth factors play a fundamental role, resulting in a selective growth advantage of the leukemic clone over the normal population. In some cases, growth autonomy resulting from an autocrine process involving GM-CSF was clearly demonstrated.

Materials and Methods

Leukemic Cells : AML patients with high leucocyte counts (> 50,000/mm^3) were selected on presentation of the disease. After informed consent, blood samples were obtained before any chemotherapy was started. Pure (>95 % by morphological and immunological criteria) populations of leukemic cells were isolated from peripheral blood by gradient centrifugation. After isolation, leukemia cells were frozen at -80 °C in RPMI medium containing 10 % (vol/vol) dimethylsulfoxide (DMSO) and 15 % (vol/vol) fetal calf serum (FCS , Flow Laboratories). Cells were then transferred to liquid nitrogen until further use. After freezing and thawing, viability assessed by trypan blue exclusion was generally greater than 80%.
Leukemia diagnosis and FAB classification (2) were assessed by the usual morphological and cytochemical criteria. Cytogenetic analysis was performed on all the specimens. Immunological phenotype using a panel of monoclonal antibodies was also determined, and compared to the morphological classification. In most cases, a good correlation between these different approaches was found, leading to the precise classification of each leukemia tested.
As controls, leukemic cells from acute lymphoid leukemias (ALL), from either T or B origin, were selected.

Proliferation of leukemic cells : Survival and proliferation were studied by cell count and incorporation of tritiated thymidine ([^{3}H] Td). In all cases, populations of leukemic cells were depleted of adherent cells by a 4-h adherence on plastic at 37°C in Iscove Modified Dulbecco Medium (IMDM) supplemented with 10 % FCS. It must be pointed out that for these studies, proliferation of leukemic cells was also assayed on random samples before and after removal of T lymphocytes by E-rosetting. The results obtained never showed any statistically significant difference (p< 0.05, using Student's T-test).
Purified leukemic cells were seeded in 96-flat-bottomed-well culture trays (Flow). Cells (10^5 / well) were allowed to grow in 200 µl IMDM supplemented with 10 % FCS, with or without CSF. Cultures were maintained over a period of 7 days at 37°C in a humidified incubator in air supplemented with 5 % CO$_2$. At days 1, 3, 5, and 7, or as otherwise stated, cells were pulsed for 4 h with 18.5 kBq of [^{3}H] Td (specific activity, 2 TBq/m mole; Amersham, UK), harvested, and counted in a liquid scintillation counter. Results were expressed as counts per minute (cpm), and stimulation index (SI) calculated as follows: SI = cpm of the test sample minus cpm of the control, divided by cpm of the control sample.

Production of Leukemia Conditioned Medium (L-CM) :
Living leukemic cells as determined by trypan blue exclusion were seeded at a concentration of 10^6/ml in 25 ml culture flasks, in IMDM supplemented with glutamine and 10 % FCS. Supernatants were harvested at days 1 and 3, aliquoted, heat-inactivated at 56°C for 5 min, filtered through 0.45 µm and frozen at -20°C until use.

Northern blot analysis of mRNA expression : Leukemic cells (1 x 10^6/ml) were cultured overnight (18 h) in IMDM + 10 % FCS in 25 ml culture flasks. Cycloheximide (Sigma) (10 µg/ml) was added to half of the flasks for 3 hours, in order to stabilize the mRNAs

(38). After 21 hours, all the cells were harvested, washed, and lyzed in 4,25 M guanidium isothiocyanate containing 0,1 M mercaptoethanol as described (22). Lyzed cells were layered over a cushion of cesium chloride, and centrifuged for 18 h at 30.000 g. The RNA pellets were dried and precipitated with ethanol. RNA concentration was determined by measuring the OD at 260 nm. Absence of RNA degradation was checked by electrophoresis in 1% agarose gels containing methylmercuric hydroxyde, followed by ethidium bromide staining as described (22). 25 µg aliquots of total RNA were electrophoresed in 1.2 % agarose gels containing 10 % formaldehyde, transferred to a Nylon sheet (Nytran, Schleicher and Schüll). Membranes were prehybridized for 4 hours at 42°C in SSPE buffer containing 50 % deionized formamide, 1% SDS, 1 x Denhardt's and 200 µg/ml denatured calf thymus DNA. They were hybridized for 24-48 h in the same buffer containing 10^7 cpm/ml of [^{32}P] labeled (9) probe. They were then washed and exposed to X ray films with intensifying screens.

Southern Blot Analysis : Total cellular DNA was prepared from leukemia cells as described (8). 10 µg aliquots were digested with the appropriate restriction enzymes, electrophoresed in agarose gels and transferred to nitrocellulose sheets. Prehybridization and hybridization were performed at 68°C in 6 x SSC buffer containing 0.1 % SDS, 1 x Denhardt's and 50 µg/ml denatured calf thymus DNA (22).

DNA probes : All the probes for CSF were human cDNA clones, from which the specific DNA segments were excised by digestion with the appropriate restriction enzymes, and purified by gel electrophoresis. The GM-CSF probe was a Dra I-Pst I fragment from recombinant plasmid pCDx (obtained from Dr N. Araï, DNAX, Palo Alto, USA, through Dr. J. Banchereau, Unicet, France). The G-CSF probe was a 700 bp Pst I fragment from a cDNA clone given by Dr L. Pelicci (Perugia,Italy). The CSF-1 probe was a 1800 bp Eco RI-Xho I fragment of the recombinant plasmid p3ACSFR1 (14), a gift from Dr. S. Clark (Genetics Institute, Cambridge, USA). The IL-3 probe was a 717 bp Xho I-Hpa I fragment excised from plasmid pCD-SR (30), a gift of Dr N. Araï. The IL-1α probe was a 1700 bp Hind III fragment excised from plasmid pGEM-IL-1, also provided by Dr N.Araï. The IL-1β probe was a 800 bp Pst I fragment excised from the IL-1-SP6 vector provided by Dr C. Vaquero (Paris, France), like the IL-6 probe, a 1200 bp Eco RI fragment excised from the T3T7-IL-6 recombinant plasmid.
Control hybridizations were performed with either a complete cDNA for the rat GAPDH (obtained from Dr J.M. Blanchard, Montpellier, France) or with a mouse β actin cDNA clone (obtained from Dr. Buckingham, Paris, France). All the probes were radioctively labeled by random priming (9) to specific activities of 3 to 4 x 10^8 cpm/µg.

Detection of GM-CSF Activity in leukemia cell culture supernatants : Identification of a GM-CSF like activity (GM-CSF.A) in supernatants harvested at days 1 and 3 was performed by two different assays : induction of CFU-GM from normal bone marrow cells and activation of neutrophils in presence of the N-formyl L- Methionyl -L-Leucyl-L-Phenylalanine (fMLP).
Clonogenic assays were carried out in triplicate using the plasma clot technique (32). Normal bone marrow cells were obtained from allogeneic bone marrow

donors after informed consent, and separated by gradient centrifugation on a Ficoll Hypaque gradient. Removal of T cells and monocytes was achieved by E rosetting and adherence on plastic, respectively.1 x 10^5 cells were plated in 25 mm culture dishes (NUNC) in the following medium : IMDM supplemented with transferrin, glutamin and sodium bicarbonate and containing 10 % FCS, 10 % deionized bovine serum albumin (Fraction V, Sigma), 10 % citrated bovine plasma (Gibco), 20 mg/ml L-asparagine (Sigma) and 34 mg/ml calcium chloride (Sigma). Sources of GM-CSF were conditioned media prepared from 5637 or GCT cell lines, human recombinant GM-CSF (DNAX) and supernatants of leukemic cell cultures. Medium alone and supernatants of ALL cell cultures were used as negative controls. 5637 and GCT conditioned media, as well as L-CM were used at a final concentration of 10 %. Colonies with > 20 cells were counted using an inverted microscope at days 7 and 14.

Activation of mature neutrophils by GM-CSF was assessed by cytochrome C reduction according to the technique of Weisbart et al (43). Briefly, normal granulocytes were isolated from samples of peripheral blood by centrifugation over a discontinuous Percoll gradient (Pharmacia). After washing, 2 x 10^6 granulocytes were incubated for 90 mn at 37°C with 100 units hr-GM-CSF or 10% L-CM. Cytochrome C (Sigma) and fMLP ($5 x 10^{-9}$ M) (Sigma) were added to the test samples. Cytochrome C reduction was determined by measuring the OD at 550 nm, by comparaison with a reference sample in which cytochrome C reduction was inhibited by addition of superoxyde dismutase (Sigma). The specificity of this assay was confirmed by inhibition with anti-GM-CSF antibodies and by testing other recombinant CSF and lymphokines. Results were expressed as the ratio of nmole O_2^- released in the presence of leukemia conditioned medium to the nmole O_2^- released in control medium. Ratio obtained with hr-GM-CSF were used to estimate GM-CSF activity in L-CM.

Detection of IL-1α activity in leukemia cell culture supernatants: IL-1α activity was demonstrated by induction of proliferation of the mouse IL-1 dependent cell line D10-G4 (a gift of M. Pierrès, Marseille,France) as described (15). In brief, suboptimal concentrations of PHA (< 1 mg/ml) were selected and L-CM were tested with and without PHA. Responses were considered as positive when [3H] Td uptake was 1.5 fold higher than the control (PHA alone). Positive controls were hr-IL-1α (DNAX) or conditioned medium from the 5637 cell line known to contain IL-1 α (1).

CSF-1 Activity in leukemia cell culture supernatants :
In the same supernatants, CSF-1 activity was assessed using the MM5 murine cell line, responsive to human CSF-1 (L. Guilbert, Edmonton, Canada). The proliferation was determined by measuring the reduction of tetrazolium blue after 2 days in culture in absence or presence of purified CSF-1, medium alone or L-CM. An increase in absorbance at 640 nm greater than 3 fold the negative control was considered as a positive result.

*Control cells :*Human tumor cell lines known to produce GM-CSF as well as other growth factors were used as positive controls for both GM-CSF gene expression and production of the factor. The myelomonocytic cell line HL 60, lymphoid leukemia cells and various other cell lines were used as potential negative controls.

Results :
Spontaneous proliferation :

Using a liquid culture system in microwell plates, and measuring proliferation by [3H] Td uptake, we observed autonomous growth of leukemic cells in 28 of the 44 (64%) cases tested (Table I). The threshold value for spontaneous proliferation was determined as 1500 cpm by comparison with values obtained for ALL, which do not grow. No spontaneous proliferation was observed in the group of acute promyelocytic leukemia (M3), even when higher cell concentrations were used (data not shown). To sum up, in proliferative leukemia cells, two distinct

TABLE 1
Spontaneous in vitro AML cell proliferation
3 (H)Td uptake

Patients :		Day 1	3	5	7
M1	VE	95	102	104	48
	011	67	89	251	9
	GR	247	958	322	417
	MIL	662	82	37	112
	016	154	88	124	561
	035	1983	639	283	99
	042	4623	6644	3591	3491
	083	3708	1683	1746	1126
	197	224	1880	979	471
	106	396	769	495	2041
	121	1613	2916	2903	2010
M2	021	331	1271	430	643
	079	111	53	39	43
	049	1827	5804	4892	2530
	050	8489	28335	29302	7739
	064	336	4531	5328	7962
	069	279	7863	17361	13293
	070	1434	17459	17673	14842
	074	443	2785	1401	821
	092	63	1225	2009	417
M3	LA	1271	1213	400	500
	IT	132	49	32	23
	GR	708	500	120	143
	MO	675	99	47	23
	029	1263	442	397	297
	108	257	520	140	25
M4	JO	152	125	50	38
	014	269	2055	3881	9690
	018	836	1695	2084	1831
	028	1888	2822	2248	6418
	037	523	8894	13523	29016
	081	2247	11375	11755	13855
	105	1134	1903	3619	1685
	160	2345	1157	3545	7546
M5	007	376	3497	6901	5635
	045	8440	10668	14780	12008
	061	5476	22071	18567	24320
	062	9082	37862	34822	37019
	090	290	1704	2161	5125
	099	1219	26305	22196	17805
	126	649	2866	2512	2393
	159	359	477	1797	3386

Values in boxes correspond to spontaneous growth.

patterns were observed : either a transient proliferation, lasting less than 7 days, and mostly observed in the M1 and M2 forms, or a continuous growth, with a high level of proliferation still detectable at day 7, mainly observed for the M4 and M5 forms (Table II).

TABLE II

SPONTANEOUS GROWTH OF AML CELLS :

FAB /n=	ABSENT	TRANSIENT	CONTINUOUS	% SPONT. GROWTH
M1/12	7	5	0	42
M2/9	2	3	4	78
M3/6	6	0	0	0
M4/8	1	2	5	88
M5/9	0	1	8	100
TOT.44	16 (36%)	11	17	
		64%		

Responses of leukemia cells to exogenous growth factors :

This panel of 44 leukemias was tested for a proliferative response after addition of the following hr-CSF: IL-3, GM-CSF, and G-CSF (Table III). Strikingly, the leukemia cells of the M3 type were all induced to proliferate by all three factors. In contrast, amplification of the spontaneous growth of the M4 and M5 leukemic cells was generally weak, due to the high level of spontaneous growth. Taken together, these results clearly demonstrate that 60% of AML cells respond to IL-3, 84% to GM-CSF, and 25% to G-CSF, either by an induction or an amplification of proliferation. We observed in previous studies that higher indexes of proliferation were obtained when conditioned medium from either of several tumor cell lines like 5637, Mo, and GCT was added to the culture.

TABLE III

Proliferative response of AML-cells to human recombinant CSF .

Induction of proliferation by :

FAB	n=	IL 3	GM-CSF	G-CSF
M1	12	11	10	2
M2	9	6	7	2
M3	6	6	6	6
M4	8	1*	6*	0
M5	9	3*	8*	1*
TOT.	44	27(61%)	37(84%)	11(25%)

* weak induction 1 ‹ SI ‹ 2

These cells are known to produce more than one factor: thus 5637 cells synthesize IL-1α, GM-CSF and G-CSF, like the GCT and Mo cells. To confirm the possible existence of synergism between different factors, cultures were set up with mixtures of hr-CSF. Synergism or additive effects were observed when 2 factors (IL-3/G-CSF, G-CSF/GM-CSF, IL-3/G-CSF) were added together, in most responsive leukemias (data not shown).

Synthesis of growth factors by leukemic cells in culture :

GM-CSF Activity: GM-CSF.A was measured by comparing the number of CFU-GM colonies obtained with L-CM, to the number obtained when 5637 conditioned medium was used as a source of CSF (TableIV). Supernatants of AML cultures were collected at days 1 and 3 in 21 cases. When normal bone marrow cells were tested for CFU-GM formation, a GM-CSF activity (GM-CSF A) was detectable in 11 out

TABLE IV

AML cell supernatants (Days 1 and 3) : CSF-Activities

	GM-CSF. A		IL1. A	CSF.1A
	CFU.GM*	Cyt C Red˙	D10.G4ᵗ	MM.5
M1				
35	nt	nt	3	+
54	73	3.1	3.4	nt
83	110	1.5	1.8	+++
121	55	1.3	2.1	nt
M2				
49	nt	nt	1.8	nt
70	nt	0.8	1.8	nt
92	72	1.1	nt	+++
M3				
124	0	1	nt	nt
108	0	1	0.5	nt
M4				
28	123	2	2	+++
37	109	0.9	3.5	+
160	nt	2	nt	nt
M5				
62	83	1.6	4.4	+++
88	75	1.5	1.9	nt
99	120	2.4	2.8	+++
103	0	1	1.8	nt
126	80	2	2	+++
159	nt	1.6	nt	nt
33	81	nt	4.5	+++
ALL				
n=2	0	1	‹1	–
hr GM	–	2.2	‹1	–
5637 CM	100	–	2,7	
Control	0	1	‹1	–

*CFU-GM = Percentage of CFU-GM colonies induced by L_CM. Comparison with 5637 conditioned medium.

˙Cyt C Reduction : Ratio of test value/control value.Positive results are considered for values › 1.3

ᵗIL.1 Assay : SI = Test value divided by control value (PHA).

of the 14 samples tested. Since it is known that CFU-GM can be induced not only by addition of GM-CSF, but also by IL-3 and G-CSF, confirmation of GM-CSF.A was obtained in most cases by measuring activation of mature granulocytes.

In presence of fMLP, L-CM were able to co-activate granulocytes in a specific manner. Except for leukemias 092 (M2) and 037 (M2), a good correlation was obtained between the CFU-GM and the cytochrome C reduction assays, thus demonstrating that these leukemias synthesize GM-CSF. Moreover, inhibition of cytochrome C reduction was observed when antibodies directed against GM-CSF were added to the reaction medium (data not shown).

IL-1 Activity: IL-1α activity, as detected by the D10-G4 assay, was present in most of the L-CM (Table IV). Results were considered as positive when Stimulation Index was superior or equal to 1.5. This activity was easily detectable at days 1 and 3, and sometimes at day 5. Except for case 99, IL-1 activity was found in supernatants of cells expressing IL-1 mRNA.

CSF-1 Activity: All of the supernatants tested were found to contain CSF-1.A as detected by the MM5 assay. Strong activity was detectable in 1 M1 (83), 1 M2 (92), 1 M4 (28) and 4 M5 (33, 62, 99, 126). In most cases CSF-1.A was detected in leukemias expressing CSF-1 RNA.

Expression of RNA specific for CSF and IL genes :

Total RNA was prepared from a series of 18 AML and 6 control ALL samples, all cultured for 21 hours with or without cycloheximide as described under Materials and Methods, and analyzed by Northern blotting and hybridization with specific probes. The results obtained are summarized in Table V.

The 2.2 kb IL-1α transcript was detected in almost all the AML samples (15/18). IL-1β was also observed in the same cases. GM-CSF RNA was detected in 14 out of the 18 cases, whereas G-CSF RNA was only detected in 9 samples. Among these 14 cases, GM-CSF mRNA was detected without the addition of cycloheximide in 6 cases only, thus demonstrating the instability (38) of the GM-CSF transcript in AML cells. In contrast, the IL-3 gene was not expressed in any of the samples, whereas the probe revealed the specific transcript in a T cell clone used as a positive control. Some AMLs expressed the IL-6 transcript.

Expression of CSF genes specific for the myeloid lineage, i.e., IL-3, GM- and G-CSF, was not found in any of the 6 control ALL samples, whereas IL-1β was detected in 4 such cases (Table V). In addition, we have reported previously that AML cells synthesize the transcript for the CSF-1 receptor(8), identified to the product of the c-*fms* proto-oncogene . Table V summarizes these results, and shows that all the AML cells tested synthesizing c-*fms* also express the CSF-1 gene, though to variable levels.

In conclusion, and except for IL-3, AML cells were able to express one or more CSF genes, the most frequently expressed being IL-1s, GM-CSF and CSF-1.

Inhibition of autonomous proliferation by antibodies directed against GM-CSF.

When a polyclonal antibody against GM-CSF (a gift of Dr S. Clark, Genetics Institute), was added to spontaneously growing leukemia cells, inhibitions of proliferation ranging from 27% to 78% were observed (Table VI). In all cases except one (92)

TABLE V

IL and CSF mRNA in AML cells

	IL-1α	IL-1β	IL3	GM	G	CSF.1	IL.6
M1:							
FA	+	+	-	+	-	+	+
54	+	+	-	+	+	+	+
83	+	+	-	+	+	+	-
121	+	NT	NT	+	NT	+	-
M2:							
70	-	NT	-	-	-	-	NT
92	+	NT	-	-	-	+	-
M3:							
124	-	NT	-	-	-	-	-
M4:							
28	+	NT	NT	+	+	+	+
37	+	NT	-	+	+	+	+
105	+	+	-	+	-	-	+
160	+	+	-	+	+	+	+
M5:							
62	+	+	-	+	+	+	+
88	+	NT	-	+	NT	+	NT
99	-	NT	-	+	+	+	-
103	+	NT	-	+	NT	+	NT
126	+	+	-	+	+	+	+
159	+	NT	NT	-	-	-	+
AB*	+	NT	-	+	+	+	+
ALL:							
48	-	+	-	-	-	-	-
91	-	+	-	-	-	NT	NT
101	-	NT	-	-	-	NT	-
140	-	-	-	-	-	NT	NT
161	-	+	NT	-	-	-	-
168	-	+	NT	-	-	-	-

*:Biphenotypic leukemia (Lymphoïd + Myeloïd)

TABLE VI

Inhibition of "autocrine" proliferation of AML cells by anti GM-CSF antibody.

Cases :	GM-CSF RNA	Spontaneous Growth*	% of inhibition : by Rabbit Ab
92 M2	-	+	55
28 M2	+	+ +	46
37 M4	+	+	78
62 M5	+	+ +	62
126 M5	+	+	74
99 M5	+	+ +	27
HL.60	-	+ + +	0

spontaneous growth and antibody inhibition were clearly associated with RNA expression and GM-CSF activity.

DISCUSSION

Cancer cells are likely not subject to the same exogeneous growth controls than their normal counterparts. It is quite clear that in some instances, the unregulated growth of tumor cells is the result of signals generated by the binding of growth factors to their receptors. Moreover it is generally accepted that paracrine mechanisms represent the most common mechanisms by which proliferative cells are put under the control of other cells able to regulate the proliferation and perhaps the differentiation of the progenitor cells. The development of the concept of autocrine secretion provided a new explanation for the ability of tumor cells to escape external controls by growth and/or differentiation factors. Some examples of growth factor secretion by tumor cells, or autocrine loops, have been reported. These studies reported the autocrine secretion of TGF α, TGF β, PDGF, and bombesin by some cell lines and human tumors (23, 40). More recently, the involvement of GM-CSF, IL-1α and β, and CSF-1 has been reported in myeloid leukemias (5, 12, 34, 46, 47). Results reported here not only confirm the previously published data, but also show that growth factor secretion, and response to growth factors is a common feature of AML cells. The first evidence comes from the data on spontaneous in vitro proliferation and response to the addition of CSF. Taken together, these data show that most if not all AML cells depend on the presence of growth factor(s) for survival and proliferation. Indeed, in leukemias with autonomous growth, which seem to proliferate in a growth factor independent way, synthesis of GM-CSF, IL-1α and CSF-1 is a common finding. As IL-1α was recently identified to the pluripotent hematopoietin H1 (26), these factors are all involved in the control of myelopoiesis. In leukemias without autonomous proliferation, high proliferative responses to the addition of GM-CSF, IL-3 and, to a lesser extent, G-CSF, were observed, thus confirming the presence of specific receptors on the surface of leukemic cells, as already reported in a few cases (17, 27).

The cases reported in this study were selected on the basis of high leukemic cell numbers allowing both to culture them and to obtain large amounts of RNA. In addition, we tried to increase the number of M3 specimen, in order to confirm the specific pattern observed in this group (32).

Our study also demonstrates clearly that expression of genes encoding CSFs and growth factors involved in the myelomonocytic differentiation is frequently observed. IL-1 expression was observed in almost all the cases, and GM-CSF and CSF-1 in over 50% of the cases tested and sometimes in association. It has been shown that CSF and interleukins can interact with each other in a complex network involving synergism, (1, 20, 37) transmodulation of receptors (44) or induction of gene expression and protein synthesis (10, 13). Thus IL-1 was shown not only to synergize with CSF-1 but also to induce adherent cells isolated from the bone marrow to produce GM-CSF (4, 12), and IL-3 and GM-CSF were reported to synergize not only on normal haemopoietic cells but also on leukemic cells (25). Therefore the synthesis and excretion of several growth factors by the leukemic clone is likely to result in a growth advantage involving both autocrine and paracrine mechanisms. Thus leukemic cells appear to be loaded with growth and/or activation factors able to either elicit a response of the leukemic clone via an autocrine loop or to trigger CSF synthesis by accessory cells. An autocrine loop involving GM-CSF was demonstrated in this study and confirmed by the growth inhibition induced by antibody against GM-CSF. Moreover, considering the constant expression of the c-fms proto-oncogene, whose product appears identical to CSF-1 receptor, in all AML cells, as we reported

previously (8), it can be postulated that an autocrine loop involving CSF-1 synthesis and CSF-1 receptor expression is present in some cases of AML. Moreover, the CSF-1, c-fms and GM-CSF genes are all located on the long arm of chromosome 5 (5q) and deleted in the 5q- syndrome (21, 28). Interestingly, this gene cluster appears to be activated quite frequently in AMLs. Although the IL-3 gene has been located in the same region, it is noteworthy to point out that IL-3 transcripts were never detected in myeloid leukemias cells. As IL-3 appears synthesized specifically by T cells, it is likely not synthesized by cells of the myeloid lineage, even leukemic.

The specificity of such growth factor gene expression in terms of leukemic versus normal cells remains to be established. From a classical point of view, CSF are synthesized by accessory cells (monocytes, macrophages, endothelial cells and fibroblasts) and not by the haemopoietic cells themselves. Expression of CSF-1, c-fms, and GM-CSF genes in leukemias with monocytic differentiation (M4 and M5), can be interpreted as differentiation markers, since these genes could be activated in normal monocytes and macrophages (33, 34, 35). However, the same pattern of expression was also observed in undifferentiated leukemias (M1). The only exception is made of the M3 group, in which gene expression, CSF activity and spontaneous proliferation were not observed. However these cells were shown to respond strongly to the addition of growth factors (32). Following this concept, growth factor gene expression and protein secretion appear quite specific of the leukemic state. Even though CSF RNA were not detected in normal myeloid cells (29), it must be demonstrated that autocrine mechanisms cannot be part, even temporarily, of the complex process of normal haemopoiesis.

Thus myeloid leukemia cells display a specific pattern, including gene expression, protein synthesis and biological response to haemopoietic growth factors. A recent report demonstrated that murine CSF-dependent cell lines can become CSF-independent, following minute rearrangements affecting the GM-CSF gene, leading to constitutive expression (41). In our study, no major gene rearrangements were observed in the AML DNAs studied. However, Chang et al reported rare examples of GM- and G-CSF genes rearrangements in AML cells (5). More sophisticated techniques will have to be used, in order to identify potential DNA alterations.

These data strongly support a role for Haemopoietic Growth Factors in the expansion of the leukemic clone without differentiation, and open possibilities for new therapeutical approaches aiming at the control of tumor proliferation either by antagonists, analogs or cytokines capable to regulate over-expression of growth factor genes and/or response to growth factor(s) by the leukemic clone.

ACKNOWLEDGEMENTS

We thank Drs M.J. Pébusque, P. Dubreuil, I. De Vries, R. De Waal for discussion throughout this work.
M.L. was supported by a Fellowship from the Ligue Nationale Française contre le Cancer (L.N.F.C.C.). This work was supported by INSERM and by grants of the L.N.F.C.C, A.R.C and Fondation contre la Leucémie.

REFERENCES

1. Bartelmez SH, Stanley R : Synergism between hemopoietic growth factors detected by their effects on cells bearing receptors for a lineage specific HGF : Assay of hemopoietin-1. J. cell. Physiol., 122, 370, 1985.

2. Bennet JM, Cathoushy D, Daniel MT, Flandrin G, Galton GA, Granlinick HR, Sultan C. Proposals for the classification of the acute leukemias. Br J Hematol., 33, 451, 1976.

3. Bos JL, Verlaan-de-Vries M, van der Eb AJ, Janssen JWG, Delwel R, Löwenberg B, Colly LP: Mutations in N-ras predominate in acute myeloid leukemia. Blood, 69, 1237, 1987.

4. Broudy VC, Kaushansky K, Harlan JM, Adamson JW : Interleukin-1 stimulates human endothelial cells to produce granulocyte-macrophage and granulocyte colony-stimulating factor. J. Immunol., 139, 464,1987.

5. Cheng GYM, Kelleher CA, Miyauchi J, Wang C, Wong G, Clark SC, Mc Culloch EA, Minden MD : Structure and expression of genes of GM-CSF in blast cells from patients with acute myeloblastic leukemia. Blood, 71, 204, 1988.

6. Clark SC, Kamen R: The Human Hematopoietic Colony-stimulating Factors. Science, 236,1229, 1987.

7. Dewel R, Dorsser L, Touw I, Wagemaker G, Löwenberg B : Human recombinant IL-3 stimulator of acute myelocytic leukemia progenitor cells in vitro. Blood, 70, 333, 1987.

8. Dubreuil P, Torrès H, Courcoul MA, Birg F, Mannoni P: c-fms expression is a molecular marker of human acute myeloid leukemias. Blood, 1988 (in press).

9. Feinberg AP, Volgelstein B: A technique for radiolabeling DNA restriction endonuclease fragments to high specific activity. Anal. Biochem, 137, 266, 1984.

10. Fibbe WE, Damme J, Biliau A, Voogt PJ, Duinkerken N, Kluck PMC, Falkenburg JHF : IL-1 induces release of GM-CSF activity from human mononuclear phagocytes. Blood, 68, 1316, 1986.

11. Griffin JD, Young F, Herrmann D, Wiper K, Wagner, Sabbath KD: Effects of recombinant human GM-CSF on proliferation of clonogenic cells in Acute Myeloblastic Leukemia. Blood, 67, 1448, 1986.

12. Griffin JC, Rambaldi A, Vellenga E, Young DC, Ostapoviez D, Cannistra SA : Secretion of Interleukin - 1 by Acute Myeloblastic Leukemia cells in vitro induces endothelial cells to secrete Colony Stimulating factors. Blood, 70, 1218, 1987.

13. Horiguchi J, Warren MK, Kufe D : Expression of the macrophage specific colony-stimulating factor in human monocytes treated with granulocyte-macrophage colony stimulating factor. Blood, 69, 1259, 1987.

14. Kawasaki ES, Ladner MB, Wang AM, Arsdell JV, Warren MK, Coyne MY, Schweickart VL, Lee MT, Wilson KJ, Boosman A, Stanley ER, Ralph P, Mark DF : Molecular cloning of a complementary DNA encoding human macrophage-specific colony-stimulating factor (CSF-1) Science, 230, 291, 1985.

15. Kaye JS, Gillis S, Mizel EM, Shevach TR, Malek C, Dinarello L, Lachmann B, Janeway CA: Growth of a cloned helper T cell line induced by a monoclonal antibody specific for the antigen receptor : Interleukin 1 is required for the expression of receptors for interleukin 2. J. Immunol. 133, 1399, 1984.

16. Kelleher C, Miyauchi J, Wong G, Clark S, Miden MD, Mc Culloch EA : Synergism between recombinant growth factors, GM-CSF, acting on the blasts cells of acute myeloblastic leukemia. Blood, 69, 1498, 1987.

17. Kelleher CA, Wong GG, Clark SC, Schendel PF, Minden MD, Mc Culloch EA : Binding of iodinated recombinant human GM-CSF to the blast cells of acute myeloblastic leukemia. Leukemia, 2, 211, 1988.

18. Lang RA, Metcalf D, Gough NM, Dunn AR, Gonda TJ : Expression of a hemopoietic growth factor cDNA in a factor dependent cell line results in autonomous growth and tumorigenicity. Cell, 43, 531, 1985.

19. Lang RA, Metcalf D, Cuthbertson RA, Lyons I, Stanley E, Kelso A, Kannourakis G, Williamson DJ, Klintworth GK, Gonda TJ, Dun AR : Transgenic mice expressing a hemopoietic growth factor gene (GM-CSF) develop accumulation of macrophages, blindness, and a fatal syndrome of tissue damage. Cell, 51, 675, 1987.

20. Leary AG, Ikebuchi K, Hirai Y, Wong GG, Yang YC, Clark SC, Ogawa M : Synergism between IL-6 and IL-3 in supporting proliferation of human hematopoietic stem cells : Comparison with IL-1 . Blood, 71, 1759, 1988.

21. Le Beau MM, Westbrook CA, Diaz MO, Larson RA, Rowley JD, Gasson JC, Golde DW, Sherr CJ : Evidence for the involvement of GM-CSF and FMS in the deletion (5q) in myeloid disorders. Science, 231, 984, 1986.

22. Maniatis T, Fritsch EF, Sambrook, J. Molecular cloning. A laboratory manual. Cold Spring Harbor, NY. Cold Spring Harbor Laboratory, 1982.

23. Mano J, Nishida K, Usuki Y, Kobayashi H, Hirai T, Okabe A, Urabe , Takaku. F: Constitutive expression of the Granulocyte-Macrophage Colony Stimulating Factor gene in human solid tumors. Jpn. Cancer Res. 78, 1041, 1987.

24. Metcalf D: The Molecular Biology and Function of the Granulocyte-Macrophage Colony Stimulating Factors. Blood, 67, 2, 1986.

25. Miyauchi J, Kelleher CA, Yang YC, Wong GG, Clark SC, Minden MD, Minkin S, Mc Culloch EA : The effects of three recombinant growth factors, IL-3, GM-CSF, and G-CSF, on the blast cells of acute myeloblastic leukemia maintained in short-term suspension culture. Blood, 70, 657, 1987.

26. Mochizuki DY, Eisenman JR, Conlon PJ, Larsen AD, Tuchinski RJ : IL-1 regulates hematopoietic activity, a role previously ascribed to hemopoietin 1. Proc. Natl. Acad. Sci. USA, 84, 5267, 1987.

27. Nicola NA, Begley CG, Metcalf D : Identification of

the human analogue of a regulator that induces differentiation in murine leukaemic cells. Nature, 314, 625, 1985.

28. Nienhuis AW, Bunn HF, Turner PH, Gopal TV, Nash WG, O'Brien SJ, Sherr C : Expression of the human c-fms proto-oncogene in hematopoietic cells and its deletion in the 5q- syndrome. Cell, 42, 421, 1985.

29. Nimer SD, Chan J, Slamon DJ, Golde DW, Gasson JD : Expression of human granulocyte-macrophage colony stimulating factor (GM-CSF). Blood, 66, 158 a, 1985.

30. Otsuka T, Miyajima A, Brown N, Otsu K, Abrams J, Saeland S, Caux C, De Malefijt R, De Vries J, Meyerson P, Kyoko Y, Gemmel L, Rennick D, Lee F, Arai N, Arai KI, Yokota T : Isolation and characterization of an expressible cDNA encoding human IL-3. J. Immunol, 140, 2288, 1988.

31. Pebusque MJ, Lopez M, Torres H, Carotti A, Guilbert L, Mannoni P: Growth response of human myeloid leukemia cells to colony-stimulating factors. Exp. Hematol., 16, 360, 1988.

32. Pebusque MJ Lafage M, Lopez M, Mannoni P: Preferential response of acute myeloid leukemias with translocation involving chromosome 17 to human recombinant G-CSF. Blood, 72, 257, 1988.

33. Rambaldi A, Young DC, Griffin JD : Expression of the CSF-1 gene by human monocytes. Blood, 69, 1409, 1987.

34. Rambaldi A, Wakamiya N, Vellenga E, Horiguchi J, Warren MK, Kufe D, Griffin JD : Expression of the macrophage colony-stimulating factor and c-fms genes in human AML cells. J. clin. Invest, 81, 1030, 1988.

35. Sariban E, Mitchell T, Kufe D : Expression of the c-fms proto-oncogene during human monocytic differentiation. Nature, 316, 64, 1985.

36. Sieff CA, Emerson SG, Donahue RE, Nathan DG, Wang EA, Wong GG, Clark SC : Human recombinant Granulocyte-Macrophage Colony-Stimulating Factor : A Multilineage Hematopoeitin. Science 230, 1171, 1986.

37. Sealand S, Caux C, Fabre C, Audry JP, Mannoni P, Pebusque MJ, Gentilhomme O, Otsuka T, Yokota T, Araï N, Araï K, Banchereau J, de Vries JE: Effects of recombinant human interleukin-3 on CD34 -enriched normal hemopoeitic progenitors and on myeloblastic leukemia cells. Blood, 1988 (in press).

38. Shaw G, Kamen R : A conserved AU sequence from the 3' untranslated region of GM-CSF mRNA mediates selective m RNA degradation . Cell, 46, 659, 1986.

39. Souza LM, Boone TC, Gabrilove J, Lai PH, Zsebo KM, Murdock DC, Chazin VR, Bruszewski, Lu H, Chen KK, Barendt J, Platzer F, Moore MAS, Mertelsmann R, Welte K : Recombinant human granulocyte colony-stimulating factor : Effects on normal and leukemic myeloid cells. Science, 232, 61, 1986.

40. Sporn MB, Roberts AB : Autocrine, paracrine and endocrine mechanisms of growth control. Cancer Surveys, 4, 627, 1985.

41. Stocking C, Löliger C, Kawai M, Suciu S, Gough N, Ostertag W: Identification of genes involved in growth autonomy of hematopoietic cells by analysis of factor-independent mutants. Cell, 53, 869, 1988.

42. Vellenga E, Young DC, Wagner K, Wiper Donald, Ostapovicz D, Griffin JD : The effects of GM-CSF in promoting growth of clonogenic cells in acute myeloblastic leukemia. Blood, 69, 1771, 1987.

43. Weisbart RH, Golde DW, Clark SC, Wong GG, Gasson JC : Human GM-CSF is a neutrophil activator. Nature, 314, 361, 1985.

44. Walker F, Nicola NA, Metcalf D, Burgess AW: Hierarchical down-modulation of hemopoietic growth factor receptors. Cell, 43, 269, 1985.

45. Ymer S, Tucker QJ, Sanderson CJ, Hapel AJ, Campbell HD, Young IG : Constitutive synthesis of interleukin-3 by leukemia cell line WEHI-3B is due to retroviral insertion near the gene. Nature, 317, 255, 1985.

46. Young DC, Griffin JD: Autocrine Secretion of GM-CSF in Acute Myeloblastic Leukemia. Blood, 68, 1178, 1986.

47. Young DC, Wagner K, Griffin JD : Constitutive expression of the granulocyte-macrophage colony-stimulating factor gene in acute myeloblastic leukemia. J. Clin. Invest, 79, 100, 1987.

21 Alterations in Dependence on Lipoxygenase Metabolism in Myeloid Leukemia

A.M. Miller, K. Cullen, S.M. Kobb, and R.S. Weiner

Products of the lipoxygenase pathway of arachidonic acid metabolism are essential for the normal proliferation and differentiation of hematopoietic cells [1-4]. Leukotriene (LT) C_4 and D_4 have been demonstrated to be essential for the colony stimulating factor (CSF) induced growth of the granulocyte-monocyte progenitor cell (CFU-GM) in semi-solid culture. Since the initial report that an intact lipoxygenase pathway was necessary for normal CSF induced myeloid colony formation [1], several investigators have looked at the role of this pathway in the growth and differentiation of leukemic cells [5-10].

Reports indicate that leukemic cells may have altered sensitivity to lipoxygenase pathway inhibitors, however both increased and decreased sensitivity have been reported [5-8]. We have previously reported that colony forming cells from a subset of patients with Acute Nonlymphoblastic Leukemia (ANLL) are more resistant to lipoxygenase pathway inhibitors than are normal marrow CFU-GM [5]. Lawrence et al however working with a series of leukemic cell lines, have demonstrated increased sensitivity to lipoxygenase inhibitors [8]. The monoblastic leukemia cell line, U937 has recently been shown to be dependent on LTC_4 for maintenance of logarithmic growth, with differentiation occurring when lipoxygenase is inhibited [9,10]. Other studies report lipoxygenase dependent growth of various malignant cell lines including neuroblastoma and mouse melanoma [11,12].

Ziboh et al have reported marked changes in the lipoxygenase products of the human leukemia cell line HL-60 when it is induced to differentiate [13]. Our recent studies demonstrate that inhibiting the synthesis of LTD_4 in these cells after induction with DMSO will cause them to differentiate to monocytes rather than the expected granulocytic phenotype [14].

In order to further define the role of the lipoxygenase pathway in myeloid leukemia cell growth and differentiation, we have examined the growth of three human myeloid cell lines, (HL-60, K562, and KG-1), as well as the bone marrow colony forming cells from a series of patients with ANLL, and compared the effects to those obtained for normal human bone marrow CFU-GM.

MATERIALS AND METHODS

Arachidonic Acid Metabolism Inhibitors

Inhibitors used in this study include: Nordihydroguaiaretic acid (NDGA) (Sigma Chemical Co., St. Louis, MO); caffeic acid (Sigma Chemical Co.), Buthionine Sulfoximine (BSo) (Chemical Dynamics, Plainfield, NJ); and Acivicin which has been provided by the Pharmaceutical Resources Branch of the National Cancer Institute. NDGA and caffeic acid were prepared in ethanol and concentrated so as not to exceed a final ethanol concentration of 2%, a concentration found to have no effect on cell growth. Acivicin and BSo were parepared in deionized, double distilled water.

Leukemic Cell Lines

Three well characterized human leukemic cell lines, HL-60 [15,16], K562 [15,17], and KG-1 [15,18] were utilized. All three lines were obtained from American Type Culture Collection (Rockville, MD) and maintained in liquid culture as described below.

Suspension Culture of Leukemic Cell Lines

HL-60 and K562 were maintained and grown in RPMI-1640 medium (GIBCO), Grand Island, NY) supplemented with 10% Fetal Bovine Serum (FBS) (Hyclone Laboratories, Logan, UT). KG-1 cells were maintained in Iscoves Modified Dulbeccos Medium (GIBCO) and supplemented with 20% FBS. For inhibitor studies log phase cells were diluted to 1×10^5/ml and appropriate inhibitors added at the time of plating.

Semisolid Culture of Leukemic Cell Lines

Cells taken from log phase cultures were suspended in MEMα medium (GIBCO) containing 0.3% agar and supplemented with 10% FBS. HL-60 and K562 were plated at 500 cells/ml and KG-1 at 1000 cells/ml. Inhibitors and CSF (KG-1 only) were added to plates prior to addition of cell suspensions. Colonies of 30 or more cells of HL-60 and K562 were counted on the 7th day of culture. KG-1 grow more slowly and form smaller clonal aggregates than the other cell lines, therefore all aggregates ≥ 10 cells were counted after 10 days.

Human Bone Marrow

Normal human bone marrow cells were obtained after informed consent from normal individuals donating marrow for allogeneic bone marrow transplantation. Marrow from patients with ANLL was obtained during the performance of diagnostic bone marrow aspiration, after obtaining informed consent. Investigations were performed under a protocol approved by the Institutional Review Board of the University of Florida, College of Medicine.

Human Bone Marrow CFU-GM and Leukemic Colony Forming Cell Assay

A modification of the procedure of Bradley and Metcalf was utilized [3,19]. Briefly, Ficoll-Hypaque (Histopaque, Sigma) interface cells were depleted of adherent cell populations and these cells suspended in MEMα with 0.3% agar containing 10% FBS. Conditioned medium from human bladder carcinoma cells (line 5637, American Tissue Type Collection) was added as a source of CSF at a concentration known to stimulate maximal colony formation in normal marrow samples. Cultures were incubated at 37°C in a humidified 5% CO_2 atmosphere. On the seventh day of culture, plates were scored for colonies (> 30 cells) and clusters (30-30 cells).

RESULTS

Effects of Lipoxygenase Pathway Inhibitors on CFU-GM and Leukemic Cell Lines

CFU-GM are inhibited in a dose dependent manner by caffeic acid, an inhibitor of lipoxygenase [20], with total inhibition at 100 μM (Figure 1). The three leukemic cell lines tested show similar inhibition of growth in suspension culture, with KG-1 somewhat less sensitive

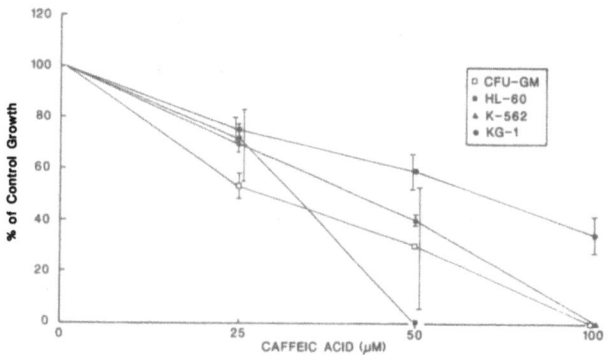

Figure 1. Effect of caffeic acid on growth of CFU-GM in semisolid culture, and leukemia cell lines in suspension culture. Results are expressed as percentage of control growth (growth of uninhibited cultures for that cell type) ± one standard deviation.

and HL-60 somewhat more sensitive than CFU-GM.

Similarly BSo, which inhibits the synthesis of LTC_4 by depleting cellular glutathione [21], causes dose dependent inhibition of CFU-GM and the leukemic cell lines (Figure 2). While CFU-GM are totally inhibited by 50 μM BSo the three leukemic cell lines continue to proliferate at higher concentrations with growth still seen at 500 μM BSo.

At all concentrations tested CFU-GM were slightly more sensitive to inhibition by Acivicin than were the three leukemic cell lines (Figure 3). Acivicin inhibits the synthesis of LTD_4 via inhibition of gamma-glutamyl transpeptidase (γGT) [22], the enzyme necessary to convert LTC_4 to LTD_4.

Effects of the Culture on BSo Inhibition of HL-60

In order to determine if the marked difference in the sensitivity of the CFU-GM and the leukemic cell lines to inhibition by BSo was due to differences in their culture conditions (semisolid vs. suspension), we grew HL-60 under both culture conditions and compared the response to BSo (Figure 4). When grown in a semisolid colony forming assay system, the HL-60 cells were notably more sensitive, than the same cells grown in suspension culture with total inhibition at 100 μM BSo.

A major difference between the two culture systems was the starting cell concentration. In the semisolid system the cells were plated at 5×10^2/ml while in suspension culture the starting concentration was 1×10^5/ml. To examine the effect of the starting concentration on BSo inhibition, we plated HL-60 in suspension culture over a one log concentration range (Figure 5). In the absence of BSo the cells doubled approximately every 1.3 days regardless of the starting cell concentration. A marked cell concentration depen-

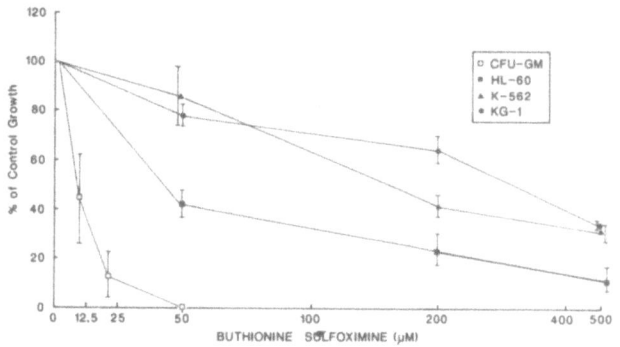

Figure 2. Effect of buthionine sulfoximine on growth of CFU-GM in semisolid culture, and leukemia cell lines in suspension culture. Results are expressed as percentage of control growth (growth of uninhibited cultures for that cell type) ± one standard deviation.

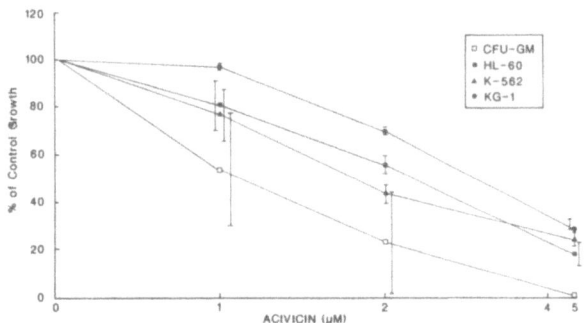

Figure 3. Effect of Acivicin on growth of CFU-GM in semisolid culture, and leukemia cell lines in suspension culture. Results are expressed as percentage of control growth (growth of uninhibited cultures for that cell type) ± one standard deviation.

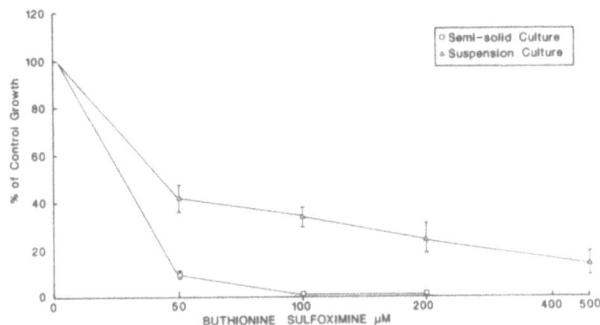

Figure 4. Effect of buthionine sulfoximine on growth of HL-60 in semisolid and suspension culture. Results are expressed as percentage of control growth (growth of uninhibited cultures for that culture condition) ± standard deviation.

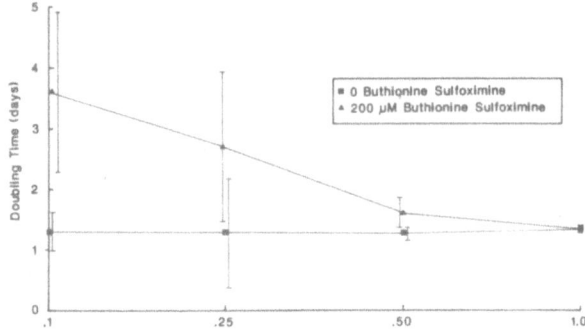

Figure 5. Effect of starting cell concentration on inhibition of HL-60 by 200 μM buthionine sulfoximine. Cells are grown in suspension culture for 4 days. Results are expressed as mean doubling time in days ± one standard deviation.

dent difference in doubling time was noted in the presence of 200 μM BSo. When the starting concentration was 1×10^4 the doubling time increased to 3.5 days, however at a starting concentration of 1×10^5 cells there was no change in doubling time as compared to controls. These data suggest that the difference in BSo sensitivity between HL-60 in semisolid culture vs. suspension are due to the relative difference in starting cell concentration.

Effect of Lipoxygenase Pathway Inhibitors on Bone Marrow CFU-GM from Patients with ANLL

We have studied the effects of NDGA, a lipoxygenase inhibitor [23], on the marrow colony forming cells of 13 patients diagnosed as having ANLL. Specimens included marrow aspirated at the time of initial presentation (6 patients), in clinical remission (9 patients), and in relapse (2 patients). Marrow from 3 of the patients was available for more than one of these clinical periods.

The marrow from several of these patients exhibited decreased sensitivity to inhibition by NDGA (Table 1). While normal marrow CFU-GM were inhibited to 9% of control in the presence of 20 M NDGA, from 5/13 patients had ≥ 40% of control growth at the same concentration. Interestingly, this lack of sensitivity to inhibition was seen in the marrow from patients in clinical complete remission as well as those with frank leukemia. The marrow from patient 1 exhibited resistance to inhibition both at initial presentation and in remission, while marrow from patient 2 was relatively sensitive both initially and in remission. None of the patients tested at different stages in their clinical course showed any change in their relative sensitivity to lipoxygenase inhibition. At the time of this writing, demonstration of

Table 1. Effect of Lipoxygenase Inhibition on Bone Marrow Colony Formation in ANLL

| | Absolute Number of Colonies + Clusters[a] (Colonies) | % of Control[b] | | |
| | | Concentration NDGA (μM) | | |
		10	20	50
Normal Individuals	246 (67)	22 $\pm$ 7	9 $\pm$ 4	0 $\pm$ 0

ANLL Patients

Initial Presentation

1	195 (0)	74	72	4
2	52 (0)	42	19	0
3	105 (0)	145	44	0
10	602 (10)	24	9	0
12	275 (47)	107	123	0
14	79 (14)	ND	28	0

Remission

1	470 (50)	62	48	2
2	199 (20)	6	6	0
4	109 (15)	56	25	0
5	30 (8)	70	40	32
6	54 (.5)	0	0	0
7	71 (6)	0	0	0
8	72 (10)	21	6	0
9	193 (8)	50	14	0
13	13 (.5)	85	54	0

Relapse

3	48 (3)	83	35	0
11	3780 (0)	34	22	3

[a]The sum of colonies (> 30 cells) and clusters (3-30 cells) per 10^5 cells plated. The number in parenthesis represents the absolute number of colonies per 10^5 cells plated.

[b]The percent of control = # of colonies + clusters in plates containing inhibitor/# of colonies + clusters in control plates for that individual. Data for normal individuals is the mean of 15 donors $\pm$ S.E.

resistance to lipoxygenase inhibitors in remission marrow has not been predictive of subsequent relapse.

Remission bone marrow cells from patient 5 demonstrated continued growth in semisolid culture at NDGA concentrations as high as 50 M. In order to determine if this resistance to inhibition was specifically to NDGA inhibition, or reflected a change in lipoxygenase pathway dependence we tested another lipoxygenase inhibitor, caffeic acid, as well as inhibitors of LTC_4 synthesis, BSo, and LTD_4 synthesis, Acivicin (Table 2). When compared to normal marrow CFU-GM marrow from patient 5 exhibited relative resistance to inhibition by each of these inhibitors. Karyotype analysis of this patients leukemic cells revealed a 48XX population, with trisomy 8 and 22. Interestingly the genes for glutathione reductase and gammaglutamyl transpeptidase, two enzymes involved in the synthesis of LTC_4 and D_4, are localized to these two chromosomes [24,25].

Table 2. Effect of Lipoxygenase Pathway inhibitors on the Bone Marrow Colony Formation of a Patient with ANLL

Inhibitor (Conc.)	Normal Individuals[a]	Patient #5
NDGA (20 μM)	9 $\pm$ 4	40
Caffeic Acid (50 μM)	18 $\pm$ 12	76
BSo (50 μM)	0 $\pm$ 0	21
Acivicin (2 μM)	12 $\pm$ 5	37

[a]Percent of control growth in the presence of the inhibitor indicated in the left column. Data for normal individuals are derived from the mean of 5-15 individual donors and represent mean $\pm$ S.E.M.

DISCUSSION

Previous reports on the role of the lipoxygenase pathway of Arachidonic Acid metabolism have reached conflicting conclusions, with both increased [8] and decreased sensitivity [5,6] reported. Our work supports altered sensitivity to lipoxygenase pathway inhibitors, primarily manifested as a decrease in sensitivity to these agents. These changes were not universal in all leukemic cell lines or patient marrows studied, and varied depending on the step in leukotriene synthesis that was inhibited. What this suggests is that a subset of myeloid leukemia manifests an alteration in the dependence on, or control of the lipoxygenase pathway.

The current studies suggest at least two plausible explanations for the discrepancies between published reports on the sensitivity of leukemic cells to lipoxygenase pathway inhibitors. The first point is that different leukemic populations may vary in their dependence on the lipoxygenase pathway. Differences in the three lines tested were noted, with KG-1 being less sensitive in liquid culture to caffeic acid than were HL-60, K562, or normal CFU-GM. In a similar manner 5/13 marrows from patients with ANLL cultured in a semi-solid system were less sensitive to lipoxygenase pathway inhibitors than normal CFU-GM, while the remaining 8 were not. Overall this suggests that different conclusions might be drawn depending on the cell population studied. Another variable in interpreting the sensitivity of leukemic cell populations to lipoxygenase pathway inhibitors is the culture system utilized. We observed large differences in the sensitivity of HL-60 cells to inhibition by BSO in semi-solid culture as compared to suspension culture. Further analysis determined that this difference was most probably due to the differing starting cell concentration used in these two systems rather than variables such as media or matrix. These experiments indicated a concentration dependent phenomena with decreasing sensitivity to lipoxygenase inhibitors at higher cell concentrations. HL-60 produce an autostimulator at high cell concentrations [26,27], and whether this or other soluble factors can protect these cells from the inhibitory effects of lipoxygenase blockers may provide insight into the survival advantages of leukemic cells.

Differences in lipoxygenase pathway metabolism between leukemic and normal populations may be exploitable in designing therapeutic strategies, but this will be a complicated process. As individual populations appear to have unique lipoxygenase pathway dependence they may be more or less sensitive than normal progenitors to lipoxygenase pathway inhibitors. Of potentially greater relevance than lipoxygenase or the lipoxygenase products themselves may be the enzymes involved in the synthesis of the peptido-leukotrienes. As was noted the leukemic cells from patient 5 had trisomy 8 and 22. The genes for two of the enzymes involved in the synthesis of LTC_4 and D_4, glutathione reductase, and γGT have been localized to chromosomes 8 and 22 respectively [24,25]. The gene for a third enzyme necessary for peptido-leukotriene synthesis, glutathione S-transferase has been localized to chromosome 11 [28]. Abnormalities of all three chromosomes, 8, 11, and 22, have been reported in association with myeloid leukemias [29]. The localization of the gene for γGT on chromosome 22 has been shown to lie within the breakpoint cluster region associated with the classic rearrangement of Chronic Myelogenous Leukemia (CML) [25] suggesting a possible association with an abnormality in this gene and CML. Ziboh et al [13] have noted a shift in the LTC_4 to LTD_4 ratio when HL-60 cells were induced to granulocytic differentiation. Our own studies demonstrate that if LTD_4 synthesis is inhibited HL-60 will differentiate to monocytes rather than granulocytes [14]. Further studies on the role of γGT and LTD_4 in leukemia are currently in progress. Studying the molecular biology of these genes in leukemic cells may provide us with new insight into the pathogenesis of myeloid leukemias and suggest new directions in treatment.

This work supported by PHS Grant #CA44838 awarded by the National Cancer Institute, DHHS, and an American Cancer Society Clinical Oncology Career Development Award (AMM). The authors wish to thank Ms. Mary Ann Gross for expert technical assistance, Mr. Jimmy Franco for preparation of figures, and Mrs. Anne Crawford for assistance in preparation of this manuscript.

REFERENCES

1. Miller AM, Ziboh VA, Yunis AA (1982) Evidence for involvement of the lipoxygenase pathway in CSF-induced human and murine myeloid colony formation. In: Powles TJ, Bockman RS, Honn KV, Ramwell P (eds) Prostaglandins and cancer: first international conference. New York: Alan J Liss, Inc, pp 482-485

2. Ziboh VA, Wong T, Wu M-C, Yunis AA (1986) Modulation of colony stimulating factor-induced murine myeloid colony formation by S-peptido-lipoxygenase products. Cancer Res 46:600-603

3. Miller AM, Weiner RS, Ziboh VA (1986 Evidence for the role of leukotriene C4 and D4 as essential intermediates in CSF stimulated human myeloid colony formation. Exp Hematol 14:760-76

4. Snyder DS, Desforges JF (1986) Lipoxygenase metabolites of arachidonic acid modulate hematopoiesis. Blood 67:1675-1679

5. Miller AM, Gross MA, Weiner RS (1986) Alterations in the requirements for leukotriene (LT) synthesis in leukemic marrow. Proceedings AACR 27:27-, 40 (abst)

6. Miller AM, Cullen MK, Kobb SM, Weiner RS (in revision) Effects of lipoxygenase and glutathione pathway inhibitors on leukemic cell line growth. J Lab Clin Med

7. Snyder DS, Desforges JF (1984) Lipoxygenase metabolites of arachidonic acid may play a role in K562 cell proliferation. Blood 64:135a (abst)

8. Lawrence HJ, Suyehira AL, Hack FM (1986) Differential sensitivity of human myeloid leukemic cells to the antiproliferative effects of 5 lipoxygenase. Blood 68:145a (abst)

9. Ondrey F, Salim N, Anderson KM, Harris J (1988) Eicosatetraynoic acid (5,8,11,14-ETYA) induces cellular differentiation in U937 cells. Proceedings of AACR 29:32 (Abst 128)

10. Harris JE, Ondrey F, Anderson KM (1988) Eicosatetraynoic acid (ETYA), a competitive inhibitor of arachidonic acid metabolism, inhibits eicosanoid and DNA biosynthesis while inducing differentiation in U937 cells: reversal of effect on DNA synthesis with leukotriene C4 (LTC4). Proceedings of AACR 29:78 (Abst 310)

11. Werner EJ, Waltenga RW, Dubowy RL, Boone S, Stuart MJ (1985) Inhibition of human malignant neuroblastoma cell DNA synthesis by lipoxygenase metabolites of arachidonic acid. Cancer Res 45:561-563

12. Honn KV, Dunn JR (1982) Nafazatrom (Bay g6575) inhibition of tumor cell lipoxygenase activity and cellular proliferation. FEBS Lett 139:65-68

13. Ziboh VA, Wong T, Wu M-C, Yunis AA (1986) Lipoxygenation of arachidonic acid by differentiated and undifferentiated human promyelocytic HL-60 cells. J Lab Clin Med 108:161-166

14. Miller AM, Kobb SM (1988) Regulation of HL-60 cell differentiation by lipoxygenase pathway metabolites. Proceedings of AARC 29:31 (Abst 122)

15. Koeffler HP, Golde DW (1980) Human myeloid leukemia cell lines: a review. Blood 56:344-350

16. Collins SJ, Gallo RC, Gallagher RE (1977) Continuous growth and differentiation of human myeloid leukaemic cells in suspension culture. Nature 27:347-349

17. Lozzio CB, Lozzio BB (1975) Human chronic myelogenous leukemic cell-line with positive Philadelphia chromosome. Blood 45:321-334

18. Koeffler HP, Golde DW (1978) Acute myelogenous leukemia: a human cell line responsive to colony-stimulating activity. Science 200:1153-1154

19. Bradley TR, Metcalf D (1966) The growth of mouse marrow cells in vitro. Aust J Exp Biol Med Sci 44:287-299

20. Koshihara Y, Neichi T, Murota S-I, Lao A-N, Fujimoto Y, Tatsuno T (1984) Caffeic acid is a selective inhibitor for leukotriene biosynthesis. Biochem Biophys Acta 792:92-97

21. Meister A (1983) Selective modification of glutathione metabolism. Science 220:472-477

22. Allison RD (1985) Gamma-glutamyl transpeptidase: kinetics and mechanism. Methods Enzymol 113:420-437

23. Chang J, Skowronek MD, Cherney ML, Lewis AJ (1984) Differential effects of putative lipoxygenase inhibitors on arachidonic acid metabolism in cell free and intact cell preparations. Inflammation 8:143-155

24. Jensen PKA, Junien C, Despoisse S, et al (1982) Inverted tandem duplication of the short arm of chromosome 8: a non-random de novo structural aberration in man. Localization of the gene for glutathione reductase in subband 8p21.1. Annales de Genetique 25:207-211

25. Bulle F, Mattei MG, Siegrist S, Pawlak A, Passage E, Chobert MN, Laperche Y, Guellaen G (1987) Assignment of the human gamma-glutamyl transferase gene to the long arm of chromosome 22. Hum Genet 76:283-286

26. Brennan JK, Abboud CN, Dipersio JF, Barlow GH, Lichtman MA (1981) Autostimulation of growth of human myelogenous leukemia cells (HL-60). Blood 58:83-812

27. Perkins SL, Androtti PE, Sinha SK, Wu M-C, Yunis AA (1984) Human myeloid leukemic cell (HL-60) autostimulator: relationship to colony stimulating factor. Cancer Res 44:5169-5175

28. Silverstein DL, Shows TB (1982) Gene
     for   glutathione   S-transferase-1
     (GST1) is on human chromosome 11.
     Somatic.   Cell Genet 8:667-675

29. Rowley JD (1984) Biological implica-
     tions of consistent chromosome re-
     arrangments in leukemia and lymphoma.
     Cancer Res 44:3159-3168

A. Hagenbeek, A.C.M. Martens, and F.W. Schultz

One of the major problems in todays leukemia treatment employing either allogeneic or autologous bone marrow transplantation (BMT) is leukemia relapse. Apparently, leukemic cells have survived high dose conditioning treatment prior to BMT. In case of autologous BMT, leukemic cells reinfused with the graft might in addition contribute to leukemia relapse after BMT.

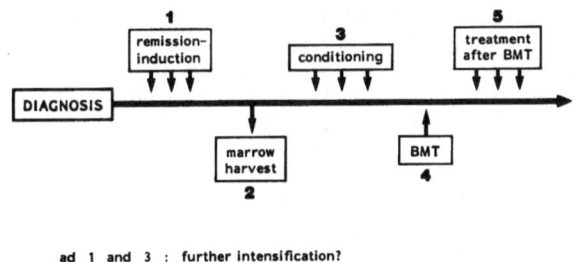

ad 1 and 3 : further intensification?
ad 2   : purging?
ad 4   : T-cell depletion?
ad 5   : low dose chemotherapy - growth factors - BRM?

Figure 1: How to prevent a leukemia relapse after bone marrow transplantation?

Figure 1 summarizes the crucial points in time during the treatment of acute leukemia at which the leukemic cell load can be reduced step by step. Some of these steps will be evaluated critically in the present paper and wherever possible data and conclusions from preclinical animal model studies will be introduced.

As a preclinical model the Brown Norway rat acute myelocytic leukemia (BNML) was used (1, 2). This chemically induced model proved to be transplantable by means of cellular transfer. After i.v. injection of leukemic cells in inbred BN rats a reproducible growth pattern is observed. In Table 1 the major characteristics of the BNML are summarized. In particular as regards its biology of growth and its sensitivity to chemotherapy, the BNML has now widely been accepted as a most relevant model for human acute myelocytic leukemia (AML). So far, 18 leukemia research centers in Europe, the U.S.A.

and Canada are presently employing the BNML in their programs. During the past years a variety of preclinical BNML studies found application in human AML as illustrated in Table 2.

Table 1

MAJOR CHARACTERISTICS OF THE BROWN NORWAY RAT ACUTE MYELOCYTIC LEUKEMIA (BNML)

- induced by DMBA
- analogy with human acute (pro)myelocytic leukemia:
  1. cytology - cytochemistry
  2. slow growth rate (10^7 cells i.v.: death at day 23)
  3. severe suppression of normal hemopoiesis
  4. diffuse intravascular coagulation (DIC)
  5. response to chemotherapy as in human AML
  6. leukemic clonogenic cells present
  7. low antigenicity
  8. no virus

Table 2

PRECLINICAL BNML STUDIES* APPLIED IN HUMAN AML

Diagnosis/detection
- monoclonal antibodies (MoAbs)
- fluorescence-activated cell sorting (FACS) (3)
- flow karyotyping (4, 5)
- heterogeneous distribution of "minimal residual disease" (6, 7)
- computer simulation of leukemia growth (8)

Treatment
- high-dose cytosine arabinoside (timed sequential treatment) (9, 10)
- pharmacokinetics of cytostatic agents (11)
- conditioning regimens prior to BMT
    - (fractionated) TBI (12)
    - supralethal chemotherapy (13)
    - combinations (14)
- in vitro chemotherapy/MoAb treatment of autologous marrow grafts (15-17)

*Radiobiological Institute TNO, Rijswijk, The Netherlands and 18 Leukemia Research Centers in Europe, United States and Canada.

In the subsequent studies two questions will be addressed to, i.e., 1) what is the contribution of leukemic cells in the autologous marrow graft to a leukemia relapse after autologous BMT?; and 2) what is the role of the so-called Graft-versus-Leukemia reaction in preventing a leukemia relapse after allogeneic BMT?

AUTOLOGOUS BONE MARROW TRANSPLANTATION

So far, clinical results show that 50-60% of patients with AML receiving an autologous marrow graft in first remission will relapse (18-20). Obviously, this may be either due to residual leukemia in the host or to leukemic cells reinfused with the graft or to a combination of both. Arguments in favor of residual disease in the host being the major determinant as regards the occurrence of a relapse are summarized in Table 3.

Table 3

ARGUMENTS IN FAVOR OF RESIDUAL LEUKEMIA IN THE HOST AS THE MAJOR CAUSE OF A LEUKEMIA RELAPSE AFTER AUTOLOGOUS BONE MARROW TRANSPLANTATION*

1. similar relapse rates after autologous - and isologous BMT

2. leukemia relapses occur at (about) the same time after autologous, isologous and allogeneic BMT (median: 9 months)

3. ED_{50} for human AML: 10^3-10^4 cells (extrapolated from preclinical data in the BNML; Leukemia Research 9, 1389-1395, 1985)

4. so far: no beneficial effect from in vitro "purging"

*first remission AML

In particular the ED_{50} concept will now be further analysed. In the BNML model it is known that 25 leukemic cells are needed to induce leukemia in 50% of normal recipient rats after i.v. injection (ED_{50}). If a total inoculum of, e.g., 1000 leukemic cells is regarded as 40 ED_{50} units, an average of 20 BNML cells would grow out in vivo. Indeed after i.v. injection of 10^7 BNML cells, 2×10^5 leukemic cells were recovered at the start of leukemia growth process in the major target organs for leukemia growth in the rat, i.e., bone marrow, spleen and liver. If an inoculum contains xED_{50} units, with each unit having a chance of 0.5 to grow out and cause overt leukemia, the chance that leukemia will not or will develop is 0.5^x or $1-0.5^x$, respectively. If these mathematics are applied to experimental data derived from experiments where the incidence of leukemia was related to the i.v. injection of graded low numbers of leukemic cells, there is a perfect agreement. With the ED_{50} model computer simulations can be performed yielding the chance that leukemia will develop from injected leukemic cells as a fraction of 1) the number of cells injected, and 2) the ED_{50} value. The ED_{50} for human AML is not known.

It seems reasonable to assume that remission-induction treatment on the average induces a 4 log leukemic cell kill. Thus, an autologous marrow graft will contain 1 leukemic cell per 10^4 normal marrow cells, or 1.5×10^6 in a graft containing a total of 1.5×10^{10} cells (2×10^8 cells/kg; 75 kg patient). In the BNML it was found that 1% of in vivo clonogenic leukemic cells survive cryopreservation. Furthermore, from in vitro studies with human AML it is concluded that only 0.1-1% of all leukemic cells can be considered to be clonogenic. Taken these two factors together, only 15-150 clonogenic leukemic cells out of 1.5×10^6 are reinfused with the marrow graft. Assuming now an ED_{50} value for human AML to be 1000 or 10,000 clonogenic cells, which seems realistic based on previous BNML studies (21, 22), it can be calculated employing the computer simulation model that the chance that reinfused leukemic cells indeed cause leukemia is 1-10% or 0.1-1%, respectively. This can be derived from the data presented in Table 4.

Thus, to prevent a leukemia relapse after autologous BMT major emphasis should be given to more effective pretreatment of the patient. From previous studies in the BNML it is known that high-dose cytosine arabinoside in combination with high-dose cyclophosphamide and total body irradiation or high-dose busulphan followed by cyclophosphamide are the most effective con-

Table 4

THEORETICAL CHANCE (%) OF LEUKEMIA DEVELOPMENT AS FUNCTION OF ASSUMED ED_{50} VALUE AND THE NUMBER OF LEUKEMIC CELLS IN THE HUMAN AUTOLOGOUS BONE MARROW GRAFT

ED_{50} value	number of leukemic cells in graft								
	10	15	100	150	500	1000	1500	5000	10000
10	50.0	64.6	99.9	100.0	100.0	100.0	100.0	100.0	100.0
100	6.7	9.9	50.0	64.6	96.9	99.9	100.0	100.0	100.0
1000	0.7	1.0	6.7	9.9	29.3	50.0	64.6	96.9	99.9
5000	0.1	0.2	1.4	2.1	6.7	12.9	18.8	50.0	75.0
10000	<0.1	0.1	0.7	1.0	3.4	6.7	9.9	29.3	50.0

The theoretical probability of leukemia development, i.e., the chance that at least one unit of ED_{50} cells yields a cell to grow out, is given by $(1-0.5^x) \cdot 100\%$, where x denotes the number of ED_{50} units injected.

ditioning regimens, both inducing a more than 10 log leukemic cell kill, the first regimen however causing a 25% treatment-related mortality (23). Based on the findings reported above, the current Dutch autologous BMT study in first remission AML employs nonpurged marrow after conditioning with busulphan-cyclophosphamide.

ALLOGENEIC BONE MARROW TRANSPLANTATION

Since the introduction of T-cell depletion a number of centers have reported an increase in leukemia relapse rate after allogeneic BMT (see, e.g., 24-26). The other way around: lower relapse rates have been observed if patients develop Graft-versus-Host disease (GvHD). Furthermore, evidence for a so-called Graft-versus-Leukemia reaction (GvLR) is derived from the observation in AML that the leukemia relapse rate after allogeneic BMT in first remission is lower than after autologous or isologous BMT.

The probabilities of leukemia cure, $Pr\{cure\}$ determined directly from Kaplan-Meier plots obtained by long term follow-up of large numbers of patients, are approximately 0.9 and 0.4 in patients with severe GvHD and without GvHD, respectively (27). In other words, GvHD apparently is related to a decrease in the number of surviving clonogenic leukemia cells. It is not known whether this antileukemia effect is merely alloreactivity or that an immunological reaction occurs specifically against leukemic cells. Experimental data indicate that the Graft-versus-Host reaction and the GvLR can be separated, although the evidence is not convincing (28).

While having a beneficial effect as far as leukemia cure probability is concerned, unfortunately, GvHD in itself causes serious morbidity and mortality, thus reducing the net success rate again. In this context it might be useful to evaluate what extra decrease, quantitatively, in survival of clonogenic leukemic cells is correlated with GvHD.

A theoretical probability of leukemia cure can be derived using either Poisson statistics or the binomial probability distribution. Let the tumor burden of a patient be M clonogenic leukemic cells. Due to treatment, let each leukemic cell have a chance, SF, to survive, and let X denote the number of surviving clonogenic leukemic cells after treatment. If the conditions for Poisson statistics are satisfied, i.e., if M is large (greater than 100) and SF is small (less than 0.1) and SF x M is constant, then the probability of cure can be written as:

$$Pr\{cure\} = Pr\{X = 0\} = e^{-SF \times M} \qquad (1)$$

Several initial tumor loads, M, and surviving fractions, SF, as listed in Table 5, were chosen such that Poisson statistics apply. The theoretical probability of cure was computed using equation (1). It decreases with either increasing original tumor load or increasing surviving fraction. Although not tabulized here, the binomial distribution too yields a decrease in theoretical probability of cure with increasing M and/or SF.

From Table 5 it is seen that a rise from about 40% to 90% cure corresponds with a differ-

ence of 1 log cell kill for several combinations of M and two subsequent SF values. In particular, this is the case when assuming - as suggested by (pre)clinical findings - that the initial leukemic population just after remission-induction therapy has a size of $M = 10^8$ cells and that the log cell kill induced by the conditioning regimen has the value eight or nine, respectively (23).

Table 5

THEORETICAL PROBABILITY OF CURE ACCORDING TO POISSON STATISTICS, AS FUNCTION OF INITIAL LEUKEMIC CELL LOAD AND "LOG CELL KILL" EFFECT

	leukemia burden				
surviving fraction SF =	$M=10^{12}$	10^{10}	$\boxed{10^8}$	10^6	10^4
10^{-1}	0	0	0	0	0
10^{-2}	0	0	0	0	4×10^{-44}
10^{-3}	0	0	0	0	5×10^{-5}
10^{-4}	0	0	0	4×10^{-44}	0.37
10^{-5}	0	0	0	5×10^{-5}	0.90
10^{-6}	0	0	4×10^{-44}	0.37	0.99
10^{-7}	0	0	5×10^{-5}	0.90	1
10^{-8}	0	4×10^{-44}	$\boxed{0.37}$	0.99	1
10^{-9}	0	5×10^{-5}	$\boxed{0.90}$	1	1
10^{-10}	4×10^{-44}	0.37	0.99	1	1

In conclusion, as the Graft-versus-Leukemia reaction at the most contributes a one log leukemic cell kill only, it is strongly advocated to seek other ways of achieving this extra kill instead of reintroducing unmanipulated allogeneic marrow grafting (i.e., going back to the Seventies) with the major drawback of Graft-versus-Host related mortality.

In summary, both in case of autologous and allogeneic BMT more effective means to reduce the leukemic cell load prior to BMT seem to be most logical. An alternative may be to apply biological response modifiers or low dose chemotherapy after BMT (Figure 1). Optimal timing employing appropriate doses were shown in the BNML model to eradicate the few logs of leukemic cells which remain after BMT without jeopardizing the graft (14, 29).

REFERENCES
1. Van Bekkum DW, Hagenbeek A: Relevance of the BN leukemia as a model for human acute myeloid leukemia. Blood Cells 3: 565, 1977.
2. Hagenbeek A, Van Bekkum DW (eds): Proceedings of a workshop on comparative evaluation of the L5222 and the BNML rat leukaemia models and their relevance for human acute leukemia. Leukemia Res 1: 75, 1977.
3. Hagenbeek A, Martens ACM: Detection of minimal residual disease in acute leukemia: Possibilities and limitations. Eur J Cancer Clin Oncol. 21: 389, 1985.

4. Arkesteijn GJA, Martens ACM, Jonker RR, Hagemeijer A, Hagenbeek A: Bivariate flow karyotyping of acute myelocytic leukemia in the BNML rat model. Cytometry 8: 618, 1987.

5. Arkesteijn GJA, Hagenbeek A: Bivariate flow karyotyping in human Philadelphia positive chronic myelocytic leukemia. Blood. In press, 1988.

6. Hagenbeek A, Martens ACM: An immunological approach to analyse the kinetics of minimal residual disease in acute leukemia. In: Hagenbeek A, Löwenberg B (eds): Minimal residual disease in acute leukemia. Dordrecht, The Netherlands, Martinus Nijhoff, p. 76, 1986.

7. Martens ACM, Schultz FW, Hagenbeek A: Nonhomogeneous distribution of leukemia in the bone marrow during minimal residual disease. Blood 70: 1073, 1987.

8. Schultz FW, Martens ACM, Hagenbeek A: Growth kinetics of minimal residual disease in the Brown Norway rat acute myelocytic leukemia. In: Hagenbeek A, Löwenberg B (eds): Minimal residual disease in acute leukemia. Dordrecht, The Netherlands, Martinus Nijhoff, p. 97, 1986.

9. Colly LP, Van Bekkum DW, Hagenbeek A: Enhanced tumor load reduction after chemotherapy-induced recruitment and synchronization in a slowly growing rat leukemia model (BNML) for human acute myelocytic leukemia. Leuk Res 8: 953, 1984.

10. Vaughan WP, Burke PJ: Development of a cell kinetic approach to curative therapy of acute myelocytic leukemia in remission using the cell cycle-specific drug 1-β-D-arabinofuranosylcytosine in a rat model. Cancer Res 43: 2005, 1983.

11. Nooter K, Sonneveld P, Deurloo J, Hagenbeek A: Tissue distribution and myelotoxicity of daunomycin in normal and leukemic rats: Rapid bolus injection versus continuous infusion. In: Harrap KR, Davis W, Calver AH (eds). Cancer chemotherapy and selective drug development. Dordrecht, Lancaster, Martinus Nijhoff, p. 546, 1985.

12. Hagenbeek A, Martens ACM: The effect of fractionated versus unfractionated total body irradiation on the growth of the BN acute myelocytic leukemia. Int J Rad Oncol Biol Phys 7: 1075, 1981.

13. Hagenbeek A, Martens ACM: High-dose cyclophosphamide treatment of acute myelocytic leukemia. Studies in the BNML rat model. Eur J Cancer Clin Oncol 18: 763, 1982.

14. Hagenbeek A, Martens ACM: The efficacy of high-dose cyclophosphamide in combination with total body irradiation in the treatment of acute myelocytic leukemia. Studies in a relevant rat model (BNML). Cancer Res 43: 408, 1983.

15. Hagenbeek A, Martens ACM: Cell separation studies in autologous bone marrow transplantation for acute leukemia. In: Gale RP (ed). Recent advances in bone marrow transplantation. UCLA Symposia on Molecular and Cellular Biology, vol. 7, New York, Alan R. Liss, p. 717, 1983.

16. Hagenbeek A, Martens ACM: Toxicity of ASTA-Z-7557 to normal- and leukemic stem cells: Implications for autologous bone marrow transplantation. Invest New Drugs: J Anticancer Agents 2: 237, 1984.

17. Sharkis SJ, Santos GW, Colvin M: Elimination of acute myelogenous leukemia cells from marrow and tumor suspensions in the rat with 4-hydroperoxycyclophosphamide. Blood 55: 521, 1980.

18. Löwenberg B, Van der Lelie J, Goudsmit R, Willemze R, Zwaan FE, Hagenbeek A, Van Putten WJL, Verdonck LF, De Gast GC: Autologous bone marrow transplantation in patients with acute myeloid leukemia in first remission. In: Dicke KA, Spitzer G, Jagannath S, Favrot M (eds). Autologous Bone Marrow Transplantation, Proceedings of the Third International Symposium. The University of Texas, MD Anderson Hospital and Tumor Institute at Houston, p. 3, 1987.

19. Burnett AK, Mackinnon S, Morrison A: Autologous transplantation of unpurged bone marrow during first remission of acute myeloid leukemia. In: Dicke KA, Spitzer G, Jagannath S, Favrot M (eds). Autologous Bone Marrow Transplantation, Proceedings of the Third International Symposium. The University of Texas, MD Anderson Hospital and Tumor Institute at Houston, p. 23, 1987.

20. Gorin NC, Laporte JP, Douay L, Lopez M, Stachowiak J, Aegerter P, Deloux J, Marin JL, Duhamel G, Laugier A, Salmon Ch, Najman A: Use of bone marrow incubated with mafosfamide in adult acute leukemia patients in remission. The experience of the Paris – Saint Antoine transplantation team. In: Dicke KA, Spitzer G, Jagannath S, Favrot M (eds). Autologous Bone Marrow Transplantation, Proceedings of the Third International Symposium. The University of Texas, MD Anderson Hospital and Tumor Institute at Houston, p. 15, 1987.

21. Hagenbeek A, Martens ACM: Reinfusion of leukemic cells with the autologous marrow graft: Preclinical studies on lodging and regrowth of leukemia. Leuk. Res. 9: 1389, 1985.

22. Hagenbeek A, Martens ACM: On the fate of leukemic cells infused with the autologous marrow graft. In: Büchner Th, Schellong A, Urbanitz D, Hiddemann W, Ritter R (eds.). Acute Leukemias. Springer Verlag, Berlin/Heidelberg. Haematology and Blood Transfusion 30: 553, 1987.

23. Hagenbeek A, Martens ACM: Conditioning regimens before bone marrow transplantation in acute myelocytic leukemia. In: Dicke KA, Spitzer G, Jagannath S, Favrot M (eds). Autologous Bone Marrow Transplantation, Proceedings of the Third International Symposium. The University of Texas, MD Anderson Hospital and Tumor Institute at Houston, p. 99, 1987.

24. Butturini A, Bortin MM, Gale RP: Graft-versus-leukemia following bone marrow transplantation. Bone Marrow Transplantation 2: 233, 1987.

25. Weisdorf DJ, Nesbit ME, Ramsay NKC et al.: Allogeneic bone marrow transplantation for acute lymphoblastic leukemia in remission: Prolonged survival associated with acute graft-versus-host disease. J. Clin. Oncol. 5: 1348, 1987.

26. Maraninchi D, Blaise D, Rio B et al.: Impact of T-cell depletion on outcome of allogeneic bone marrow transplantation for standard risk leukemias. The Lancet i: 175, 1987.

27. Weiden PL, Sullivan KM, Flournoy N, Storb R, Thomas ED and The Seattle Marrow Transplant Team: Antileukemic effect of graft-versus-host disease: Contribution to improved survival after allogeneic marrow transplantation. N. Engl. J. Med. 304: 1529, 1981.

28. Meredith RF, O'Kunewick JP: Possibility of graft-versus-leukemia determinants indepen-dent of the major histocompatibility complex in allogeneic marrow transplantation. Trans-plantation 35: 378, 1983.

29. Hagenbeek A, Martens ACM: BCG treatment of residual disease in acute leukemia. Studies in a rat model for human acute myelocytic leukemia (BNML). Leuk Res 7: 547, 1983.

Part VI.   Bone Marrow Transplantation

Chairperson: K.A. Dicke

23   Allogeneic Marrow Transplantation

R. Storb

INTRODUCTION

Allogeneic marrow transplantation has been increasingly applied to the treatment of patients with various hematological diseases (1-6). A recent survey by the International Bone Marrow Transplantation Registry estimated the number of transplants carried out through the year 1987 to be on the order of 20,000, more than 10,000 of these during the years of 1985 through 1987 (7). More than 80% of marrow transplants have been for therapy of malignant hematological diseases. Approximately 10% of transplants have been for the treatment of patients with acquired or inherited marrow dysfunction, e.g. aplastic anemia, and 5-6% of marrow grafts have been used to treat congenital defects of the hematopoietic and immune systems, e.g. thalassemia major, severe combined immunodeficiency disease, and other inborn errors.

There have been remarkable advances in marrow transplantation since 1970 when the technique was restricted to patients with advanced hematologic malignancies, and disease-free survival on the order of 15% was observed (8). Recent studies in patients with acute non-lymphoblastic leukemia (ANL) in first chemotherapy-induced remission have shown superior survival in patients undergoing marrow transplantation compared to those given chemotherapy (50% versus 20% actuarial survival with the longest patient at 10 years). Patients with acute lymphoblastic leukemia (ALL) given grafts in 2nd or subsequent remission have shown disease-free survival of approximately 35% whereas similar patients undergoing chemotherapy have all died of recurrent disease within 3-1/2 years of the initiation of therapy. Patients with chronic myelocytic leukemia (CML) cannot be cured with chemotherapy alone and disease-free survival of approximately 50-60% has been obtained with marrow grafts in the chronic phase. In patients with aplastic anemia survival with marrow grafting has improved from 45% to 60-80% compared with only 20% survival with supportive therapy and 40-50% survival with immunosuppression by antithymocyte globulin. Marrow grafting has produced 70% disease-free survival in patients with thalassemia major (9) whereas approximately 50-60% of patients given marrow grafts for severe combined immunodeficiency disease and other inborn errors have become long-term survivors (10).

Despite impressive improvements in the results of marrow transplantation, major problems and complications remain (1-6,8, 10-23). These are illustrated in Table 1. Relapse of leukemia has accounted for 17-75% of treatment failures in patients grafted for leukemia, whereas graft rejection has resulted in death of 5-12% of patients grafted for aplastic anemia. Significant acute graft-versus-host disease (GVHD) is seen in 18-45% of patients and it has a case fatality rate of approximately 50%. It is responsible for 10-25% of treatment failures. Conditioning regimen related toxicity or bacterial or fungal infection during the early period of neutropenia result in 5-10% of deaths. Fatal interstitial pneumonias are often associated with acute GVHD or may be the result of drug and radiation toxicity. For results of marrow transplantation to improve, progress in each of these problem areas is needed.

PREVENTION OF GVHD

A critical issue is the prevention of GVHD (reviewed in 15,22). This is customarily involved postgrafting immunosuppression. In many patients, immunosuppressive therapy can be discontinued by 3 to 6 months after transplantation when a stable state of graft-host tolerance has been achieved. Omission of immunosuppression in patients given unmanipulated marrow has caused an unacceptably high incidence of acute GVHD and transplantation-related death. Controlled randomized trials have shown methotrexate and cyclosporine to

Table 1.   COMPLICATIONS AND SURVIVAL AFTER HLA-IDENTICAL
MARROW TRANSPLANTATION (1-6,8,10-23)

Disease [a]	AA	ALL				ANL			CML		
Disease phase [b]	severe	1st CR	2nd CR	3rd CR	2nd+ Rel	1st CR	2nd+ CR	1st Rel	1st CP	AP	BC
% 5 yr disease-free survival	60-80	54	35	30	18	50	25	34	58	30	20
% Relapse	---	35	45	58	75	22	45	31	17	45	70
% Grade II-IV Acute GVHD [c]	18-35	35				40-45			45		
% Chronic GVHD	30-45	25				25-35			35		
% Interstitial Pneumonia [d]	5-15	15				15-35			22		
% VOD [e]	<1	7				28			25		
% Bacterial + Fungal Infections During first 3 months Before Engraftment After Engraftment After first 3 months	15 12 20					20 12 20					
% Graft Failure	5-17					<1					
% Secondary Malignancies	1					5					

[a]   AA = aplastic anemia;   ALL = acute lymphoblastic leukemia;   ANL = acute nonlymphoblastic leukemia;   CML = chronic myelocytic leukemia.   Patients with AA are usually
conditioned with cyclophosphamide alone or in combination with total lymphoid or
thoracoabdominal irradiation; patients with leukemia are usually given cyclophosphamide
and total body irradiation.

[b]   CR = complete remission;   Rel = relapse;   CP = chronic phase;   AP = accelerated phase;
BC = blast crisis.

[c]   GVHD = graft-versus-host-disease.

[d]   Includes both idiopathic & cytomegalovirus interstitial pneumonias.

[e]   VOD = veno-occlusive disease of the liver.   VOD is rare in patients conditioned only
with cyclophosphamide.

be equivalent in their ability to prevent
GVHD (reviewed in 24).   A combination of
methotrexate and cyclosporine is significantly better than either drug alone in preventing acute GVHD and leads to increased
survival (25,26).   However, chronic GVHD
continues to be seen (27).

Another way to prevent GVHD has been the
removal of T cells from the marrow by immunological or mechanical means (reviewed in
6,28,29).   With any of the techniques used,
there is a reduction in the number of infused
T cells by 1 to 3 logs.   In this manner, most
differentiated immune cells causing GVHD
would be eliminated and the immune system
returned to an early prenatal state.   Any new
stem cell-derived T cells would accept the
host's antigenic environment as "self" and
become tolerant to it.   T cell depletion has
worked well in rodent models but has been
less convincing in random bred large animals
where the problem of graft failure was first
noted.   Graft failure seems to be the result
of host immune cells which have survived the
conditioning program and whose continued survival is assured through the absence of GVHD.
Nearly all clinical studies have shown a significant reduction in acute GVHD in patients
given T cell depleted marrow grafts (Table
2).   This result provides convincing evidence
for a favorable effect of T cell depletion on
GVHD.   However, the reduction in acute GVHD
was achieved at the price of substantial
increases in graft rejection and leukemic
relapse (Tables 2 and 3).   Thus, the overall
incidence of graft rejection in HLA-identical
marrow transplant recipients increased from
1% to 12% and in HLA-nonidentical recipients

Table 2. T-CELL DEPLETION, GVHD AND GRAFT FAILURE IN PATIENTS GRAFTED FOR LEUKEMIA (28)

Source of Marrow	T-deple-tion	% Acute GVHD	% Graft Failure
HLA-identical	+	11	12
	-	45	1
HLA-nonidentical	+	31	32
	-	75	5

Table 3. T-CELL DEPLETION AND LEUKEMIC RELAPSE (IBMTR data reviewed in ref. 28)

Disease[a]	No. of pts.	% Relapse	
		Untreated Marrow	T-depleted Marrow
CML - CP	309	8	43
ANL - 1st CR	538	18	45
ALL - 1st CR	205	25	35
ALL - 2nd CR	179	55	75

[a]CML - CP = chronic myelocytic leukemia in chronic phase
ANL - CR = acute nonlymphoblastic leukemia in complete remission
ALL - CR = acute lymphoblastic leukemia in complete remission

from 5% to 32%. Additionally, relapse rates in patients with leukemia increased significantly, most impressively in patients grafted for CML in chronic phase (Table 3). High graft failure rates with T cell depleted marrow grafts have also been seen in patients with aplastic anemia.

Given that graft rejection and leukemic relapse almost uniformly result in death, an improvement in survival has not been realized in patients given T-depleted marrow transplants. Nevertheless, the significant decrease in the incidence of acute GVHD suggests that the technique of T cell depletion is promising provided that the risk of graft rejection and relapse can be lessened. To achieve this aim, two different methods can be envisioned. One includes improvement of pretransplant conditioning programs which better eradicate immune cells of host type as well as malignant cells. As discussed below, this aim may be achievable through better use of currently available chemoradiation therapy and through innovative approaches using antibody isotope conjugates in addition to chemoradiation therapy.

The other method is based on the hope that T cells causing GVHD are distinct from those enhancing allogeneic marrow engraftment and causing graft versus leukemia effects. A better understanding of the precise role of lymphocytes in mediating these diverse immune functions is needed. This might result in the development of strategies to eliminate GVHD without affecting engraftment and the graft versus leukemia effect.

GRAFT REJECTION IN APLASTIC ANEMIA

Graft failure has been a common problem in patients given HLA identical marrow grafts for the treatment of severe aplastic anemia after conditioning with high-dose cyclophosphamide (3,5,21,30). This problem was seen in 30-60% of patients treated by marrow grafts in the early 1970's. Two factors were associated with rejection: first, positive in vitro tests of cell mediated immunity, indicating reaction of host lymphocytes against antigens on donor cells before transplantation; and secondly, a low number of transplanted marrow cells ($< 3 \times 10^8$ cells/kg). Immunity of recipient against donor was thought to be the result of transfusion-induced sensitization, a speculation which is amply supported by studies in experimental animals. Studies in dogs have indicated that dendritic mononuclear cells in the transfused blood products lead to sensitization of the recipient against minor antigens of the donor which may not be suppressed by the immunosuppressive conditioning programs (31). When transplants are carried out in patients who have not received preceding transfusions, graft failure is the exception. Eighty percent of untransfused patients are alive with functioning grafts, suggesting that immunological mechanisms involved in graft failure are, for the most part, iatrogenic, e.g. induced by previous blood transfusions.

Many programs to avoid rejection in multiply-transfused patients are being used, mainly involving more intensive immunosuppression. All programs include cyclophosphamide, but other feature vary, including the use of total body irradiation (TBI), total lymphoid irradiation, total nodal irradiation and thoracoabdominal irradiation. The Seattle team has employed viable donor buffy coat cells along with the marrow graft, since the donor's peripheral blood is a potential source of hemopoietic stem cells and/or lymphoid cells capable of abrogating rejection. With all of these conditioning programs, the rejection rates have decreased, and survival has increased with most centers now reporting survivals between 60% and 70% in multiply-transfused patients.

Most of the conditioning programs have associated risks. The addition of buffy coat cells may lead to an increase in chronic GVHD. Regimens involving irradiation carry the potential risk for late cancer. Because of these associated risks and the still existing mortality from rejection, emphasis should be placed on measures that prevent rather than overcome sensitization by blood transfusions. This is best done by early transplantation before transfusion. In case transfusions are needed, buffy coat-poor red blood cells and platelets should be used. Recent data in dogs have shown that sensitization can be prevented if blood transfusion products are exposed to ultraviolet light irradiation (32).

CONDITIONING REGIMENS TO REDUCE GRAFT FAILURE AND RELAPSE IN PATIENTS WITH LEUKEMIA

The most commonly used conditioning agents for patients with leukemia have involved a combination of cyclophosphamide and TBI (1-4,8,11-13). In efforts to reduce the leukemic recurrence rates, numerous therapeutic reagents have been used in addition to or in lieu of cyclophosphamide, including etoposide, high-dose cytosine arabinoside, piperazinedione, BCNU and others. For the past decade, fractionated TBI has slowly replaced single dose TBI. A prospective comparison of these two radiation schedules showed fractionated TBI to be better tolerated and to result in fewer long-term complications without any apparent increase in postgrafting relapse rates (33). Hyperfractionated TBI followed by cyclophosphamide has been used by the Sloan-Kettering team with apparently superior results in patients with ALL in second or subsequent remission (11). The Johns-Hopkins team has used busulfan combined with cyclophosphamide without TBI, and they reported very low leukemic recurrence rates in patients with ANL in first remission while relapse rates in patients with more advanced ANL appeared not to be different from those seen after cyclophosphamide/TBI regimens (14). This latter result is in apparent contrast to results reported by the Ohio State team suggesting that relapse rates are low not only in patients with ANL in first remission but also in patients with advanced ANL or ALL even though the doses of busulfan and cyclophosphamide have been reduced (34). It appears, however, that for all approaches involving systemic chemotherapy and TBI, the limits of nonhemopoietic toxicity have been reached and no substantial improvements in relapse rates and survival can be expected using these methods.

Ideally, the most efficient method to eradicate cancer would be to use agents designed to interact specifically with malignant cells. The method approaching this ideal most closely is the use of monoclonal antibodies directed against tumor-associated antigens. It is known that monoclonal antibodies injected in vivo can concentrate on tumor cells; however, the anti-tumor effect is limited, in part due to the fact that some tumor cells lack target antigens, and in part because some cells, though coated by antibody, may not be killed by it. Attempts are being made to link antibodies to toxins such as the Ricin A chain for more effective tumor cell kill. Also in progress are studies attaching monoclonal antibodies to short-lived radioactive isotopes which deposit most of their energy within a 1-2 mm radius. This way, cells expressing the target antigens will be killed as will also be neighboring cells which may be antigen negative. In the case of hematologic malignancies, this ap-

proach would ablate normal marrow cells, and subsequent marrow "rescue" would be needed. Initial experiments in a canine model of marrow transplantation have shown appropriate antibody isotope conjugates to localize preferentially in the marrow and spleen and to a lesser extent also in lymph nodes (35,36). The amount of isotope in the marrow compared to other organs achieves a ratio of 5:1 or better. The marrow aplasia caused by radiolabeled antibodies can be reversed by infusion of cryopreserved autologous marrow eight days later at a time when very little radioactivity is left. Various combinations of chemotherapy, TBI, and radiolabeled antibodies are being explored for their ability to prepare dogs for T cell depleted marrow grafts. It is probable that refinements of this approach, particularly the use of high energy beta emitting isotopes with short linear energy transfer, will lead to less toxic but more efficient conditioning programs, not only providing better elimination of malignant disease but also ameliorating the problem of graft failure.

Radiolabeled antibodies might be useful not only in transplantation of patients with malignant hematological diseases but also of patients with aplastic anemia, thalassemia, and other genetic diseases. Radiolabeled antibodies might allow one to reduce the dose of cyclophosphamide needed for engraftment in patients with aplastic anemia and to replace busulfan in patients with thalassemia major.

PREVENTION OR TREATMENT OF INTERSTITIAL PNEUMONIA

Among the most serious complications during the first 3-4 months after transplantation are interstitial pneumonias (reviewed in 23,37,38). These are less frequent in patients grafted for aplastic anemia following cyclophosphamide than in patients with leukemia whose conditioning regimen includes TBI or busulfan. Pneumocystis carinii infection, formerly the cause of about 10% of all interstitial pneumonias, is now being prevented by prophylactic trimethoprim sulfamethoxazole. By far the most critical infection is cytomegalovirus (CMV) infection. Evidence of CMV activation is seen in about 3/4 of all patients with positive antibody titers to CMV before transplant. While often asymptomatic and manifested only by viral excretion in the urine or by increasing antibody titers, CMV activation can develop into a serious complication in the form of CMV pneumonia. CMV pneumonia has a case fatality rate of approximately 85%. Patients who are CMV seronegative before transplant can be protected from infection by the use of CMV seronegative blood products after transplant. If possible, only CMV negative blood products should be given to any CMV negative patient who is a potential transplant candidate. Immunoprophylaxis using CMV immunoglobulin has been controversial. There is no current effective proven therapy for established CMV infection. The use of an acyclovir derivative, dihydroxymethylethoxymethylguanine, is not effective in treating CMV pneumonia although it's use has significantly reduced the amount of virus in the lung tissues. It may be beneficial when given along with CMV immunoglobulin in treating established CMV pneumonia. Also, the drug might be useful in prophylactic trials. Idiopathic interstitial pneumonia was seen in approximately 13% of patients given single dose TBI, but the incidence has declined to 3% with the use of fractionated TBI.

CONDITIONING RELATED TOXICITY AND EARLY INFECTIONS

Conditioning regimen related toxicity can only be influenced by the development of less toxic regimens, as described above. It is possible that the use of certain recombinant human hematopoietic growth factors, such as IL-1, IL-3, G-CSF, GM-CSF, etc., might shorten the period of granulocytopenia or thrombocytopenia after grafting, reducing the incidence of early infection and resulting in a modest improvement of survival, on the order of 5%.

CONCLUSIONS

In the early 1970's marrow transplants were restricted to patients who had advanced acute leukemia, severe aplastic anemia, and severe combined immunodeficiency diseases. Since then, the technique has been shown to be beneficial and even curative for patients with many different hematological conditions. In younger patients, marrow grafting is the treatment of choice for aplastic anemia, immunodeficiency disease, certain genetic disorders of hemopoiesis, any leukemia which has relapsed at least once, ANL in first remission and CML. For patients who have thalassemia major, CML in chronic phase and ANL in first remission, the risk of early death from transplant complications must be weighed against the benefit of long-term cure.

Despite impressive advances in transplant results, major problems remain. These include recurrence of leukemia, graft failure in patients given T depleted or HLA-nonidentical grafts, acute and chronic GVHD, infections associated with prolonged immunodeficiency, and late occurring complications resulting from the conditioning programs. Major advances must come in the area of more effective and less toxic conditioning regimens. The use of monoclonal antibodies linked to short-lived radioactive isotopes with short linear energy transfer holds much promise in this regard. It is likely that more effective conditioning programs will reduce drastically the problems of recurrence of leukemia and of graft failure. They may permit a broader application of T cell depletion to prevent acute and chronic GVHD, thus extending marrow grafting to include more HLA-nonidentical and unrelated patients. As combinations of immunosuppressive agents such as methotrexate and cyclosporine have already decreased the incidence of acute GVHD, the use of recombinant hemopoietic growth factors

might reduce the risk of early infections. The problem of CMV infection in seropositive recipients will remain until effective antiviral drugs have been developed.

ACKNOWLEDGEMENTS

This work was supported in part by grant HL 36444 from the National Heart, Lung and Blood Institute, and by grants CA 18029, CA 18221, CA 31787, CA 18105 and CA 15704 awarded by the National Cancer Institute, National Institutes of Health, DHHS.

REFERENCES

1. Gratwohl A, Hermans J, Barrett AJ, Ernst P, Frassoni F, Gahrton G, Granena A, Kolb HJ, Marmont A, Prentice HG, Speck B, Vernant JP, Zwaan FJ (1988) Allogeneic bone marrow transplantation for leukaemia in Europe: Report from the Working Party on Leukaemia, European Group for Bone Marrow Transplantation. Lancet i: 1379-1382

2. Ringden O, Zwaan F, Hermans J, Gratwohl A, for the Leukemia Working Party of the European Group for Bone Marrow Transplantation (1987) European Experience of Bone Marrow Transplantation for Leukemia. Transp Proc 19: 2600-2604

3. Champlin R, for the Advisory Committee of the International Bone Marrow Transplant Registry (1987) Bone marrow transplantation for acute leukemia: A preliminary report from the International Bone Marrow Transplant Registry. Transplant Proc 19: 2626-2628

4. Gluckman E (1987) Current status of bone marrow transplantation for severe aplastic anemia: A preliminary report from the International Bone Marrow Transplant Registry. Transplant Proc 19: 2597-2599

5. Storb R, Doney K, Thomas ED, Anasetti C, Appelbaum F, Beatty P, Bensinger W, Buckner CD, Clift R, Fefer A, Hansen J, Hill R, Martin P, McGuffin R, Sanders J, Singer J, Stewart P, Sullivan K, Witherspoon R (1988) Allogeneic and syngeneic marrow transplantation for aplastic anemia: Overview of Seattle results. In: Baum SJ, Santos GW, Takaku F, eds. Recent Advances and Future Directions in Bone Marrow Transplantation (Experimental Hematology Today - 1987). New York: Springer-Verlag, 119-124

6. Storb R (1987) Critical issues in bone marrow transplantation. Transplant Proc 19: 2774-2781

7. Bortin MM, for the Advisory Committee of the International Bone Marrow Transplant Registry (1988) Key results from recent analyses: A report from the International Bone Marrow Transplant Registry. Proceedings of the Seventeenth

Annual Meeting of the International Society for Experimental Hematology, Houston, TX. In press.

8. Thomas ED, Storb R, Clift RA, Fefer A, Johnson FL, Neiman PE, Lerner KG, Glucksberg H, Buckner CD (1975) Bone-marrow transplantation. N Engl J Med 292: 832-843, 895-902

9. Lucarelli G, Galimberti M, Polchi P, Giardini C, Politi P, Baronciani D, Angelucci E, Manenti F, Delfini C, Aureli G, Muretto P (1987) Marrow transplantation in patients with advanced thalassemia. N Engl J Med 316: 1050-1055

10. O'Reilly RJ (1983) Allogeneic bone marrow transplantation: Current status and future directions. Blood 62: 941-964

11. Brochstein JA, Kernan NA, Groshen S, Cirrincione C, Shank B, Emanuel D, Laver J, O'Reilly RJ (1987) Allogeneic bone marrow transplantation after hyperfractionated total-body irradiation and cyclophosphamide in children with acute leukemia. N Engl J Med 317: 1618-1624

12. Goldman JM, Apperley JF, Jones L, Marcus R, Goolden AWG, Batchelor R, Hale G, Waldmann H, Reid CD, Hows J, Gordon-Smith E, Catovsky D, Galton DAG (1986) Bone marrow transplantation for patients with chronic myeloid leukemia. N Engl J Med 314: 202-207

13. Dinsmore R, Kirkpatrick D, Flomenberg N, Gulati S, Kapoor N, Shank B, Reid A, Groshen S, O'Reilly RJ (1983) Allogeneic bone marrow transplantation for patients with acute lymphoblastic leukemia. Blood 62: 381-388

14. Santos GW, Tutschka PJ, Brookmeyer R, Saral R, Beschorner WE, Bias WB, Braine HG, Burns WH, Elfenbein GJ, Kaizer H, Mellits D, Sensenbrenner LL, Stuart RK, Yeager AM (1983) Marrow transplantation for acute nonlymphocytic leukemia after treatment with busulfan and cyclophosphamide. N Engl J Med 309: 1347-1353

15. Gale RP, Bortin MM, van Bekkum DW, Biggs JC, Dicke KA, Gluckman E, Good RA, Hoffmann RG, Kay HEM, Kersey JH, Marmont A, Masaoka T, Rimm AA, van Rood JJ, Zwaan FE (1987) Risk factors for acute graft-versus-host disease. Br J Haematol 67: 397-406

16. Thomas ED, Clift RA, Fefer A, Appelbaum FR, Beatty PG, Bensinger WI, Buckner CD, Cheever MA, Deeg HJ, Doney K, Flournoy N, Greenberg P, Hansen JA, Martin P, McGuffin R, Ramberg R, Sanders JE, Singer J, Stewart P, Storb R, Sullivan K, Weiden PL, Witherspoon R (1986) Marrow transplantation for the treatment of chronic myelogenous leukemia.

Ann Intern Med 104: 155-163

17. Thomas ED, Buckner CD, Clift RA, Fefer A, Johnson FL, Neiman PE, Sale GE, Sanders JE, Singer JW, Shulman H, Storb R, Weiden PL (1979) Marrow transplantation for acute nonlymphoblastic leukemia in first remission. N Engl J Med 301: 597-599

18. Thomas ED, Sanders JE, Flournoy N, Johnson FL, Buckner CD, Clift RA, Fefer A, Goodell BW, Storb R, Weiden P (1979) Marrow transplantation for patients with acute lymphoblastic leukemia in remission. Blood 54: 468-476

19. Appelbaum FR, Dahlberg S, Thomas ED, Buckner CD, Cheever MA, Clift RA, Crowley J, Deeg HJ, Fefer A, Greenberg P, Kadin M, Smith W, Stewart P, Sullivan KM, Storb R, Weiden P (1984) Bone marrow transplantation or chemotherapy after remission induction for adults with acute nonlymphoblastic leukemia: A prospective comparison. Ann Intern Med 101: 581-588

20. Clift RA, Buckner CD, Thomas ED, Kopecky KJ, Appelbaum FR, Tallman M, Storb R, Sanders J, Sullivan K, Banaji M, Beatty P, Bensinger W, Cheever M, Deeg J, Doney K, Fefer A, Greenberg P, Hansen JA, Hackman R, Hill R, Martin P, Meyers J, McGuffin R, Neiman P, Sale G, Shulman H, Singer J, Stewart P, Weiden P, Witherspoon R (1987) The treatment of acute nonlymphoblastic leukemia by allogeneic marrow transplantation. Bone Marrow Transplant 2: 243-258

21. Storb R, Thomas ED, Buckner CD, Appelbaum FR, Clift RA, Deeg HJ, Doney K, Hansen JA, Prentice RL, Sanders JE, Stewart P, Sullivan KM, Witherspoon RP (1984) Marrow transplantation for aplastic anemia. Semin Hematol 21: 27-35

22. Storb R, Thomas ED (1985) Graft-vs-host disease in dog and man: The Seattle experience. In: Moller G, ed. Immunological reviews, No. 88. Copenhagen: Munksgaard, 215-238

23. Meyers JD (1988) Prevention and treatment of cytomegalovirus infection after marrow transplantation. Bone Marrow Transplant 3: 95-104

24. Storb R, Deeg HJ, Fisher LD, Appelbaum F, Buckner CD, Bensinger W, Clift R, Doney K, Irle C, McGuffin R, Martin P, Sanders J, Schoch G, Singer J, Stewart P, Sullivan K, Witherspoon R, Thomas ED (1988) Cyclosporine versus methotrexate for graft-versus-host disease prevention in patients given marrow grafts for leukemia: Long-term follow-up of three controlled trials. Blood 71: 293-298

25. Storb R, Deeg HJ, Whitehead J, Appelbaum

F, Beatty P, Bensinger W, Buckner CD, Clift R, Doney K, Farewell V, Hansen J, Hill R, Lum L, Martin P, McGuffin R, Sanders J, Stewart P, Sullivan K, Witherspoon R, Yee G, Thomas ED (1986) Methotrexate and cyclosporine compared with cyclosporine alone for prophylaxis of acute graft versus host disease after marrow transplantation for leukemia. N Engl J Med 314: 729-735

26. Storb R, Deeg HJ, Farewell V, Doney K, Appelbaum F, Beatty P, Bensinger W, Buckner CD, Clift R, Hansen J, Hill R, Longton G, Lum L, Martin P, McGuffin R, Sanders J, Singer J, Stewart P, Sullivan K, Witherspoon R, Thomas ED (1986) Marrow transplantation for severe aplastic anemia: Methotrexate alone compared with a combination of methotrexate and cyclosporine for prevention of acute graft-versus-host disease. Blood 68: 119-125

27. Sullivan KM, Witherspoon R, Storb R, Appelbaum F, Beatty P, Bensinger W, Bigelow C, Buckner CD, Cheever M, Clift R, Doney K, Fefer A, Greenberg P, Hansen J, Martin P, Matthews D, McDonald G, Meyers J, Petersen FB, Sanders J, Shulman H, Singer J, Stewart P, Thomas ED (1988) Chronic graft-versus-host disease: Pathogenesis, diagnosis, treatment and prognostic factors. In: Baum SJ, Santos GW, Takaku F, eds. Recent Advances and Future Directions in Bone Marrow Transplantation (Experimental Hematology Today - 1987). New York: Springer-Verlag, 150-157

28. Butturini A, Franceschini F, Gale RP (1988) Critical analysis of T-cell depletion in man. In: Martelli MF, Grignani F, Reisner Y, eds. T-cell Depletion in Allogeneic Bone Marrow Transplantation. Rome: Ares-Serono Symposia, 1-13

29. Goldman JM, Gale RP, Horowitz MM, Biggs JC, Champlin RE, Gluckman E, Hoffmann RG, Jacobsen SJ, Marmont AM, McGlave PB, Messner HA, Rimm AA, Rozman C, Speck B, Tura S, Weiner RS, Bortin MM (1988) Bone marrow transplantation for chronic myelogenous leukemia in chronic phase: Increased risk for relapse associated with T-cell depletion. Ann Intern Med 108: 806-814

30. Gordon-Smith EC (1987) Recent advances and future trends in bone marrow transplantation for severe aplastic anemia. In: Baum SJ, Santos GW, Takaku F, eds. Recent Advances and Future Directions in Bone Marrow Transplantation (Experimental Hematology Today - 1987). New York: Springer-Verlag, 125-129

31. Deeg HJ, Aprile J, Storb R, Graham TC, Hackman R, Appelbaum FR, Schuening F (1988) Functional dendritic cells are required for transfusion-induced sensitization in canine marrow graft recipients. Blood 71: 1138-1140

32. Deeg HJ, Aprile J, Graham TC, Appelbaum FR, Storb R (1986) Ultraviolet irradiation of blood prevents transfusion-induced sensitization and marrow graft rejection in dogs. Concise Report. Blood 67: 537-539

33. Thomas ED, Clift RA, Hersman J, Sanders JE, Stewart P, Buckner CD, Fefer A, McGuffin R, Smith JW, Storb R (1982) Marrow transplantation for acute non-lymphoblastic leukemia in first remission using fractionated or single-dose irradiation. Int J Radiat Oncol Biol Phys 8: 817-821

34. Tutschka PJ, Copelan EA, Klein JP (1987) Bone marrow transplantation for leukemia following a new busulfan and cyclophosphamide regimen. Blood 70: 1382-1388

35. Appelbaum FR, Badger C, Deeg HJ, Nelp WB, Storb R (1987) Use of Iodine-131-labeled anti-immune response-associated monoclonal antibody as a preparative regimen prior to bone marrow transplantation: Initial dosimetry. NCI Monographs 3: 67-71

36. Appelbaum FR, Brown PA, Graham TC, Sandmaier BM, Schuening FW, Storb R (1988) Characterization of malignant lymphoma in dogs and use as a model for the development of treatment strategies. In: Baum SJ, Santos GW, Takaku F, eds. Recent Advances and Future Directions in Bone Marrow Transplantation (Experimental Hematology Today - 1987). New York: Springer-Verlag, 31-35

37. Meyers JD, Flournoy N, Thomas ED (1982) Nonbacterial pneumonia after allogeneic marrow transplantation: A review of ten years' experience. Rev Infect Dis 4: 1119-1132

38. Winston DJ, Ho WG, Champlin RE, Gale RP (1984) Infectious complications of bone marrow transplantation. Exp Hematol 12: 205-215

24 Autologous Bone Marrow Transplantation for Acute Leukemia in Remission: Fifth European Survey. Results of 1021 Cases

N.C. Gorin, P. Aegerter, and B. Auvert

The European bone marrow transplantation group (EBMTG) which was born from the initiative of a few individuals in 1974 has since these early days considerably increased to a point where it now embraces a total of 600 physicians and scientists from 105 centres throughout Europe, who report annually their data for computer analysis. Scientific analysis within the EBMTG are conducted by 5 separate working parties : aplastic anemia, leukemias, immunology and bone marrow transplantation (BMT) inborn errors, and autologous bone marrow transplantation (ABMT). In the past decade, the EBMTG has published several interesting analysis (1,2) on the value of allogeneic BMT in aplastic anemia (AA) (3,4), acute leukemias (AL) (5,6,7) chronic myelocytic leukemia (CML) (8), lymphomas (9), myelomas (10) and immune deficiencies (ID). These analysis have produced results similar to those of the international bone marrow transplant registry (IBMTR) which combines data from Europe and the USA. (11,12,13,14,15,16).

Acknowledgment : Miss Patricia PALUT for help in computer analysis and preparation of manuscript.

Supported by EBMTG, grant 87750 from European Communities, grant 161 187 from Ministère de l'emploi et des Affaires sociales- Direction générale de la Santé , and grant 6113 from Association pour la recherche contre le cancer - Villejuif - France.

Correspondence : Pr. N.C. GORIN, service des maladies du sang - Hôpital Saint-Antoine - 184, rue du Fbg Saint-Antoine - 75571 PARIS CEDEX 12 - FRANCE.

The Working party on ABMT was initiated in 1981 ; its major aim was to collect information on individual patients treated with ABMT in Europe, for computer analysis, in an effort to answer several important questions such as the respective value of different pretransplant regimens, the validity of marrow purging, and most of all the efficacy of ABMT per se for the treatment of acute leukemias in various stages of the disease. It has also been its responsibility to make recommendations on methodology and if possible to build randomized trials to answer those questions. Previous reports and annual surveys of the EBMTG on ABMT have successively pointed out

1)   The high rate of remission obtained with ABMT in AL in relapse, unfortunately associated with the absence of reported long term disease free survival (17,18)

2)   In contrast the value of ABMT delivered as a high dose consolidation regimen with marrow rescue in patients in remission (CR) bearing only minimal residual disease (19,20,21)

3)   The differences existing in the kinetics of recovery of hemopoiesis between AML an ALL, with slower engraftment in AML on all 3 parameters:

PMN >. 5 $10^9/l$ (30 days vs 21, p<0.01), reticulocytes >.1% (24 days vs 15, p <0.01) and platelets > 50 $10^9/l$ (82 days vs 38, p<0.0001), and long lasting thrombocytopenias (> 1 year) only observed in AML whatever the nature of the autograft, purged or unpurged. (20,21)

4)   The possible importance of multiple courses of conventional chemotherapy for

consolidation prior to marrow collection, considered as in vivo purging. (21)

5)   Better results of ABMT in acute myelocytic leukemia (AML) over acute lymphoblastic leukemia (ALL) with the exception of the myelomonocytic (M4) subcategory yielding poorer results. (19,21)

The present report is based on 1021 patients with AL in CR autografted since 1981, who were analized for the 5th European survey. While this survey was conducted as the previous ones, a special effort was made to study the possible impact of intervals pretransplant and in vitro marrow purging on the final outcome.

I - MATERIAL AND METHODS :

In June 1987, all previously reporting teams received a complete print out of their own data for verification and necessary corrections. Modifications were entered in September 1987. New questionnaires (new patients and follow up forms) were sent to 54 teams of whom 50 had appropriately answered by December 15, 1987. Forms have been reviewed up to January 30, 1988 and further clarification requested for 10%. A considerable effort has been made to ensure the accuracy of the present database prior to the introduction of the new homogenized EBMTG computer program. This analysis has been done on 1021 cases from 50 reporting institutions (see appendix). Patients were classified as high risk (hr) if they had anyone of the following characteristics

(a) Leukocyte count > 100 10^9 at presentation

(b) Acute leukemia secondary to treatment for other malignancy

(c) Neuroleukemia

(d) Presence of a Philadelphia chromosome. Patients without any high risk feature were classified as having " standard risk " (std).

Distribution according to diagnosis and status was the following : ALL : 436, CR std : 138, pr : 63, CR2 std : 138, pr : 31. Median age of the population 17 y (1-54), children (<15y) : 43%, adult >45y : 3%. AML : 540, CR1 std : 353, pr : 40, CR2 std : 108, pr : 6. Median age of the population 32y (1-67), children : 15%, adults>45y 15%.
Others : 45.
Median interval diagnosis-ABMT : 200 days (55-1290). Among all complications (infections : bacterial 42%, viral 18%, fungal 13%, pneumonitis 9.5%, cardiac failure 3.6%, liver V.O.D. 2.9%), only liver veino occlusive disease was reported with a significantly higher incidence in AML (3.9% vs 1.6% in ALL, p<0.05).

Populations studied were stratified in subcategories as follows : ALL, AML and according to the FAB classification, CR1 and CR2, standard and poor risk, adults and children, all grafted patients (median follow up : 700 days) and populations grafted prior to January 1987 (follow up 826 days) and prior to January 1986 (follow up 1026 days). Different pretransplant regimens were studied but TBI was chosen for subsequent studies as the reference. The impact of the intervals pretransplant was studied by comparing patients who were grafted with intervals CR-ABMT of <3 months, 4 to 6 months, 6 to 9 months and >9 months. In vitro treatments considered were : mafosfamide globally (dose ≥50μg/ml) and mafosfamide dissected in 5 different laboratory techniques including treatment of ficoll purified mononuclear cells, buffy coat and treatment adjusted to individual patients (22) ; monoclonal antibodies (ALL only) and other chemotherapies (highly heterogenous). While studying the value of marrow purging in all possible situations as mentionned above, the following working hypothesis were made : purging efficacy might appear particularly in situations of real minimal residual disease (CR1, post TBI) and/or conversely in the absence of effective previous in vivo purging (interval CR-ABMT <3 months). This hypothesis apparently turned out to be fruitful.
Fig 1 and tables 1 to 5 indicate the annual reported numbers of ABMT (fig.1), the age distribution of the patients (table 1), the different pretransplant regimens used (table 2), the incidence of reported toxic complications (table 3), and the proportions of ABMT done with unpurged and purged marrow in ALL (table 4) and AML (table 5).

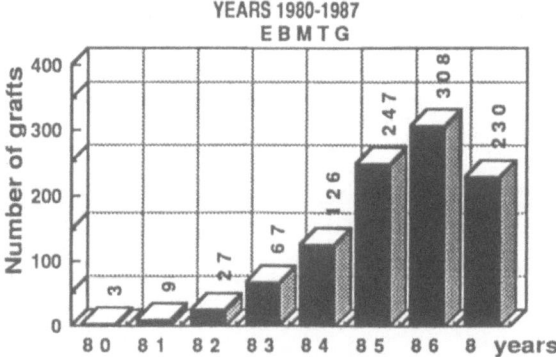

NUMBER OF A B M T FOR ACUTE LEUKEMIA
YEARS 1980-1987
E B M T G

ABMT IN AL
EBMTG SURVEY 4/88
AGE DISTRIBUTION

	ALL	AML
Median y	17	32
Range y	0 - 54	0 - 67
Children % < 15y	43	15
Adult % 15 - 45	52	69
Adult % > 45	3	15

TABLE 1

A B M T in ACUTE LEUKEMIA
EBMTG SURVEY 4/88
PRETRANSPLANT REGIMENS %

	ALL (n=436)			AML (n=540)		
	CR1 std	CR1 poor risk	CR2	CR1std	CR1 poor risk	CR2
TBI	41	36.5	24	33	35	13
FTBI	40	57	54	23	17.5	37
BU + CY	4.5	1.5	6.5	9	0	16
BACT	1.5	1.5	0.7	3	12.5	1
BAVC	1.5	0	3.5	11	7.5	15
UCH	8	0	6	7.5	12.5	4.5

TABLE 2

ABMT FOR AL : EBMTG SURVEY
April 88
COMPLICATIONS REPORTED
INCIDENCE (%)*

NATURE OF COMPLICATIONS	ALL	AML	TOTAL
BACTERIAL INFECTIONS	43.1	40.6	41.9
VIRAL INFECTIONS	17.9	17	17.7
FUNGAL INFECTIONS	14.2	12.8	13.3
PNEUMONITIS	8.7	10	9.5
CARDIAC FAILURE	3.9	3.3	3.6
LIVER V.O.D.	1.6 **	3.9 **	2.9

*Number of patients with indicated
complication/total number of patients autografted
** p < 0.05

TABLE 3

ABMT in AL
EBMTG SURVEY 4/88
MARROW PURGING IN ALL (n=436) %

	CR1 Std	CR1 poor risk	CR2 Std	CR2 poor risk
NONE	46	13	22.5	16
MAFOSFAMIDE	18	42.8	31	55
OTHER CHEMO.	11	9.5	16	16
MoAb	24	35	13	13

There is a tendency to purge
and to use Mafosfamide
in poor risk ALL

TABLE 4

A B M T in AL
EBMTG SURVEY 4/88
MARROW PURGING IN AML % (n=540)

	CR1 Std	CR1 poor risk	CR2
NONE	71	65	53
MAFOSFAMIDE	22	27.5	28.5
OTHER CHEMO.	5.5	7.5	16.5
MoAb	1.5	0	2

TABLE 5

II - RESULTS IN ALL

1) The DFS for patients autografted in CR1 std (more than 80% received TBI) was 40% at 60 months. For patients grafted in CR2, results were better in children (40% at 45 months : 103 patients) than in adults (18% at 20 months : 52 patients). (Fig. 2,3).

2) Long intervals pre ABMT were associated with better DFS in several situations . For the whole population of patients autografted in CR1 and those with standard risk factors only, the DFS was better for long intervals CR1-ABMT (p = 0,008 and p = 0,009). In patients grafted in CR2, both the durations of intervals CR2-ABMT and CR marrow collection had an impact on the outcome (table 6). For instance, in ALL CR1 std DFS were respectively 30% and 68% for delays CR1-ABMT ≤ 6 months and ≥7 months (p<0,01) (Fig.4). In ALL CR2, the DFS was 68% at 45 months in patients whose marrows had been collected ≥6 mo post CR2 as compared to 25% at 20 months for smaller intervals or ABMT done with marrow collected in CR1. (Fig. 5).

3) There was a trend in favour of marrow purging with mafosfamide in the following situations :
CR1 poor risk : - Mafosfamide vs monoclonal antibodies : DFS 63% at 21

months vs 39% (p = 0,06). - Mafosfamide individually adjusted vs Mo Ab 82% at 20 months vs 40% (p<0,05). (Fig.6 and 7).
<u>CR1 with interval CR1-ABMT ≤3 months :</u>
Mafosfamide vs Mo Ab 51% at 22 months vs 44% (p = 0.15).
However in opposition to AML (see below) the numbers of patients in the groups studied were small and all other comparisons have not shown any advantage of marrow purging.

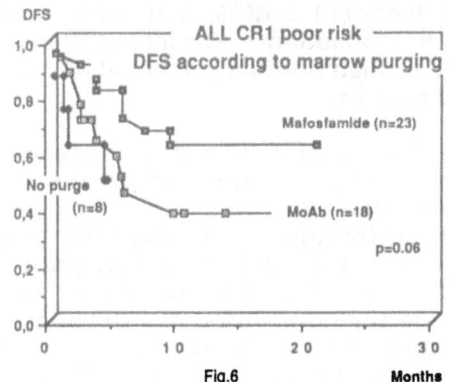

Fig.6

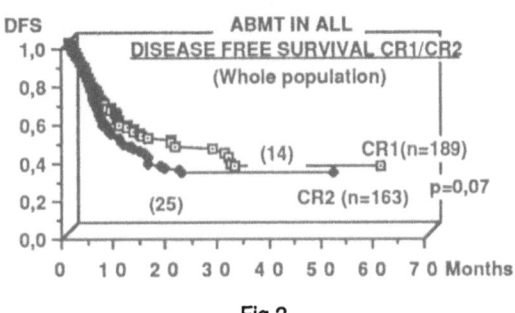

Fig.2

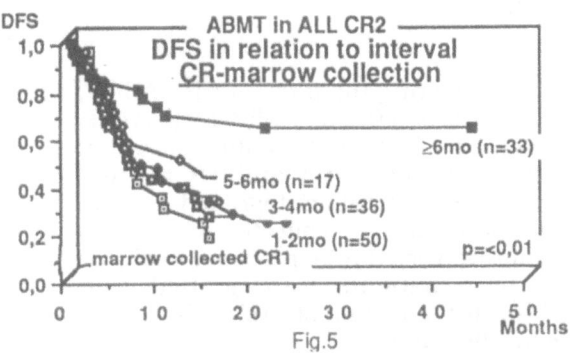

Fig.5

AUTOLOGOUS BONE MARROW TRANSPLANTATION IN ALL
INFLUENCE OF INTERVALS PRE TRANSPLANT

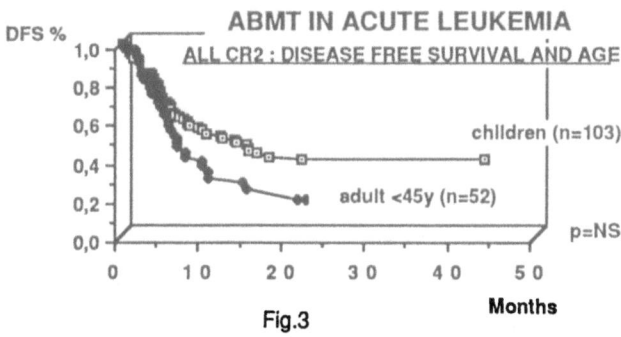

Fig.3

POPULATION STUDIED (n)	INTERVAL	STATISTICAL SIGNIFICANCE (P)*
CR1 (178)	CR1-ABMT	0,008
	CR1-Collection	0,24
	Coll-ABMT	0,32
CR1 Std (124)	CR1-ABMT	0,009
	CR1-Collection	0,1
	Coll-ABMT	0,11
CR1 poor (54)	CR1- ABMT	0,75
	CR1-Collection	0,68
	Coll-ABMT	0,005 (?)
CR2 (160)	CR2-ABMT	0,028
	CR1/CR2-Collection	0,009
	Coll-ABMT	0,77
CR2 Std (130)	CR2-ABMT	0,16
	CR2-Collection	0,055
	Coll-ABMT	0,47

* Mantel Cox

TABLE 6

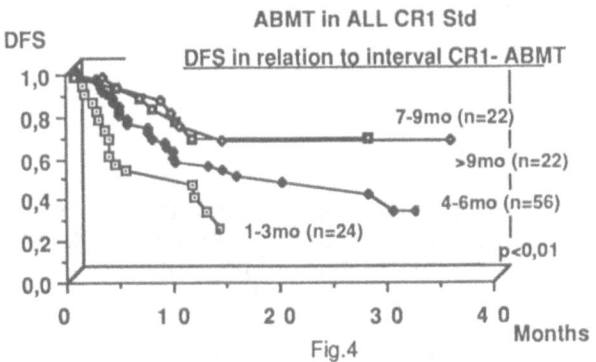

Fig.4

III - RESULTS IN AML

1) The disease free survival (DFS) in the whole heterogenous group of patients autografted in CR1 was 34% at 75 months comparing with 32% at 40 months for patients grafted in CR2. (Fig.8). According to the FAB classifications, DFS were : M1 : 38%, M2 : 39%, M3 : 45%, M4 : 30%, M5 : 53% (p = NS). For AML CR1 std, DFS in relation to pretransplant regimens were :

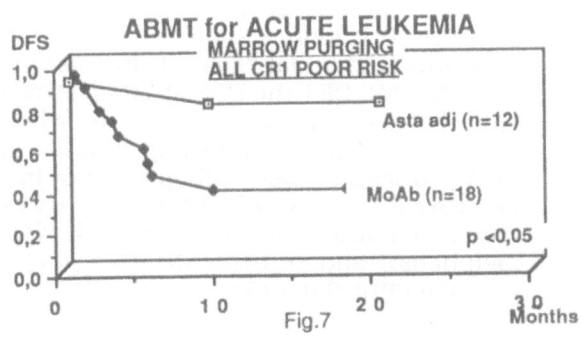

Fig.7

UCH (London) : 60%, TBI 45%, at 50 months, Busulfan + Cyclophosphamide, BACT or high dose melphalan alone <30% at 20 months.

2) As in ALL, there was an impact of the pretransplant intervals on the final outcome (table 7) :
Long intervals pretransplant were associated with better DFS in patients grafted in CR1. For instance, when considering the interval CR1-marrow collection, DFS at 50 months were 38% below 6 months and 60 % above 6 months (p = 0.02).

When considering the interval CR1-ABMT, DFS were 26%, 40% and 65% respectively for intervals of <3 mo (63 patients), 4 to 9 months (222 patients) and >9 mo (41 patients). For patients autografted in CR2, utilization of marrow collected in CR1 did not improve the DFS (one reason might be that these patients were grafted very quickly after induction of CR2 with no time for consolidation).

3) The impact of marrow purging with mafosfamide was demonstrated or suggested in the following situations (Table 8).
CR1 std : DFS if marrow purged by mafosfamide 52% at 43 months, vs 30% in the absence of purge (p = 0.10).
CR1 autografted after TBI : DFS 56% at 43 months vs 32% (p = 0.02) (Fig.9).
CR1 autografted less than 3 months post CR1 : 40% at 23 months vs 20% (p = 0.06).
CR1 std : marrow purged with mafosfamide individually adjusted vs no purge : 64 % at 38 mo vs 35% (p = 0.009).
However this last comparison remains highly questionable since individual adaptation of the dose of mafosfamide for marrow purging was done in only 2 institutions (Paris Saint-Antoine and Tours, France).

AUTOLOGOUS BONE MARROW TRANSPLANTATION IN AML
INFLUENCE OF INTERVALS PRE TRANSPLANT

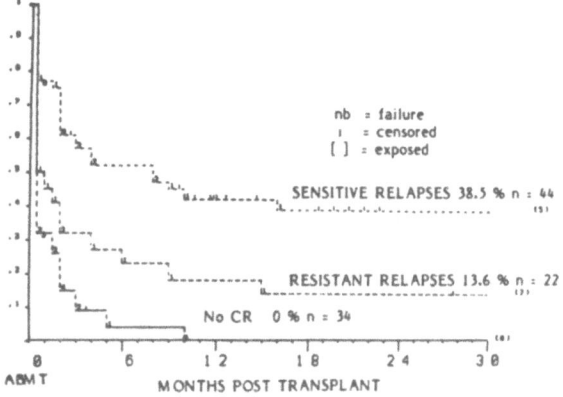

MANTEL COX   Table 7

EFFICACY OF MARROW PURGING IN ABMT FOR AML

RESULTS

DFS (months)

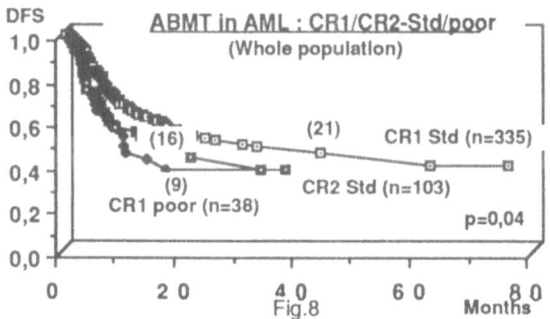

POPULATION (number of patients)	MAFOSFAMIDE	NO PURGE	P *
CR1 Std (69/237)	52 % (43)	30 % (75)	0.10
CR1 TBI (57/138)	56 % (43)	32 % (59)	0.02
CR1 CR-ABMT<3mo (12/47)	40 % (23)	20 % (16)	0.06
CR1 Std (27/237)	ADJUSTED 64 % (38)	35 % (78)	0.009

* MANTEL COX

Table 8

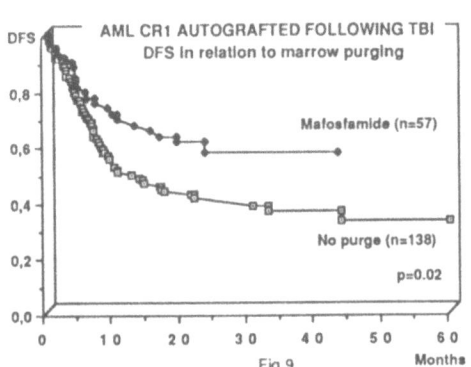

Fig.8

Fig.9

IV - DISCUSSION

The data of the EBMTG registry on 1021 AL consolidated by ABMT in CR, give several informations of potential value :
In ALL CR1, the DFS at 60 months is 38% when considering the whole population, 40% when restricting the study to CR1 std receiving total body irradiation. It rises to 68% when considering only the minority of patients grafted late, more than 7 months post initial diagnosis (n = 44).
In AML CR1 the DFS at 70 months is only 34% when considering the whole population of patients, mixing together all FAB subcategories, all pretransplant regimens including even monochemotherapy (such as high dose melphalan) and transplants with purged and non purged marrow.

It however rises to 44% when taking into account only standard risk patients receiving TBI. The DFS seems better in the group of patients receiving TBI and autografted with marrow purged by mafosfamide (58%). In the minority (n = 41) autografted late, more than 9 months post initial diagnosis, the DFS reaches a plateau at 64% from 35 to 55 months. The best DFS (68% at 40 months is observed in patients with std CR1 receiving TBI pretransplant and marrow purged by mafosfamide at the individually CFUGM LD 95-90 level. This finding remains however highly questionable since it concerns only a group of 27 patients treated in 2 institutions so that a selection bias cannot be excluded.

ABMT in CR2 is associated with DFS as high as 32% both in AML and ALL. However in AML the follow up remains still short (32% DFS at 40 months) and in ALL, the DFS in fact can be subdivided in 40% at 45 months for children in contrast to 18% only at 20 months for adults.

Two important findings of this study were the influence of the pretransplant intervals and marrow purging on DFS :

a) In both ALL and AML CR1 std, long intervals CR1-marrow collection and CR1-ABMT were signifcantly associated with better DFS. The cut-off points for CR1-marrow collection and CR1-ABMT were respectively 5 and 7 months for ALL and 6 and 9 months for AML. As mentionned above the DFS in patients autografted more than 7 months post diagnosis in ALL and more than 9 months post diagnosis in AML were high, 68% and 64% respectively.

In contrast patients grafted early (less than 3 months post obtention of CR1) had the worst DFS (ALL : 20% at 15 months ; AML 26% at 30 months). Patients grafted late were a minority in AML where they represented 12.6% of the whole population (41/326) and this minority was somehow balanced by the population grafted early which represented 19% (63/326). In fact global results in AML were close to results observed in the intermediate population autografted 4 to 6 and 7 to 9 months post CR (34% at 70 months and 40% at 60 months), an observation which is consistent with the observed median duration of the interval CR-ABMT of 134 days. Therefore, it seems reasonable to assume that the population of patients grafted late did not significantly modify and/or improve the overall DFS in AML. In contrast in ALL, where the population autografted late represented a higher percentage (35% : 44/124), not totally

balanced by the population transplanted early (19% : 24/124), such an influence cannot be ruled out. The influence of the delay CR-marrow collection was also found to influence DFS in patients with ALL autografted in second CR : while the DFS was only 20% at 20 months in those grafted before 6 months, it was 64% in 33 patients grafted later. Such a relation could not be observed in AML CR2. The finding of a relationship between DFS and intervals pretransplant with a favourable impact of long intervals was not unexpected : 2 explanations may be proposed : the first is a selection bias: patients transplanted late are more likely to be cured already by chemotherapy at the time when they are transplanted and therefore the relapse rate is smaller. Comparison of these patients with similar groups treated by conventional chemotherapy with DFS curves initiated only with those still in CR, 7 and 9 months post induction, would be one way to test this hypothesis. An alternative explanation can be that patients transplanted late have received more courses of consolidation chemotherapy in the pretransplant period. Whether these additional courses have had a direct impact by reducing the tumor load in the patient, or an indirect one by introducing a reduction in the contamination of the marrow collected (in vivo purging) or even both cannot be sorted out. It is tempting to postulate that patients with ALL in CR2 who were autografted with marrow collected in CR1 had a poor DFS because they were transplanted early and received less consolidation courses than those who received their marrow collected in CR2.

b) While the study on the efficacy of marrow purging was done in all possible subpopulations of patients, we felt that we should select TBI as being the most homogeneous tumoricidal pretransplant regimen. Then from preclinical animal studies (23,24,25) we postulated that purging efficacy might be easier to demonstrate in situations of real minimal disease, such as CR1, post TBI. Also if effective in vivo purging (i,e numerous consolidation courses) might mask the possible benefit of in vitro purging (which would then be unnecessary), conversely the efficacy of in vitro purging would rather appear in its absence or insufficiency such as in the group of patients grafted early. These working hypothesis turned out to be fruitful, most of all in AML : In patients autografted in CR1 following TBI, the DFS was 56% in the group receiving marrow purged with mafosfamide versus 32% in the group receiving non purged marrow (p = 0.02). In patients with standard risk AML the value of marrow purging was suggested in those grafted less than 3 months post CR (40% vs 20% at 23 months p = 0.06). There was still a trend in favour of marrow purging for those grafted 4 to 9 months post CR but results were identical for longer intervals. This demonstration or suggestion of the efficacy of marrow purging in AML is obtained after 2 unfructuous previous analysis of the EBMTG (1985,1986) in which no difference could be shown whether marrow was purged or not. As a consequence of these previous reports, many teams throughout Europe have given up purging in AML and several multicentric randomized studies comparing conventional CT to allogeneic and autologous BMT have selected unpurged marrow for ABMT, leading to a considerable reduction in the numbers of patients autografted with purged marrow. Data presented here in favour of purging, now obtained with more patients, a longer follow up and maybe a more complete methodology may modify the whole therapeutic strategy.

Trends in favour of marrow purging were similarly observed in ALL but only in particular situations : poor risk ALL in CR1, and interval CR1-ABMT ≤ 3 months.

Finally, while the use of mafosfamide adjusted at an individual level of sensitivity defined as CFUGM LD-90-95% was associated with the best observed DFS both in AML and ALL, these data still remain controversial since coming from only 2 institutions, which therefore cannot rule out some selection bias.

IV - CONCLUSIONS

1) Autologous bone marrow transplantation using marrow purged by mafosfamide, following total body irradiation, produces DFS of 58% in AML CR1, 40% in ALL CR1, 32% in AML CR2 and 40% in childhood ALL CR2.

2) Marrow purging by Mafosfamide now appears to be effective or likely so, in selected situations : such as CR1, post TBI, presence of poor risk factors, interval CR-ABMT<3 months, with results in AML being more suggestive. We propose that these results in these particular situations reflect the efficacy of marrow purging when the residual disease is really minimal and/or the absence of a previous sufficient in vivo purging renders in vitro purging necessary/effective and evaluable. Efforts should now be conducted towards the generation of randomized protocols studying the value of marrow purging by mafosfamide, in AML CR2 and possibly AML CR1 autografted after TBI and poor risk ALL CR1.

Outside such randomized studies, we feel that abstention of marrow purging should not any longer be approved/accepted in all situations.

3) Recommendations of today may include :

TBI as the standard pretransplant regimen (BU + Cy not yet evaluable).
Delay CR-ABMT not <3 months, to allow minimum in vivo purging.
In vitro purging with Mafosfamide.

APPENDIX
AUTOLOGOUS BONE MARROW TRANSPLANTATION
FOR ACUTE LEUKEMIA
LIST OF INSTITUTIONS REPORTING DATA

TEAM	COORDINATOR	NUMBER OF PATIENTS
ROMA Univ. La Sapienza, ITALY	MELONI	93
PARIS ST ANTOINE, FRANCE	GORIN	79
HEIDELBERG, WEST GERMANY	KORBLING	69
BESANCON, FRANCE	HERVE	68
BLOOMSBURY, ENGLAND	GOLDSTONE	60
MARSEILLE, FRANCE	MARANINCHI	52
GLASGOW, UK	BURNETT	39
GENOVA, ITALY	CARELLA	38
LEIPZIG, DDR	HELBIG	38
LYON, FRANCE	SOUILLET	30
UPPSALA, SWEDEN	SIMONSSON	30
UTRECHT, THE NETHERLANDS	VERDONCK	25
PARMA, ITALY	RIZZOLI	25
TOURS, FRANCE	COLOMBAT	21
PESARO, ITALY	PORCELLINI	20
BRUXELLES, BELGIUM	FERRANT	20
WESTMINSTER, ENGLAND	BARRET-POYNTON	18
BORDEAUX, FRANCE	REIFFERS	17
ROTTERDAM, THE NETHERLANDS	LOWENBERG	17
NANTES, FRANCE	HAROUSSEAU	16
NIJMEGEN, THE NETHERLANDS	DE WITTE	16
PESCARA, ITALY	TORLONTANO	16
MILANO, ITALY	POLLI	15
PAVIA, ITALY	ALESSANDRINO	15
LEIDEN, THE NETHERLANDS	DE PLANQUE	14
ULM, WEST GERMANY	WIESNETH	12
TRIESTE, ITALY	ANDOLINA	12
TORINO, ITALY	AGLIETTA	12
BIRMINGHAM, ENGLAND	FRANKLIN	11
NEWCASTLE, ENGLAND	PROCTOR	11
PARIS HOTEL DIEU, FRANCE	ZITTOUN	11
NANCY, FRANCE	WITZ	10
PADOVA, ITALY	COLLESELLI	10
BARCELONA, SPAIN	BRUNET	9
GENEVA, SWITZERLAND	CHAPUIS	8
BOLOGNA, ITALY	VISANI-TURA	8
WIEN, AUSTRIA	HINTERBERGER	7
ROMA S. CAMILLO, ITALY	DE LAURENZI	7
VALL D'HEBRON, SPAIN	ORTEGA	7
NICE, FRANCE	GRATECOS	6
PARIS, LA PITIE, FRANCE	LEBLOND	6
BERN, SWITZERLAND	BRUN DEL RE	6
INNSBRUCK, AUSTRIA	HUBER	5
BOLZANO, ITALY	COSER	4
ST ETIENNE, FRANCE	FREYCON-FRAPPAZ	3
LONDON, UK	PRENTICE	2
PARIS, COCHIN , FRANCE	BELANGER	2
MILANO-HOSP. NIGUARDA, ITALY	DE CATALDO	2
ROTONDO S.GIOVANNI, ITALY	GRECO	2
	TOTAL	1021

REFERENCES

1. Proceedings of the XIIth annual meeting of the European cooperative group for bone marrow transplantation : Bone Marrow Transplantation, sup 1, 1, 1986.

2. Proceedings of the XIIIth annual meeting of the European cooperative group for bone marrow transplantation : Bone Marrow Transplantation, sup 2, 2, 1987.

3. BACIGALUPO A., VAN LINT M.T., CONGIN M., PITTALUGA P.A., OCCHINI D., MARMONT A.M. : Treatment of SAA in Europe 1970-1 1985 : A report of the SAA working party, Bone Marrow Transplant. 1, sup 1,19, 1986.

4. GLUCKMAN E., MARMONT A., SPECK B., GORDON-SMITH E.C. : for the Working party of severe aplastic anemia of the European group for bone marrow transplantation : seminars in Hemat, 21,11-19, 1984.

5. FRASSONI F., BARRETT A.J., GRANENA A., ZWAAN F., GRATWOHL A. : for the leukemia working party of the European Group for bone marrow transplantation " relapse after allogeneic bone marrow transplantation for acute leukaemia: A survey by the EBMT of 117 cases. Submitted for publication.

6. GRATWOHL A., HERMANS J., BARRETT A.J., ERNST P., FRASSONI F., GAHRTON G., GRANENA A., KOLB H.J., MARMONT A., PRENTICE H.G., SPECK B., VERNANT J.P., ZWAAN F.J. : Allogeneic bone marrow transplantation for leukemia in Europe. Report from the leukemia working party of the European group for bone marrow transplantation, Lancet 1988, in Press.

7. RINGDEN O., ZWAAN, HERMANS J., GRATWOHL A. : for the leukemia Working party of the European group for bone marrow transplantation: European Experience of bone marrow transplantation for leukemia. Transpl. Proceedings, 19, 2600-2604, 1987.

8. GRATWOHL A., GOLDMAN J.M., GLUCKMAN E., ZWAAN F. : Effect of splenectomy before bone marrow transplantation on survival in chronic granulocytic leukaemia., Lancet, 2, 1290, 1985.

9. ERNST P., MARANINCHI O., JACOBSEN N., KOLB H.J., BORDIGONI P., LJUNGMAN P., BANDINI G., PARKER A.C., VOLIN L., POWLES R., GORIN N.C., RIO B. : Marrow transplantation for non Hodgkin's lymphoma : a multi centre study from the European Cooperative bone marrow transplant group : Bone Marrow Transplantation, 1, 81-86, 1986.

10. GAHRTON G., TURA S., FLESCH M., GRATWOHL A., GRAVETT P., LUCARELLI G., MICHALLET M., REIFFERS J., RINGDEN O., VAN LINT M.T., VERNANT J.P. and ZWAAN F. : Bone marrow transplantation in multiple myeloma. Report from the European Cooperative group for bone marrow transplantation, Blood 69, 1262-1264, 1987.

11. BORTIN M.M. : Bone Marrow Transplantation : The difficulties ahead. In :Experimental Hematology Today - 1987, BAUM S.J. SANTOS G.W., TAKAKU F., (Eds.), SPRINGER-VERLAG, New York, 134- 140, 1987.

12. BORTIN M.M., GALE R.P. : Current status of allogeneic bone marrow transplantation : A report from the International bone Marrow Transplant Registry. In : Clinical Transplants, 1987, (Terasaki PI, Ed), UCLA Tissue Typing Laboratory, Los Angeles, California, 127- 140, 1987.

13. GALE R.P., BORTIN M.M., VAN BEKKUM D.W., BIGGS J.C., DICKE K.A., GLUCKMAN E., GOOD R.A., HOFFMANN R.G., KAY H.E.M., KERSEY J.H., MARMONT A., MASAOKA T., RIMM A.A., VAN ROOD J.J., ZWAAN F. : Acute graft-vs-host disease following bone marrow transplantation in humans : Assessment of risk factors. Br J Haematol, 67, 397-406, 1988.

14 GOLDMAN J.M., BORTIN M.M., BIGGS J.C, CHAMPLIN R;E., GALE R.P., HOFFMANN R.G., JACBOBSEN S.J., MARMONT A.M., MESSNER H.A., McGLAVE P.B., RIMM A.A., ROZMAN C., SPECK B., TURA S., WEINER R.S.: Bone Marrow Transplantation for chronic myelogenous leukemia in chronic phase : Increased risk of relapse associated with T cell depletion. Annals of Intern Med, in press, 1988.

15. SPECK B., BORTIN M.M., CHAMPLIN R.,GOLDMAN J.M., HERZIG R.H., McGLAVE P.B., MESSNER H.A., WEINER R.S., RIMM A.A. : Allogeneic bone-marrow transplantation for chronic myelogenous leukemia, Lancet 1, 665-668, 1984.

16. WEINER R.S., BORTIN M.M., GALE R.P., GLUCKMAN E., KAY H.E.M., KOLB H.J., RIMM A.A. : Risk factors associated with interstitial pneumonitis following allogeneic bone marrow transplantation for leukemia. Transplant. Proc.17,470-474, 1985.

17. GORIN N.C., HERVE P., PHILIP T. : High dose therapy and autologous bone marrow transplantation in France : in Bone Marrow Transplantation in Europe 2 : Symposia Fondation Merieux 6:TOURAINE J.L., GLUCKMAN E., GRISCELLI C., Eds, Excerpta Medica Amsterdam - Oxford-Princeton, 42- 45, 1981.

18. GORIN N.C., AEGERTER P., PARLIER Y. : For the Working party on autologous bone marrow transplantation of the EBMTG : Autologous bone marrow transplantation for acute leukemia in remission : 2nd European survey: Exp Hematol 13, sup 17,18-19, 1985.

19. GORIN N.C., HERVE P., AEGERTER P., GOLDSTONE A., LINCH D., MARANINCHI D., BURNETT A., HELBIG W., MELONI G., VERDONCK L.F., DE WITTE T., RIZZOLI V., CARELLA A., PARLIER Y., AUVERT B. and GOLDMAN J. : for the Working party on autologous bone marrow transplantation of the EBMTG : Brit J. Haemat, 64, 385-395, 1986.

20. GORIN N.C., AEGERTER P. : For the working party on autologous bone marrow transplantation of the European bone marrow transplantation group : Autologous bone marrow transplantation for acute leukaemia in remission : 3rd European survey : March 1986. Bone Marrow Transpl 1, sup 1, 255-258, 1986.

21. GORIN N.C., AEGERTER P. : For the working party on autologous bone marrow transplantation of the EBMTG : Autologous bone marrow transplantationfor acute leukaemia in Europe : 4th European survey. Bone Marrow Transpl, 2, sup 1, 320-322, 1987.

22. GORIN N.C., DOUAY L., LAPORTE J.P., LOPEZ M., MARY J.Y., NAJMAN A., SALMON C., AEGERTER P., STACHOWIAK J., DAVID R., PENE F., KANTOR G., DELOUX J., VAN DEN AKKER J., GEROTA J., PARLIER Y., and DUHAMEL G. : Autologous Bone Marrow Transplantation Using Marrow Incubated with Asta Z 7557 in Adult Acute Leukemia . Blood, 1367-1376, 1986.

23. KROLICK K.A., UHR J.W., VITETTA E.S. : Selective killing of leukaemia cells by antibody-toxin conjugates : implications for autologous bone marrow transplantation, Nature 295, 604-605, 1982.

24. SHARKIS S., SANTOS G.W., COLVIN M. : Elimination of acute myelogenous leukemic cells from marrow and tumor suspensions in the rat with 4 hydroperoxycyclophosphamide, Blood, 55,521-523, 1980.

25. SIEBER F., SIEBER-BLUM M. : Dye mediated photosensitization of murine neuroblastoma cells : Cancer Research, 46, 2072-2076, 1986.

J.A. Spinolo, K.A. Dicke, S. Jagannath, L.J. Horwitz, F. Dunphy, J. Yau, and G. Spitzer

The field of high dose chemotherapy with autologous bone marrow transplantation (ABMT) is in constant evolution. New indications, approaches and methodologies are constantly being introduced. Three new trends will be discussed that are influencing the way in which we use ABMT and which influence survival after transplant.

TANDEM DELIVERY OF HIGH DOSE THERAPY

The standard way to apply the dose/response principle to ABMT is to give very high doses of cytoreductive therapy, which are escalated up to the point of maximal tolerable extramedullary toxicity. Usually, this type of bolus treatment can be given only once, since mortality rates of 15-20% are not unusual and severe toxicity is common. Moreover, this single treatment approach can be less advantageous in tumors with low growth fraction, in which many cells may be in an insensitive phase of the cell cycle at the time of treatment.

The concept of tandem delivery involves the administration of very high doses of cytoreductive therapy that are given in two fractions within a short time. The total dose is significantly higher than in standard bolus therapy, but each fraction is equivalent to 85-90% of the dose given in most preparative regimens. This modest decrement in each individual dose markedly reduces extramedullary toxicity. The drug combinations utilized do not have significant cumulative non-hemopoietic toxicity, and the two fractions are given at a time interval of approximately 3-4 weeks, upon recovery of blood counts. If the therapeutic efficacy is maintained, and morbidity and mortality are reduced, the therapeutic ratio is increased. Avoidance of severe toxicity may also increase cost effectiveness, even when the more prolonged hospitalization is taken into account.

We have used this concept for the last eight years in the treatment of solid tumors. Our initial experience in small cell bronchogenic carcinoma (SCBC) [1] showed that this approach is feasible, toxicity is minimal, and therapeutic efficacy is not compromised. Our current protocol for treatment of hormone resistant metastatic adenocarcinoma of the breast consists of two cycles of high dose CVP at a 28 day interval [2]. Current doses are: cyclophosphamide 1.75 g/m^2 days 1, 2 and 3; cis-platinum 60 mg/m^2 days 1, 2 and 3, and VP-16 400 mg/^{2} days 1, 2 and 3. These doses are slightly below the level of drugs used in bolus programs for adenocarcinoma of the breast (Table I); however, when compared to the doses used in the bolus treatments, the sum of the two fractions is 187% higher for cyclophosphamide, and 218% higher for cisplatinum.

Forty-six patients up to the age of 62 have been treated, of which 36 are evaluable for response. Complete remission rate is 67%, and partial remission rate is 28%. Ten patients have been in CR for more than a year, of which 3 are beyond 2 years. Three patients died of treatment related complications (mortality: 6.5%). This mortality is low when compared to the 10-20% reported in other studies using bolus programs [4,5]. The response rate is similar; it is too early to compare the progression free survival in the various programs. Interpretation of comparative results is limited since differences might be due to patient selection. Be that as it may, a low mortality in a high dose chemotherapy program is essential, especially when such programs will be moved to earlier stages of the disease in which

TABLE 1
Individual Drug Doses in High Dose Combination Therapy Programs

Institution	CTX	Cis-DDP	BCNU
M. D. Anderson*	10,500	360	**
Duke	5,625	165	600
Dana-Farber	5,625	165	600

Note: Doses are in mg/m^2.
Abbreviations: CTX, cyclophosphamide; Cis-DDP, Cis-platinum; VP-16, etoposide.

* Total tandem dose over two fractions.
** Replaced by VP-16 2400 mg/m^2

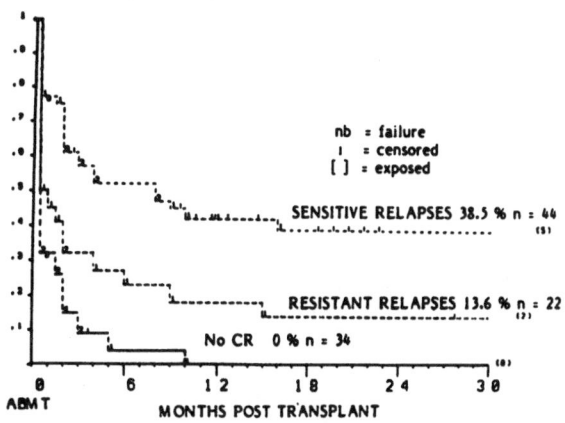

Figure 1.

the life expectancy is longer and conventional alternative treatments are not totally ineffective.

Chemoresponsiveness is a valuable prognostic indicator of survival after high dose chemotherapy and ABMT. The use of two cycles is also beneficial in that the response and tolerance to the first cycle can be used to decide whether the patient receives the second one. The tandem approach can also be tailored so that two non cross-resistant combinations are used sequentially to achieve better cytoreduction. In this approach, the least toxic regimen would be used first. We are currently evaluating a second drug combination of mitoxantrone, VP-16 and thiotepa, to be used as the second part of the tandem.

PATIENT SELECTION BY CHEMOSENSITIVITY

The efficacy of high dose chemotherapy with ABMT varies in heterogeneous patient cohorts. The response status of the patient at the time of treatment is one of the most important clinical variables related to response and survival after high dose cytoreductive therapy. In acute leukemia, the long term survival rate after allogeneic BMT is highest when transplanted in first remission, and lowest in relapse [5]. In chronic myelogenous leukemia, the results in blast crisis are dismal compared to those in first and second chronic phase of the disease [6]. This does not hold for allogeneic transplantation alone; our experience with ABMT in SCBC showed that long term survivors are almost exclusively those patients who achieved a CR with initial conventional therapy and were consolidated with ABMT. In contrast, all but one of the patients with initial PR who converted to CR after high dose therapy eventually relapsed [1].

The question whether the chemotherapy sensitivity status at the time of autologous BMT plays a role in the long term disease free survival rate after ABMT has been investigated by Philip et al [7]. They did a retrospective analysis of 100 transplant patients with intermediate and high grade non-Hodgkin's lymphoma treated in various institutions around the world with high dose therapy and ABMT (Figure 1). They found that the worst disease free survival rate was in the patient population most resistant to chemotherapy, namely patients who showed early progressive disease on their initial therapy. In this group, all patients died within a year after BMT. In contrast, patients who achieved a CR and then relapsed had a better prognosis after ABMT. Patients treated in relapse who had an objective response to conventional dose salvage therapy ("sensitive relapse") had a 36% 3 year disease free survival, whereas that rate was only 14% in those with chemotherapy resistant relapse. Further analysis demonstrated that chemosensitivity of the disease was an independent risk factor, whereas tumor bulk was not. Thus, it appears that chemosensitivity is not a mere expression of residual tumor bulk, but a separate prognostic factor in itself.

In relapsed Hodgkin's disease, a retrospective analysis of our results with high dose chemotherapy with CBV and ABMT

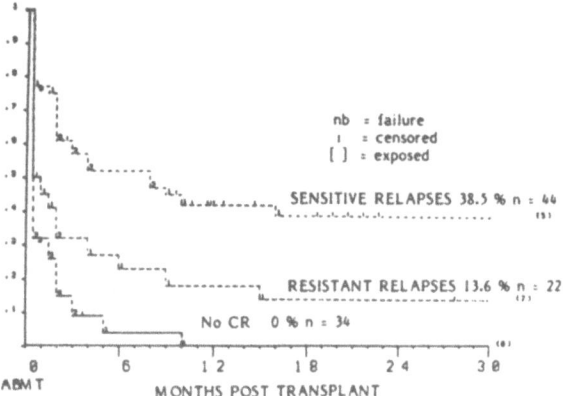

Figure 3. Disease free survival (DFS) is significantly better in the purged group. *(Used by permission of Dr. N. C. Gorin.)*

showed that the 12 patients with sensitive relapse had a 2 year survival rate of 83%, whereas the rate for the 33 patients with resistant relapse was only 48% (Figure 2) [8]. This difference is significant at a P level of 0.15 (Generalized Wilcoxon test). Larger numbers of patients are needed to prove whether this trend is as significant for Hodgkin's disease as for intermediate and high grade non-Hodgkin's lymphomas.

The above retrospective findings allow us for more rational application of ABMT. All these studies suggest that the use of conventional dose therapy prior to ABMT acts as an *in vivo* test of the tumor's sensitivity, and that the response to that therapy predicts for survival after BMT. The Dana Farber group has selected patients for their lymphoma study based on chemoresponsiveness [9]; only patients whose disease was reduced to minimal bulk were accepted for this protocol (about one third of the patients with relapsed lymphoma referred to that center). This selection by chemosensitivity undoubtedly contributed to their excellent results; this group of patients had a 65% disease free survival at a median of 11+ months.

Currently, two international cooperative studies use chemosensitivity as a prerequisite for acceptance: the Parma Study, for patients with relapsed intermediate and high grade lymphoma, and the U.S.A.-Netherlands study for Hodgkin's disease. In the former one, patients are required to have an objective response to two cycles of DHAP (dexamethasone, high dose ARA-C and cis-platinum), and are subsequently randomized to treatment with high dose BEAC (BCNU, etoposide, ARA-C and cyclophosphamide) and ABMT or with continued DHAP. In the latter study, patient's sensitivity to DHAP is again required to then receive treatment with high dose CBV (cyclophosphamide, BCNU and VP 16) and ABMT. The application of chemosensitivity to patient selection allows for better allocation of resources, and reserves this toxic and expensive mode of treatment for the groups of patients who are more likely to benefit from it.

PRE-BMT INTENSIFICATION IN ACUTE MYELOGENOUS LEUKEMIA

High dose cytoreduction with autologous bone marrow rescue has been extensively used to intensify first remission in acute myelogenous leukemia [10-15]. The bone marrow is collected after achievement of remission. However, results have been variable, and it is unclear whether they are better than those of conventional chemotherapy alone. In the Dutch Cooperative study [10], the long term disease free survival is 30%. These relatively disappointing results were confirmed by a review of the pooled results of the European Bone Marrow Transplantation Study Group by Gorin [16] (Figure 3). Patients treated with a total body irradiation (TBI) containing regimen and unpurged marrow had a 32% disease free survival. However, patients treated with the same regimen and marrow purged with mafosfamide had a three year disease free survival (DFS) of 56%. The results were even better for patients whose

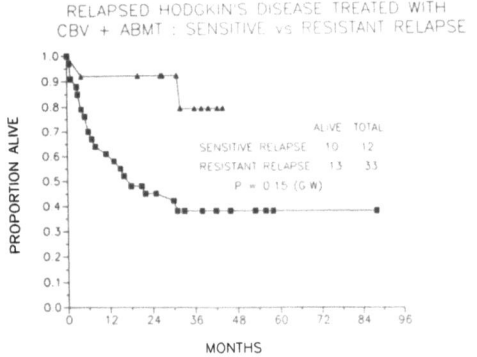

Figure 2. A trend for improved survival is noted in chemosensitive patients.

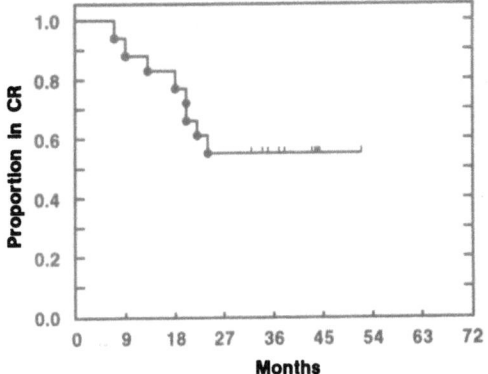

Figure 4.

mafosfamide dose was adjusted to obtain optimal cytoreduction, who had a DFS of 64%. This seems to indicate that, even with comparable remission induction regimens and pre-BMT preparative regimens, the quality of the infused bone marrow has a significant impact in disease free survival.

In our unpurged autologous BMT study for patients with AML in first remission [17], the disease free survival rate is 56% (Figure 4), similar to the results with bone marrow purging reported by Gorin. The conditioning regimen used by us was CBV (cyclophosphamide 1.5 g/m^2 days 1-4, BCNU 300 mg/m^2 day 1, and VP-16 125 mg/m^2 q 12 hrs days 1-3). This regimen is most likely not superior to the TBI containing programs in the studies reported by Gorin and Lowenberg. The major difference between our study and the former studies resides in the use of an amsacrine + high dose ARA-C intensification shortly after induction of CR and prior to bone marrow harvest. It is known that high dose ARA-C intensification of first remission is associated with improved long term disease free survival in AML [18,19]. This is most likely due to improved cytoreduction in comparison to conventional doses of ARA-C and anthracycline. In our study, such pre-BMT intensification may have reduced the leukemic cell contamination in the marrow at the time of harvest. Thus, these patients would have the equivalent of an *"in vivo* purging" of their bone marrow by the use of this therapy. In addition, if the time interval between pre-BMT intensification and high dose cytoreduction is limited, the leukemic cell burden in the recipient would still be low at the time of ABMT, and the patient would undergo a possibly synergistic double intensification.

Another potential benefit of the pre-BMT intensification would be faster decrease of the total tumor burden. According to Goldie and Coldman [20], the likelihood of emergence of drug resistant cell clones is related to tumor bulk and to the number of cell division that the original malignant clone goes through. Consequently, the early use of a pre-BMT intensification with high dose ARA-C may also prevent emergence of resistant cell clones.

CONCLUSIONS

High dose chemotherapy with autologous bone marrow transplantation has proven to be successful in defined subgroups of patients with malignant diseases. The split administration of high doses of chemotherapy may result in a decrease of transplantation associated mortality, making the procedure safer. Patient selection on the basis of sensitivity to conventional doses of salvage chemotherapy will lead to increased response rates to the higher dose regimens. The use of high dose ARA-C prior to bone marrow harvest in acute myelogenous leukemia may lead to more favorable results in autologous bone marrow transplantation.

Further improvement of bone marrow transplantation technology and of the scheduling of drug administration may lead to better results in this field.

BIBLIOGRAPHY

1) Spitzer G, Farha P, Valdivieso M, Dicke KA, Zander A, Vellekoop L, et al. (1986) High dose intensification therapy with autologous bone marrow support for limited small cell bronchogenic carcinoma. J Clin Oncol 4:4-13

2) Dunphy FR, Spitzer G, Dicke KA, Buzdar AU, Hortobagyi GN (1988) Tandem high-dose chemotherapy as intensification in stage IV breast cancer. In: Gale RP, Champlin R (eds). Bone Marrow Transplantation: Current Controversies. UCLA Symposia on Molecular and Cellular Biology, New Series; Vol. 87. New York: Alan R. Liss, Inc, (in press)

3) Antman K, Gale RP (1988) Advanced breast cancer: high dose chemotherapy and bone marrow autotransplants. Ann Int Med 108:570-574

4) Peters WP, Eder JP, Henner WD, Schryber S, Wilmore D, Finberg R, et al. (1986) High dose combination alkylating agents with autologous bone marrow support. A Phase I trial. J Clin Oncol 4:646-654

5) Champlin R, Gale RP (1987) Bone marrow transplantation for acute leukemia: Recent advances and comparison with alternative therapies. Sem Hematology 24:55-67

6) Goldman JM, Apperley JF, Jones L, Marcus R, Goolden AWG, Batchelor R, et al. (1986) Bone marrow transplantation for patients with chronic myeloid leukemia. N Engl J Med 314:202-207

7) Philip T, Armitage JO, Spitzer G, Chauvin F, Jagannath S, Cahn JY, et al. (1987) High-dose therapy and autologous bone marrow transplantation after failure of conventional chemotherapy in adults with intermediate-grade or high-grade non-Hodgkin's lymphoma. N Engl J Med 316:1493-1498

8) Jagannath S, Armitage JO, Dicke KA, Tucker SL, Velasquez WS, Smith K, et al. (1988) Prognostic factors for response and survival after high dose cyclophosphamide, carmustine and etoposide with autologous bone marrow transplantation for relapsed Hodgkin's disease. J Clin Oncol (in press)

9) Takvorian T, Canellos GP, Ritz J, Freedman AS, Anderson KC, Mauch P, et al. (1987) Prolonged disease-free survival after autologous bone marrow transplantation in patients with non-Hodgkin's lymphoma with a poor prognosis. N Engl J Med 316:1499-1505

10) Lowenberg B, van der Lelie J, Goudsmit R, Willemze R, Zwaan FE, Hagenbeek A, et al. (1987) Autologous bone marrow transplantation in patients with acute myeloid leukemia in first remission. In: Dicke KA, Spitzer G, Jagannath S (eds): Autologous Bone Marrow Transplantation. Proceedings of the Third International Symposium. Houston, University of Texas, pp 3-7

11) Burnett AK, Mackinnon S, Morrison A (1987) Autologous transplantation of unpurged bone marrow during first remission of acute myeloid leukemia. In: Dicke KA, Spitzer G, Jagannath S (eds): Autologous Bone Marrow Transplantation. Proceedings of the Third International Symposium. Houston, University of Texas, pp 23-29

12) Cahn JY, Flesch M, Plouvier E, Pignon B, Behar C, Audhuy B (1987) Autologous bone marrow transplantation for acute myelogenous leukemia: 24 unselected patients in first remission. In: Dicke KA, Spitzer G, Jagannath S (eds): Autologous Bone Marrow Transplantation. Proceedings of the Third International Symposium. Houston, University of Texas, pp 31-33

13) Carella AM, Meloni G, Mangoni L, Porcellini A, Alessandrino EP, Visani G, et al. (1987) Massive therapy (MT) and autologous BMT (ABMT) in ANLL. An update of the Italian Study Group (IABMTG). Bone Marrow Transplantation 2(suppl.1):45

14) Goldstone AH, Anderson CC, Linch DC, Franklin IM, Boughton BJ, Cawley JC and Richards JDM (1986) Autologous bone marrow transplantation following high dose chemotherapy for the treatment of adult patients with acute myeloid leukemia. Br J Haematol 64:529-537

15) Gorin NC, Laporte JP, Douay L, Lopez M, Stachowiak J, Aegerter P, et al. (1987) Use of bone marrow incubated with mafosfamide in adult acute leukemia patients in remission: The experience of the Paris Saint-Antoine

transplant team. In: Dicke KA, Spitzer G, Jagannath S (eds): Autologous Bone Marrow Transplantation. Proceedings of the Third International Symposium. Houston, University of Texas, pp 15-22

16) Gorin NC, Aegerter P (1988) Autologous bone marrow transplantation in acute leukemia in remission. Fifth European Survey of the European Bone Marrow Transplantation Group. Evidence in favor of marrow purging and influence of intervals pretransplant. Proceedings of the Meeting on Autologous Bone Marrow Transplantation, Chamonix, France, April 10-14, 1988

17) Spinolo J, Dicke KA, Horwitz LJ, Jagannath S, Spitzer G (1988) Autologous bone marrow transplantation (ABMT) for remission intensification in acute myelogenous leukemia (AML): Long term follow-up. Experimental Hematology 16:487

18) Wolff SN, Marion J, Stein RS, Flexner JM, Lazarus HM, Spitzer TR, et al. (1985) High-dose cytosine arabinoside and daunorubicin as consolidation therapy for acute nonlymphocytic leukemia in first remission: A pilot study. Blood 65:1407-1411

19) Preisler HD, Raza A, Early A, Kirshner J, Brecher M, Freeman A, et al. (1987) Intensive remission consolidation therapy in the treatment of acute nonlymphocytic leukemia. J Clin Oncol 5:722-730

20) Goldie JH, Coldman AJ (1979) A mathematical model for relating the drug sensitivity of tumors to the spontaneous mutation rate. Cancer Treat Rep 63:1727-1733

26 High Dose Therapy and Autologous Bone Marrow Transplantation in Non Hodgkin's Lymphoma and Neuroblastoma

T. Philip

INTRODUCTION

The importance of dose intensity in determining outcome of chemotherapy was extensively reviewed in the 1987 ASCO Educational Session on Dose Intensity (Educational booklet 1987 pages 23 to 38). The Rigdway sarcoma model in dogs had also shown (Figure 1) that small modification of dose intensity can modify tumor regression from partial response to complete response with subsequent relapse and finally at the maximum dosage to complete response with cure (10).

In the field of solid tumors and lymphoma a graft versus tumors effect of bone marrow transplantation is not demonstrated and every published comparison had shown no difference in this setting between allogeneic and autologous bone marrow transplantation (7). High dose therapy is then the major objective in the field of autologous bone marrow transplantation. This strategy had already produced very promising results and we will focus today in non Hodgkin's lymphoma and childhood neuroblastoma as models of high dose therapy with bone marrow rescue.

NON-HODGKIN'S LYMPHOMA

A retrospective study of data from "France Autogreffe" Study group and London was performed in July 1983 (6). From these data as shown in Figure 1A, some conclusions could be drawn on the following points: - **Real resistant relapse** (i.e., patients still progressing at time of ABMT) will usually not be cured by massive therapy. However responses were observed in 73% of the cases but as shown in Figure 1A, only 1/16 patients is still alive 3 years post-ABMT (i.e., 6%). - **Non-resistant relapse** (i.e., all other patients excluding stable disease and minor response on rescue protocol) could be long-term survivors in approximately half of the cases. Our initial data concerned 15/19 patients who had relapsed on therapy (Figure 1A) (i.e., it was clearly a group with a very bad prognosis) and showed no difference between PR or CR prior to ABMT if CR is obtained after ABMT.

The conclusion was that patients in relapse have to be separated into 2 groups: resistant relapse (patients with progressive, stable or minor response on salvage chemotherapy) and non-resistant relapse or sensitive relapse (patients achieving partial or complete response within the first 2 courses of salvage chemotherapy). Two major criticisms have been addressed to this study: this retrospective analysis included both adults and children (2/3 adults, 1/3 children) and there was a high proportion of childhood non-Hodgkin's Lymphoma of the Burkitt type (12/42 BL; 15 intermediate grade; and 15 high grade non-BL)(4,8,9). As shown Figures 1B and 1C, a European study by Goldstone (Reference in 7) later concluded that no statistical difference was observed either between adults and children (Figure 1B) or between Burkitt and non-Burkitt (Figure 1C). The concept of resistant and non-resistant relapse has been subsequently confirmed in:

- A review of 42 cases of Burkitt's Lymphoma in France (Figure 1D)

- A review of 42 cases of Adult Diffuse Lymphoma from France and England

- A review of 39 cases of NHL from Houston and Omaha (11).

In 1986, data from bone marrow transplant centers in Europe and America were pooled to determine the outcome of ABMT in adult patients with relapsed diffuse intermediate or high grade NHL (excluding Burkitt lymphoma), and to identify the prognostic significance of response to therapy preceding the bone marrow transplant procedure (5). Patients were treated with high-dose chemotherapy alone (61 patients) or high dose chemotherapy plus total body irradiation (TBI) (39 patients). The median age was 35 years and the median Karnofsky performance score was 80%. Thirty-four patients had disease that was primarily refractory to chemotherapy (i.e., never achieved complete remission) and had progressive disease (no CR). Sixty-six patients achieved a complete remission (CR) with primary chemotherapy but later relapsed. After receiving further chemotherapy (salvage) at "traditional" doses 22 patients had no response or disease progression (i.e., resistant relapse - RR) and 44 patients responded with partial or complete responses to salvage chemotherapy (i.e., sensitive relapse - SR). The actuarial 3-year disease-free survival for the entire group was 19%, with the last death at 31 months and a median observation time of 40 months (5). Disease-free survival was significantly related to previous response to chemotherapy. The two-year disease-free survival was 0% in the no CR group, 14% in the RR group, and 36% in the SR group (5). Patients who had achieved a CR to initial chemotherapy had a superior disease-free survival after ABMT when compared to patients never achieving a CR (p<.0001) (5). Patients with SR had a better disease-free survival than patients with RR (p<,001). Outcome was not affected by treatment regimen and histologic grade. Whether relapse occurred on or off therapy was also not of significance, but the probability of SR was significantly higher for relapses off-therapy. In conclusion, it appears that prior response to chemotherapy is an important prognostic variable in patients with intermediate or high grade NHL undergoing ABMT. These results and others (1, 2, 3) were the background for the current international randomized study and explained why patients with no previous CR and RR were excluded from the randomized study. The question of whether

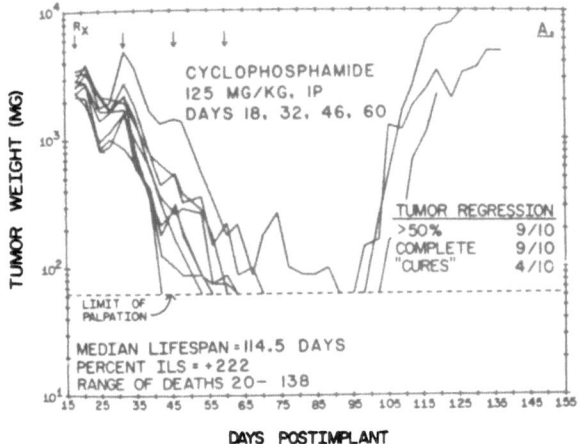

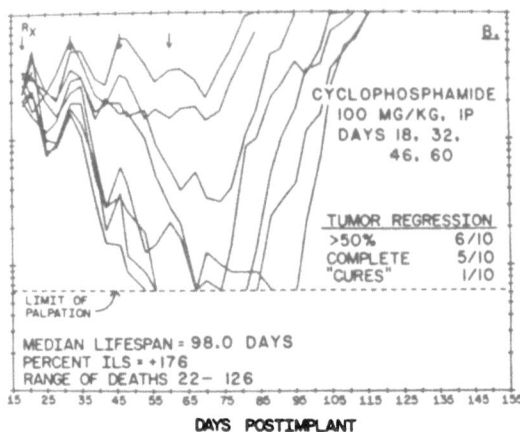

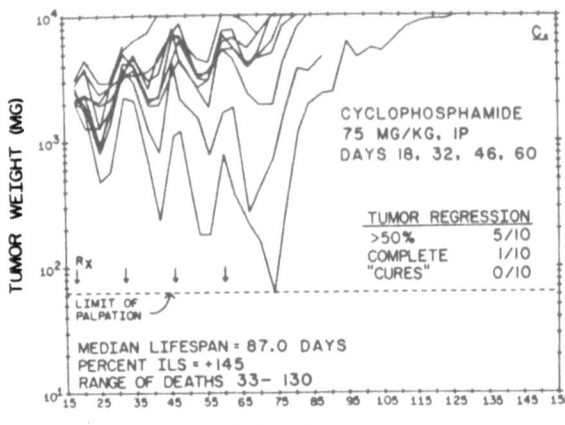

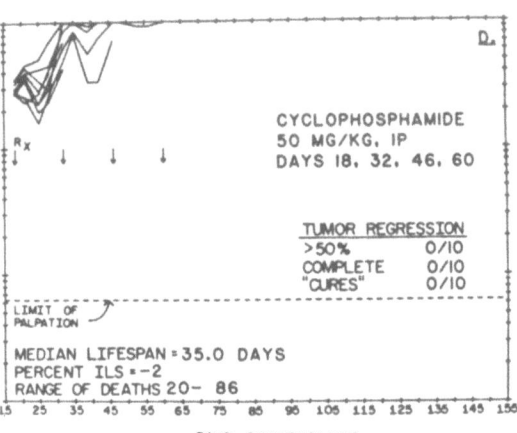

Figs. 1A–1D. Variable regression responses of ROS. Treatment with CPA. Reprinted with permission from Schabel FM Jr, Griswold DP Jr, Corbett TH, Laster WR Jr, Mayo JG, Lloyd HH. Testing therapeutic hypotheses in mice and man: Observations on the therapeutic activity against advanced solid tumors of mice treated with anticancer drugs that have demonstrated or potential clinical utility for treatment of advanced solid tumors of man. In: De Vita V Jr, Busch, eds. Cancer Drug Development, Part B: Methods in Cancer Research, Vol. 17. New York: Academic Press, 1979;3-51.[2]

cures can be obtained with conventional salvage regimens without ABMT remains, as there are reports of occasional long-term survivors after relapses treated with MIME (Methyl Gag, Ifosfamide, Methotrexate, Etoposide) or DHAP (Dexamethasone, Aracytine, Platinum)(3). A randomized study was urgently needed in 1987 to compare both treatment modalities. This randomized study is currently activated among 52 centers around the world. Two courses of the DHAP rescue regimen allowed selection of sensitive patients. These patients are randomized between ABMT and continuation of DHAP. The study is expected to be completed in 3 years.

NEUROBLASTOMA

Neuroblastoma is the second commonest solid childhood tumor and the outlook in limited disease (25% of the cases) is good. However, with more advanced or metastatic disease, complex and demanding combinations of surgery, chemotherapy and radiotherapy are required. Even with these treatment modalities the elusive goal of cure remains a frustrating distant objective in many patients. The outcome with surgical treatment alone is good in the rare stage 1 patient and also in stage 2 cases where there is no associated lymph node involvement. Chemotherapy is highly effective in remaining stage 2 patients with long term survival in excess of 75%. With more advanced initially unresectable disease the outcome appears to depend to at least some extent upon the completeness of eventual surgical resection after primary chemotherapy. The wide variation in long term survival for this stage (40% to 70%) thus reflects both the activity of the chemotherapy and surgical expertise. In stage 4 patients, age remains an important prognostic factor and for patients under 1 year, even with metastatic disease, cure rates may exceed 60%. Between 1 and 2 years the prognosis is intermediate but for patients over 2 years at diagnosis there is almost universal agreement that the likelihood of long term survival with standard chemotherapy is around 10%. The initial response to chemotherapy is often encouraging and with most modern regimens up to 80-90% will achieve at least a partial response with clearing or improvement of metastatic disease. Unfortunately, even with surgery and radiotherapy this is rarely converted to a completed response and disease progression usually occurs within the following year.

The dose effect concept is widely used in neuroblastoma with the administration of single or combination therapy at doses limited only by nonhemopoietic tissue tolerance using bone marrow rescue procedures. Because of the limited availability of matched related donors in this very young group of patients experience with autologous bone marrow is greater than with allogeneic. Moreever, unlike leukemia, neuroblastoma only involves marrow secondarily which may be cleared of tumor by effective chemotherapy prior to marrow harvest. Various purging techniques have been developed which in vitro are capable of removing residual tumor cells from marrow. The need for such purging procedures in practice remains a contentious issue as the clonogenic potential of reinfused neuroblastoma cells is difficult to demonstrate. The main limiting effect in the use of megatherapy procedures remains the ability of such therapy to completely ablate the malignant cell population. It is difficult to summarize the state of the art due to the lack of published data. However, in the recent Houston Autologous Bone Marrow Transplantation Meeting (December, 1986, published in 1988, Karel Dicke, Editor) all the major teams have reported current results.

In relapsed highly selected, patients 20% progression free survival at 2 years is reported when TBI was used in the conditioning regimen and no progression free survivors reported when a non TBI regimen was used.

With or without TBI, with single or double graft procedures, 40 to 45% progression free survivors at 2 years were reported in patients grafted in complete remission. No advantage was shown at two years in this group for the TBI containing regimens but the mortality rate is higher (20% versus 10%). It is clear that in all studies no plateau has yet been reached on the curves after 2 years. It is also clear that patients alive disease free at 24 months are not cured and that at least 50% will relapsed before 48 months post BMT.

The timing of relapse post BMT (i.e. between 3 to 48

months post graft for the majority of the cases) has kindled interest in the concept of a double graft procedure in this disease. Selected patients in partial remission after induction therapy receiving a double graft without TBI were reported with no survivors at 2 years. The Lyon Marseille Curie Group have early data using a double graft, including TBI for the second conditioning regimen, in relapsed patients, with 50% survivors up to 2 years.

Phase II studies with these high dose regimens have been encouraging although their construction has to a large extent been empirical. An exception to this is high dose melphalan. The European Neuroblastoma Study Group (ENSG) has carried out a prospective randomized study in state 3 and 4 patients who achieved at least partial remission after a standardized platinum containing induction regimen. There was a significant advantage for the melphalan group both in terms of median survival and disease progression free duration. This provides a rational basis for the inclusion of this agent at high dose in future protocols. TBI is still included in many schedules because of in vitro radiosensitivity data and early clinical experience. However, because of concern about the contribution of this modality to both short and long term morbidity its inclusion requires prospective evaluation. It is possible that the substitution of other drugs at high dose might provide a better therapeutic index.

In conclusion, although advanced neuroblastoma remains a difficult and challenging disease there is now a significant percentage of long term survivors who may be cured by the application of intensive strategies. With careful monitoring long term sequelae should be minimal and are essentially platinum related even in the absence of high dose alkylating agents or TBI. Late effects such as second malignancies are beginning to appear in the literature and are obviously being born in mind in the design of new regimens. However, for the present the first concern is to achieve prolonged good quality remission in these patients thus providing a realistic chance of long term survival.

REFERENCES

1. Appelbaum FR, Deisseroth AB, Graw RG, Levine AS, Herzig GP, Ziegler JL (1978) Prolonged complete remission following high-dose chemotherapy of Burkitt's lymphoma in relapse. Cancer 1978, 41:1059-1063

2. Appelbaum FR, Thomas ED (1983) Review of the use of marrow transplantation in the treatment of non-Hodgkin lymphomas. Clin Oncol 7:440-447

3. Cabanillas F, Hagemeister FB, Riggs S, Salvador P, Velasquez WS, McLaughlin P, Smith T (1985) Results of ifosfamide-etoposide combinations for patients with recurrent or refractory aggressive lymphoma. In: Malignant Lymphomas and Hodgkin's Disease: Experimental and Therapeutic Advances. Martinus Nijhoff Publishing. Boston, Cavalli F (ed), p485-492

4. Philip I, Philip T, Favrot MC, Vuillaume M, Fontaniere B, Chamard D, Lenoir GM (1984) Establishment of lymphomatous cell lines from bone marrow samples from patients with Burkitt's lymphoma. J. Natl Cancer Insti, 73:835-841

5. Philip T, Armitage JO, Spitzer G, Chauvin F, Jagannath S, Cahn JY, Colombat P, Goldstone AH, Gorin NC, Flesh M, Laport JP, Maraninchi D, Pico J, Bosly A, Anderson C, Schots R, Biron P, Cabanillas F, Dicke KA (1987) High dose therapy and autologous bone marrow transplantation after failure of conventional chemotherapy in adults with intermediate grade or high grade non hodgkin's lymphoma. N. Engl J Med 316:1493-1498.

6. Philip T, Biron P, Maraninchi D, Gastaut JL, Herve P, Flesh Y, Goldstone AH, Souhami RH (1984) Role of massive chemotherapy and ABMT in non-Hodgkin's malignant lymphoma. Lancet ii, 391

7. Philip T, Pinkerton R (1987) Very high dose therapy in lymphomas and solid tumors. New Direction in Cancer Treatment, Y. Magrath Ed., (in press).

8. Philip T, Pinkerton R, Hartmann O, Patte C, Philip I, Biron P, Favrot MC (1986) The role of massive therapy with autologous bone marrow transplantation in Burkitt's lymphomas. Clinics in Hematol, 15:205-217

9. Pinkerton R, Philip T, Bouffet E, Lashford L, Kemshead J (1986) Autologous bone marrow transplantation in paediatric solid tumors. Clinic in Haematol. 15:187

10. Shabel FM Jr (1975) Animal models as predictive systems. In: Cancer Chemotherapy Year Book Medical, Publisher Chicago, pp. 323-355

11. Spitzer G, Jagannath S, Dicke KA, Armitage J, Zander AR, Vellekoop L, Horwitz LJ, Cabanillas F, Zagarse GK, Velasquez WS (1986) High-dose melphalan and total body irradiation with bone marrow transplantation for refractory malignancies. Europ J Cancer and Clin Oncol 22:677-684

R.P. Gale and A. Butturini

Exposure to total body radiation results in dose dependent suppression of hematopoiesis (reviewed in 1-3). Variables influencing the extent of bone marrow suppression include total dose, dose rate, schedule, shielding, dose uniformity, as well as source-term parameters. Single-dose total body radiation at doses ≥ 1 Gy and dose rates ≥ 1 cGy per minute produce granulocytopenia and thrombocytopenia. Doses > 2 Gy can cause death from infection and bleeding. The 50% lethal dose (LD_{50}) in humans is presumed to be 4-5 Gy based on data in animals. Higher radiation doses carry an increasing risk of death from bone marrow suppression; survival is unlikely after doses > 8-10 Gy. Doses > 15-20 Gy results in death from toxicity to other tissues such as the gastrointestinal tract or central nervous system.

There are several approaches to preventing death from bone marrow suppression following accidental radiation exposure (reviewed in 4-6). One involves supportive measures designed to prevent infection or bleeding while permitting spontaneous bone marrow recovery. Another approach is to replace the damaged hematopoietic system via a bone marrow transplant (reviewed in 5,7). A third alternative, subject of this commentary, are attempts to stimulate recovery of endogenous hematopoiesis using molecularly cloned hematopoietic growth factors.

Two considerations underly this approach. First is the notion that some hematopoietic stem cells survive even the highest doses of radiation. Because radiation induced cell damage is associated with a fractional log-cell kill and because only a small proportion of hematpoietic stem cells are dividing at any time, it is unlikely that any dose of radiation will destroy all hematopoietic stem cells. The second notion is that only one or a very few stem cells are needed for hematopoietic reconstitution. This concept is supported by experimental data in mice using cytogenetically or retrovirus marked bone marrow cells (8,9).

If these concepts are correct, it should be possible to prevent deaths from bone marrow suppression and to increase survival if these residual hematopoietic stem cells can be stimulated to divide. In this commentary we consider experimental data which support this approach using molecularly cloned hematopoietic growth factors. We also cite results of preliminary trials in humans following a recent radiation accident in Goiania, Brazil (10).

Several years ago technics were developed to study hematopoietic cells (reviewed in 11,12). In these assays stem cells give rise to clonal progeny when cultured in semi-solid matrices in vitro or when introduced into the spleen of radiated mice in vivo. One by-product of studies of in vitro colony-forming cells was the realization that in vitro growth was not autonomous, but required exogenous growth factors; these were termed colony-stimulating factors. Eventually these colony-stimulating factors were biochemically purified. Most recently they have been isolated by molecular cloning. In humans three colony-stimulating factors have been identified: multi-CSF, GM-CSF and G-CSF (for reviews see 13-15). Multi-CSF stimulates growth of pluripotent stem cells; it is also referred to as interleukin-3. GM-CSF stimulates growth of progenitors of granulocytes and macrophages in vitro. G-CSF stimulates granulocyte progenitors. There are also other hematopoietic growth factors. Some like erythropoietin, have been isolated and molecularly cloned. Others, like M-CSF, are being investigated. Colony-stimulating factors are glycoproteins. Multi-, GM, and G-CSF share structural homology with one another but not with other known proteins, growth factors or oncogenes. They are the only known molecules which directly stimulate myeloid progenitor cells and bind specifically to receptors on these cells (for reviews see 13-15). In addition to their effect on hematopoietic stem cells, the colony-stimulating factors also effect some functions of mature cells such as chemotaxis or oxidative burst of neutrophils.

Because hematopoietic growth factors in general and colony stimulating factors specifically stimulate proliferation of hematopoietic stem cells, they have been studied as potential therapeutic agents following radiation exposure. Considerable in vitro data indicate that colony-stimulating factors can increase the number (cloning efficiency) and size of myeloid stem cells in semisolid matrices. Dogs and monkeys exposed to high doses of total body radiation have more recovery of hematopoiesis following treatment with molecularly cloned hematopoietic growth factors (15,18). Different factors have different effects. For example, treatment with GM-CSF increases predominatly granulocytes, macrophages, and eosinophils, whereas use of multi-CSF also increases platelets.

Preliminary data suggest that combinations of molecularly cloned hematopoietic growth factors, such as multi-CSF followed by GM-CSF, results in more rapid and complete recovery of hematopoiesis than either used singly. Similar data are available in man. Persons with AIDS related bone marrow failure respond to therapy with GM-CSF (19). Similar observations apply to cancer patients receiving chemotherapy and to autotransplant recipients (20-23).

Recently we used molecularly cloned GM-CSF to treat eight victims of a 137-cesium radiation accident in Goiania, Brazil. Details have been reported (24). Briefly, all seven evaluable persons responded rapidly to GM-CSF administration; data are reviewed in Figure 1. The tempo of recovery observed differed considerably from individuals not receiving GM-CSF (Figure 2). These and other data, such as a dose response relationship led us to conclude that GM-CSF treatment was responsible for the accelerated recovery of granulocytes observed in these persons. Furthermore, this approach seemed to prevent infections in two individuals treated before severe granulocytopenia developed.

The likelihood of death from bone marrow failure following radiation exposure is correlated with severity and duration of granulocytopenia and thrombocytopenia. Measures that decrease either severity or duration of hematopoietic suppression must therefore improve survival unless they produce undesirable consequences (vide infra).

Therapy with molecularly cloned hematopoietic growth factors may have other beneficial effects. As indicated, neutrophil function is increased following exposure to GM-CSF. Whether this is clinically important is unknown.

One important question is whether treatment with molecularly cloned hematopoietic growth factors can influence the likelihood of bone marrow recovery following radiation exposure by directly affecting the probability of survival of hematopoietic stem cells. Recent data in mice suggest that in vitro treatment of bone marrow cells with GM-CSF followed by transplantation into radiated recipients results in a higher survival than untreated bone marrow cells (25). This is consistent with the notion of an operational increase in the efficiency of hematopoietic stem cells repopulating capacity. Whether such an effect occurs in humans is unknown; if present, this would shift the radiation survival curve favorably.

Treatment with molecularly cloned hematopoietic growth factors is not without potential adverse effects. The most serious conceptual problem is that treatment might disturb the delicate stem cell balance between self-replication and differentiation or maturation. A consequence could be to deplete the stem cell pool thus the initial more rapid recovery of hematopoiesis would be followed by hematopoietic failure. Although this is an interesting possibility there are little experimental data that treatment with most molecularly cloned hematopoietic growth factors results in late bone marrow failure. This has also not been observed in humans treated with GM-CSF or in the Brazil radiation victims. In our opinion, this complication is unlikely since only a few stem cells are needed to restore normal hematopoiesis following radiation exposure. Furthermore, most of the available molecularly cloned hematopoietic growth factors act on relatively differentiated hematopoietic stem cells and not on pluripotent stem cells. Nevertheless, it may be prudent to use molecules such as GM-CSF rather than multi-CSF so as to preserve less mature stem cells. This notion is supported by recent data in dogs suggesting that treatment of radiated dogs with multi-CSF results in an early granulocytosis followed by delayed or absent bone marrow recovery (MacVittie T., Personal Communication). Other potential adverse consequences of molecularly cloned hematopoietic growth factors can result from increased levels or function of neutrophils or platelets. For example, at very high leukocyte levels agglutination can occur in the pulmonary vasculature. The likelihood of this complication may be increased in tissues damaged by radiation. However, most of these effects can be controlled by appropriate dose reductions.

In summary, the use of molecularly cloned hematopoietic growth factors is likely to be important in treating persons with bone marrow failure resulting from radiation accidents. Some effects, such as increases in granulocytes or platelets, are of certain therapeutic benefit. Other effects, such as direct action on survival of hematopoietic stem cells and increased granulocyte function, may also improve survival but have not yet been convincingly demonstrated in humans. Many other important areas remain to be studied including which molecularly cloned hematopoietic growth factor to use and the optimal dose and timing of administration. Some of these issues can be addressed in clinical trials; others will require animal models. Much progress has been made but much remains to be done as we enter a new era in treatment of radiation accidents.

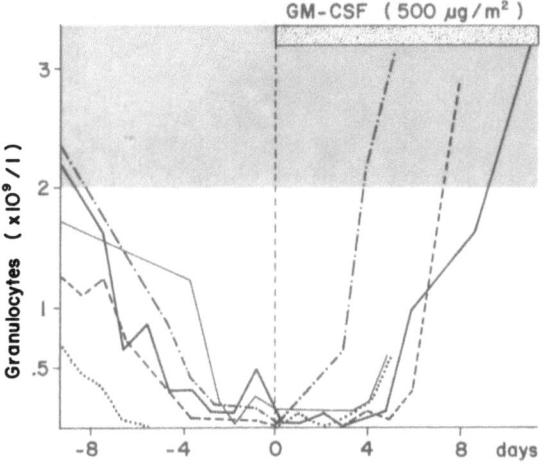

FIGURE 1. Response of 5 radiation victims to treatment with rHuGm-CSF.

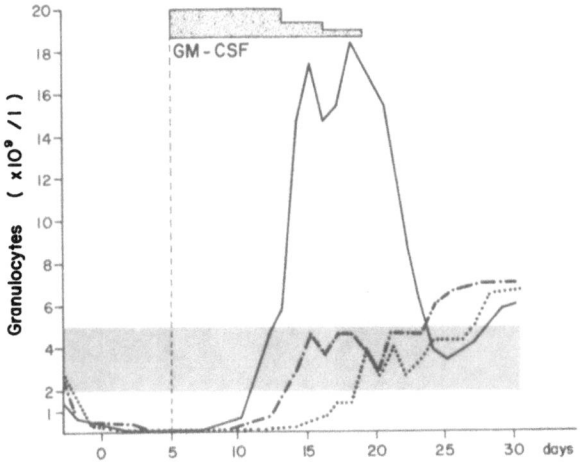

FIGURE 2. Comparison of 1 victim receiving rHuGM-CSF with 2 unrelated victims.

REFERENCES

1. UNSCEAR 1982. Ionizing radiation: sources and biologic effects. Report to the General Assembly, Vienna and New York.
2. Medical Research Council Committee on Effects of Ionizing Radiation. A forum on lethality from acute and protracted radiation exposure in man. Intl J Radiat Biol 1984; 46:209-17.
3. Mettler FA Jr, Moseley RD Jr: Medical Effects of Ionizing Radiation. New York Grune and Stratton. 1985.
4. Gale RP. Immediate medical consequences of nuclear accidents: Lessons from Chernobyl. JAMA 1987; 258:625-8.
5. Gale RP. The role of bone marrow transplantation following nuclear accidents. Bone Marrow Transplantation 1987; 2:1-6.
6. Gale RP. The medical response to radiation and nuclear accidents: Lessons for the future. J Natl Cancer Inst Submitted
7. Gale RP, Reisner Y. Are bone marrow transplants effective after nuclear accidents? Lancet 1988; i:923-5.
8. Abramson S, Miller RG, Phillips RA. The identification in adult bone marrow of pluripotent and restricted stem cells of the myeloid and lymphoid systems. J Exp Med 1977; 145:1567-79.
9. Lemischka IR, Raulet DH, Mulligan RC. Developmental potential and dynamic behavior of hematopoietic stem cells. Cell 1986; 45:917-27.
10. Roberts L. Radiation accident grips Goiania. Science. 1987; 238:1028-31.
11. Golde DW, Takaku F (eds). Hematopoietic Stem Cells. New York, Marcel Dekker, Inc. 1985.
12. Wright DG, Greenberger JS (eds). Long-Term Bone Marrow Culture. New York, Alan R. Liss, Inc. 1984.
13. Metcalf D. The molecular biology and functions of the granulocyte-marcophage colony-stimulating factors. Blood 1986; 67:257-67.
14. Sieff CA. Hematopoietic growth factors. J Clin Invest 1987; 79:1549-57.
15. Clark SC, Kamen R. The human hematopoietic colony stimulating factors. Science 1987; 236:1229-37.
16. Donahue RE, Wange EA, Stone DK, et al. Stimulation of haematopoiesis in primates by continuous infusion of recombinant human GM-CSF. Nature 1986; 321:872-5.
17. Welte K, Platzer E, Lu L, et al. Purification and biological characterization of human pluripotent hematopoietic colony stimulating factor. Proc Natl Acad Sci 1985; 82:1526-30.
18. Donahue RE, Seehra J, Norton C, et al. Hematologic effects of recombinant human interleukin-3 (rhIL-3) and granulocyte/macrophage colony-stimulating factor (rhGM-CSF) in primates. Proceedings of ASCO, Vol 7, March 1988, p 162.

19. Groopman JE, Mitsuyasu RT, DeLeo JM, Oette DH, Golde DW. Effect of recombinant human granulocyte-macrophage colony-stimulating factor on myelopoiesis in the acquired immunodeficiency syndrome. N Engl J Med 1987; 317:593-8.

20. Brandt SJ, Peters WP, Atwater SK, et al. Effect of recombinant human granulocyte-macrophate colony-stimulating factor on hematopoietic reconstitution after high dose chemotherapy and autologous bone marrow transplantation. N Engl J Med 1988; 318:870-6.

21. Antman K, Griffin J, Elias A, et al. Use of rGM-CSF to ameliorate chemotherapy induced myelosuppression in sarcoma patients. Blood 1987; 70:Suppl 1:129a.

22. Antin JH, Smith BR, Rosenthal DS, et al. Phase I/II study of recombinant human granulocyte-macrophage colony-stimulating factor (GM-CSF) in bone marrow failure. Blood 1987; 70:Suppl 1:129a.

23. Morstyn G, Souza LM, Keech J, et al. Effect of granulocyte colony stimulating factor on neutropenia induced by cytotoxic chemotherapy. Lancet 1988; i:667-71.

24. Butturini AB, De Souza PC, Gale RP, et al. Use of recombinant GM-CSF in the Brazil radiation accident. Submitted

25. Blazar BR, Widmer MB, Soderling, et al. Augmentation of donor bone marrow engraftment in histoincompatible murine recipients by granulocyte-macrophage colony-stimulating factor. Blood 1988; 71:320-8.